Biomimetic Scaffolds for Hard Tissue Surgery

Biomimetic Scaffolds for Hard Tissue Surgery

Editors

Ryszard Uklejewski
Mariusz Winiecki

MDPI

Basel • Beijing • Wuhan • Barcelona • Belgrade • Novi Sad • Cluj • Manchester

Ryszard Uklejewski
Kazimierz Wielki University
Bydgoszcz
Poland

Mariusz Winiecki
Kazimierz Wielki University
Bydgoszcz
Poland

Editorial Office
MDPI AG
Grosspeteranlage 5
4052 Basel, Switzerland

This is a reprint of the Special Issue, published open access by the journal *Biomimetics* (ISSN 2313-7673), freely accessible at: https://www.mdpi.com/journal/biomimetics/special_issues/3902Y6X7SE.

For citation purposes, cite each article independently as indicated on the article page online and using the guide below:

Lastname, A.A.; Lastname, B.B. Article Title. *Journal Name* **Year**, *Volume Number*, Page Range.

ISBN 978-3-7258-1577-7 (Hbk)
ISBN 978-3-7258-1578-4 (PDF)
https://doi.org/10.3390/books978-3-7258-1578-4

Cover image courtesy of Ryszard Uklejewski and Mariusz Winiecki
Intraoperative image of implantation in the swine knee of the prototype biomimetic multi-spiked connecting scaffold for a new generation of resurfacing endoprostheses (partial knee endoprosthesis) – a task of research grant of National Science Centre Poland (No. N518412638, head: R. Uklejewski)

Contents

About the Editors

Ryszard Uklejewski

Ryszard Uklejewski, PhD Eng., PhD MD is a Professor of Biomedical Engineering. He received an MSc at the Poznan University of Technology, Poland (1974) and a PhD at the Lodz University of Technology (1979), and he studied mathematics (1976–1980) at the University of Lodz, Poland. He is also a qualified medical doctor (1996, Poznan University of Medical Sciences, Poland) and received a PhD in Medicine (in Endocrinology) at the University of Medical Sciences in Poznan (2010). In 1995, he received a Habilitation in Biomedical Engineering at the Institute of Biocybernetics and Biomedical Engineering, Pol. Acad. Sci. with a thesis on the bioelectromechanics of porous bone. Since 1997, he has worked at the Kazimierz Wielki University in Bydgoszcz (Poland) as the Head of the Department of Medical Bioengineering Fundamentals (until 2021) and currently as the Chair of Constructional Materials and Biomaterials. His main research areas are the structural–biomechanical biocompatibility of the bone–implant interface, bone tissue biomechatronics, and growing human skeleton endocrinology. He is the author of more than 180 research publications, was the head of two three-university research projects of the National Science Centre Poland on the development of prototypes of non-cemented resurfacing arthroplasty endoprostheses, and is a co-author of the monograph "Prototype of a Biomimetic Multi-Spiked Connecting Scaffold for a New Generation of Resurfacing Endoprostheses", Taylor & Francis/CRC (2024). He is a member of American Biographical Institute Research Association, European Federation for Research in Rehabilitation, and Polish Society for Biomaterials and a former member of AAAS and NYAS (for at least 10 years), and he is included in the list of "2000 Outstanding Scientists of the 20th Century" published by the International Biographical Centre, Cambridge, UK.

Mariusz Winiecki

Mariusz Winiecki, PhD Eng., is an Assistant Professor in the Department of Constructional Materials and Biomaterials at the Faculty of Materials Engineering at the Kazimierz Wielki University, Bydgoszcz, Poland. In 2001, he received an MSc in Mechanical Engineering (specialization in Mechatronics) at the Poznan University of Technology, Poland. In 2006, he received his PhD in Mechanical Engineering at the Poznan University of Technology, Poland for his thesis on the investigation of the micro-geometric constructional properties of porous intraosseous implants and the influence of these properties on the strength of the bone–implant model fixation. In 2007, he finished his Postgraduate Studies in Biomaterials Engineering at the AGH University of Science and Technology, Cracow, Poland, and since 2007, he has been a member of the Polish Society for Biomaterials. His main research areas in the field of biomaterials engineering are the engineering of bone–implant interfacing and implant surface functionalization towards improving the conditions for osseointegration. His collaborative interdisciplinary research group has designed, prototyped, and manufactured in selective laser melting technology. Furthermore, through bioengineering research, his group developed the essential innovation in the fixation of components of a new generation of entirely cementless resurfacing hip arthroplasty endoprostheses in the subchondral trabecular bone by means of the biomimetic multi-spiked connecting scaffold. He was a key investigator in two research projects of the Polish Ministry of Science and the National Science Centre Poland and is co-author of the scientific monograph "Prototype of a Biomimetic Multi-Spiked Connecting Scaffold for a New Generation of Resurfacing Endoprostheses", Taylor & Francis /CRC (2024).

biomimetics

MDPI

Editorial

Advances in Biomimetic Scaffolds for Hard Tissue Surgery

Ryszard Uklejewski * and Mariusz Winiecki *

Department of Constructional Materials and Biomaterials, Faculty of Materials Engineering, Kazimierz Wielki University, Jan Karol Chodkiewicz Street 30, 85-064 Bydgoszcz, Poland
* Correspondence: uklejew@ukw.edu.pl (R.U.); winiecki@ukw.edu.pl (M.W.)

1. Introduction

Hard tissues are living mineralized tissues that possess a high degree of hardness and are found in organs such as bones and teeth (enamel, dentin, and cementum). The ultimate goal of bone and joint surgery, craniomaxillofacial surgery, oral/dental surgery or, in general, hard tissue surgery is reconstruction via the implantation of a biomaterial or a device to replace bones and/or joints affected by various diseases, traumatic damages, or deformities. The reconstruction of critical-sized loss or defects caused by trauma, tumor excision, osteoarthritis, and other bone-resorption-related diseases or disorders remains a significant challenge [1–7]. However, three-dimensional biomaterial scaffolds (produced by means of engineering or tissue engineering technologies) have emerged as relatively novel tools used to repair such damaged hard tissues [8–11]. Biomimetic scaffolds are designed and generated as biomaterial architectures that promote the regeneration of native tissue [12–16]. Hard tissue surgery scaffolds require mechanical stability in order to support the needed geometry of tissue loss or defects and facilitate external loading. Such scaffolds should provide internal microarchitecture to the tissue that is to be regenerated with an internal, interconnected porous network of effective space for the infiltration, growth, and differentiation of bone marrow mesenchymal stem cells, vasculature ingrowth, and new tissue growth, with the aim of ensuring a channel of material exchange with the external environment (delivering oxygen and other nutrients to the cells, in addition to waste removal) [17–20]. Thus, the design of such scaffolds is extremely important to the success of clinical outcomes in hard tissue surgery. The newest trend in this field is the viable bioinspired structural and functional design of tissue-mimicking 3D-printed (composite or hybrid) scaffolds with interconnected pore structures of controlled and often gradual porosity of implants, with the synergistic functions of promoting bone regeneration (often seeded with mesenchymal stromal cells and involving biomolecules and growth factors) and reducing local bacterial infections (intrinsically antimicrobial or loaded with antibiotics, peptides, antimicrobial metallic ions, and/or nanoparticles, anticancer drugs, etc.) [21–30].

This Special Issue aims to exhibit and discuss the latest advances in biomimetic scaffolds for hard tissue surgery, and it includes contributions on potential topics including, but not limited to, (1) biomimetic design strategies for scaffolds, (2) techniques for fabricating biomimetic scaffolds, (3) novel biomaterials for biomimetic scaffolds, (4) the biodegradability design of biomimetic scaffolds, (5) the surface functionalization of biomimetic scaffolds, and (6) clinical applications of biomimetic scaffolds.

This Special Issue presents eleven contributions submitted by scholars with renowned backgrounds in scaffolds design, fabrication, functionalization, or clinical applications [31–50]; four of these contributions are valuable review articles, and seven are original new research articles. Thematically, according to the area of application, the contributions can be grouped into articles concerning the following:

- Biomimetic scaffolds for bone and joint surgery;
- Biomimetic scaffolds for maxillofacial surgery and oral surgery;
- Biomimetic scaffolds for general surgery (i.e., hard- and soft tissue surgeries).

Citation: Uklejewski, R.; Winiecki, M. Advances in Biomimetic Scaffolds for Hard Tissue Surgery. *Biomimetics* **2024**, *9*, 279. https://doi.org/10.3390/biomimetics9050279

Received: 27 March 2024
Accepted: 6 May 2024
Published: 8 May 2024

2. Overview of the Published Papers

Within the topic of biomimetic scaffolds for bone and joint surgery, two review articles highlight the advances that have been made in biomimetic scaffolds. The first review [Contribution 1] regards biomimetic silk-based materials and scaffolds for tissue engineering, with a focus on skeletal tissues, whereas the second review [Contribution 2] explores the background of currently used designs in resurfacing arthroplasty (RA) endoprostheses and presents a new approach for their component fixation via a biomimetic multi-spiked prototype scaffold.

Branković et al. (2024) [Contribution 1] review the latest research related to the potential applications of spider silk and silk-based biomaterials in reconstructive and regenerative medicine and tissue engineering, with a focus on musculoskeletal tissues. The structure and properties of spider silk, along with natural spider silk synthesis and the further advanced recombinant production of spider silk proteins, are reviewed. Research insights into possible spider silk structures, such as fibers (1D), coatings (2D), and 3D constructs, including porous structures, hydrogels, and organ-on-chip designs, are presented, with a review of applications of silk-based materials and scaffolds in musculoskeletal tissue engineering. These include bone and cartilage in addition to muscle and tendon, as well as the advanced design and precise engineering of artificial skin and vascular tissues and the design of bioactive silk-based biomaterials for smart medical implants and controlled drug delivery systems.

Uklejewski et al. (2024) [Contribution 2] take the over-hundred-year-long history of RA into scope and discuss how the designs of RA endoprostheses have evolved via a variety of designs of endoprosthesis components, different choices of materials used, and changes in methods of fixation in bone. The milestones of past design generations of RA endoprostheses are chronologically discussed along with critical insight into contemporary hip RA endoprostheses designs and their failure scenarios. As pointed out by the authors, coupled with technological advancements is the need for innovations directed towards more biomimetic designs that have materialized with the first biomimetic fixation for the RA endoprostheses being introduced. This new design type of completely cementless and stemless RA endoprostheses of knee joints and hip joints (and other diarthrodial joints), where endoprosthesis components are embedded in the surrounding bone via the prototype biomimetic multi-spiked connecting scaffold (MSC-Scaffold), initiates the first at-all generations of biomimetic endoprostheses of diarthrodial joints [44].

Of the research articles published in Part I, on biomimetic scaffolds for bone and joint surgery, two [Contributions 3 and 4] concern biomimetic design strategies for scaffolds, as well as techniques for fabricating biomimetic scaffolds, and three are concerned with novel biomaterials for biomimetic scaffolds. The personalized biomimetic scaffold for the restoration of long-bone segmental defects is the subject of Contribution 3, and the drug-releasing biomimetic scaffold for bone tissue repair is the subject of Contribution 4. Both articles present the development and characterization of 3D constructs with biological verification. Among the novel materials for biomimetic scaffolds, the three consecutive original research articles present new composite material mimicking the natural bone structure for bone scaffolds [Contribution 5], advances in ceramic biocomposites of good biomimetics of human bone composition and the means by which they are applicable in bone scaffolds [Contribution 6], and a novel biomimetic highly porous material as a scaffold with osteointegration potential [Contribution 7].

Popkov et al. (2023) [Contribution 3] propose a method for the restoration of long-bone segmental defects with the use of a bioactive degradable 3D-printed scaffold. The porous implant of a gyroid-like triply periodic minimal surface (TPMS) cellular structure was designed, fabricated via the fused deposition modeling (FDM) additive technology of ε-polycaprolactone (PCL), and coated with hydroxyapatite (HA). These implants were experimentally implanted in laboratory sheep to fill a 20 mm long segmental tibial defect, and radiological examinations demonstrated evident reparative bone tissue regeneration occurring from the proximal and distal bone fragments in the third week after surgery, the ingrowth of bone tissue into the cylindrical PCL-HA implant from the adjacent bone ends, and periosteal structures from day 7 of the study, in addition to the implant cellular structure filled with newly formed bone tissue on day 30 of the postoperative day. This in vivo study proved that the personalized biomimetic scaffold proposed provides stimulation of reparative osteogenesis and osseointegration in a single-implant bone block and is suitable for the regeneration of long-bone segmental defects.

Ensoylu et al. (2023) [Contribution 4] present the results of producing a borate-based 13-93B3 bioactive glass composite scaffold mimicking native bone tissue. In their work, hexagonal boron nitride hBN nanoparticles were included directly inside the bioactive glass matrix, and dense three-dimensional scaffolds were fabricated using the polymer foam replication method. The structural, mechanical, and biological performance of the scaffolds was investigated, and the drug delivery properties of the scaffolds loaded with gentamicin and fluorouracil were explored. The results indicate that the hBN nanoparticles, up to a certain concentration in the glass matrix, improved the mechanical strength of the glass scaffolds, while their presence enhanced the in vitro hydroxyapatite-forming ability of bioactive glass composites and accelerated the drug release rates of the system. The authors conclude that bioactive glass/hBN composite scaffolds that mimic native bone tissue could be used for bone tissue repair and regeneration applications.

Matos et al. (2023) [Contribution 5] investigated the potential of composites produced from polyvinylpyrrolidone (PVP) nanofibers containing mesoporous bioactive glass (MBG) 80S15 nanoparticles by the electrospinning technique to be used in bone tissue engineering. These polymeric scaffolds revealed the absence of cytotoxic effects on Saos-2 cells and enhanced bioactivity considering the rapid formation of hydroxycarbonate apatite (HCA) when exposed to simulated body fluid (SBF). Their degradation and swelling assays showed an ability to tailor their properties by varying the amount of MBG powder incorporated or the cross-linking properties applied, which also translates into the bioactivity (bone-bonding potential) of the composites. The authors conclude that, considering their inherent properties, the composites produced can be used for bone scaffolds because they not only reveal a high level of biocompatibility but also a swelling capacity suitable for further development, including drug delivery.

Ferro et al. (2023) [Contribution 6] present their advancements in the development of metal ion-doped ceramic biocomposites with a high level of similarity in composition to human bone and a bone-like morphology. In the study, tricalcium phosphate-based biocomposites were designed and sintered, both dense and poly(methyl methacrylate (PMMA) induced porous, doped with combinations of metal ions of magnesium (Mg^{2+}), manganese (Mn^{2+}), zinc (Zn^{2+}), and iron (Fe^{3+}); and simultaneously reinforced (for strengthening) with tetragonal zirconia (t-ZrO_2) and cubic zirconia (c-ZrO_2). A detailed evaluation of their physical, mechanical, and microstructural properties was performed with an evaluation of cytocompatibility in a human osteoblast (hOB) culture. The presented results led the authors to conclude that the addition of tetragonal and cubic zirconia resulted in a significant improvement in strength—up to 22% and 55%, respectively—and the addition of PMMA-generated porosity resulted in an improvement of strength up to 30% and an improvement in interconnectivity, with excellent hOB cellular viability achieved for all biocomposites produced.

Tavarro et al. (2024) [Contribution 7] propose an innovative bioactive aerogel-based composite with piezoelectric properties to assist bone regeneration. Aerogels of hydroxyapatite (HA) nanowires with barium titanate (BT, $BaTiO_3$) particles were synthesized and characterized for their physical and chemical properties, bioactivity, and in vitro cytotoxicity. The results demonstrated that the HA/BT aerogel, characterized by good bioactivity and biocompatibility, constitutes a new biomaterial with osteointegration potential, which combines the advantages of a highly porous structure such as a cell scaffold (providing osteoconductivity) and can accelerate osteogenesis and osteoinduction by the presence of surface charges induced by piezoelectric BTs; thus, it could be suitable for non-load-bearing applications, such as cavity filling.

The group of papers published in Part II, titled 'Biomimetic Scaffolds for Maxillofacial Surgery and Oral Surgery', includes one review article [Contribution 8] discussing the recent advances in biological scaffolds for bone formation as a new tissue engineering technique for maxillofacial surgery and one new research article [Contribution 9] on the development of a 3D biomimetic scaffold to serve as a biomechanical model of fibrous periodontal ligament behavior.

Ramezanzade et al. (2023) [Contribution 8] provide a systematic review of the current literature on the angle of reconstruction of critical-sized maxillofacial defects resulting from trauma or a benign pathologic disease, using composite allogeneic tissue engineering. This technique uses an allogenic graft as a biologic scaffold in conjunction with harvested mesenchymal stem cells and recombinant human bone morphogenic protein-2 (rhBMP-2) to create, as a custom-made graft, a favorable microenvironment for new bone formation. The discussion of its reliability and efficacy allows one to see the potential of using large-scale transplantable, vascularized, and customizable bone to reconstruct large maxillofacial bony defects as a promising alternative to current therapeutic clinical options that include extensive autogenous bone harvesting and many patient morbidities.

In their work, Gauthier et al. (2023) [Contribution 9] focus on the development of a 3D fibrous scaffold with biomechanical properties representative of those of the periodontal ligament (PDL), the fibers of which would be capable of transmitting mechanical loading to ligament cells, as it is carried out in vivo by collagen bundles. Three-dimensional fibrous polycaprolactone (PCL) scaffolds were synthesized by electrospinning and seeded with human periodontal ligament cells (PDLCs), and the behavior of the cells was observed in terms of their cellular organization and signaling under static and sinusoidal axial compressive loads. The results highlight the finding that electrospun fibrous PCL 3D scaffolds mimic the in vivo mechanical deformation of PDL collagen bundles and the transmission of load to PDLCs; therefore, they might represent an interesting (suitable) experimental model for investigating PDL mechanobiology and analyzing PDLC mechanobiological behavior.

Under the topic of scaffolds for general surgery (in Part III), one review paper [Contribution 10] discusses the challenges of mechanical property adaptation in tissue-engineered scaffolds for clinical applications, and one research article [Contribution 11] presents a novel biomimetic biomaterial for tissue engineering.

Based on a study of all the relevant papers published between the years 2021 and 2023, Johnston and Callanan (2023) [Contribution 10] provide a review of the current techniques by which the mechanical properties and biological compatibility of tissue-engineered grafts (or bioscaffolds)—promoting the repair of damaged soft and hard tissues—are enhanced via hybrid material usage, multi-layer scaffold designs, and surface-modified-type scaffolds, as well as clinically translated designs, whose uses and outcomes were published within the above-given years. From their studies, the authors draw a notable observation—that accounting for the complex range of mechanical properties present in biological tissues is generally beyond the scope of a single-material-based design, and a multifaceted approach consisting of multiple material types, layers, or surface treatment methods working in tandem is likely to yield the most successful designs. They underline the fact that, in terms of translation to clinical use, the time-sensitive nature of surgical treatment suggests that 3D printing, as a fabrication method, is a strong contender for the successful realization of these designs.

Dey et al. (2023) [Contribution 11] present results on the design and development of new gelatin-based electroconductive hydrogel scaffolds and examine the angle of their potential applications for tissue engineering scaffolds. Gelatin/poly(ethylene glycol)diglycidyl ether/chitosan (G/PEG/CH) nanocomposite hydrogels with the incorporated nanosized carbon black (CB) were prepared via a mild processing condition that included aqueous media, various polymer assembly, and cross-linking chemistry, facilitating gelation with CB nanomaterial. The synthesized nanocomposite hydrogels and underwent structural, mechanical, and thermal property characteristics. As a result, the authors concluded that these nanocomposite hydrogels, which are still in the development and optimization stages, are compositionally, morphologically, mechanically, and electrically similar to native extracellular matrixes of many tissues and show great promise for use as conducting substrates for the growth of electroresponsive cells in tissue engineering.

3. Conclusions

These contributions published in this Special Issue on Biomimetic Scaffolds for Hard Tissue Surgery (i.e., four valuable review articles and seven new research articles) comprehensively summarize the latest achievements in the field, with a discussion of prospects and challenges for further research. Along with different proposed and developed novel biomaterials and 3D constructs of specific complementary properties that, through their biomechanical, biostructural, biochemical, biomimetic, and functional properties, possess great potential for assisting local tissue regeneration, the contributions of this Special Issue demonstrate remarkable advancements in this interesting and dynamically developing field.

Author Contributions: R.U.: drafted, edited, and approved the manuscript; M.W.: drafted, edited, and approved the manuscript. All authors have read and agreed to the published version of the manuscript.

Funding: This research received no external funding.

Institutional Review Board Statement: Not applicable.

Data Availability Statement: No new data were created or analyzed in this study. Data sharing is not applicable to this article.

Conflicts of Interest: The authors declare no conflicts of interest.

List of Contributions with Thematic Division

Part I. Biomimetic Scaffolds for Bone and Joint Surgery

1. Branković, M.; Živić, F.; Grujovic, N.; Stojadinovic, I.; Milenkovic, S.; Kotorcevic, N. Review of Spider Silk Applications in Biomedical and Tissue Engineering. *Biomimetics* **2024**, *9*, 169. https://doi.org/10.3390/biomimetics9030169.
2. Uklejewski, R.; Winiecki, M.; Dąbrowski, M.; Rogala, P. Towards the First Generation of Biomimetic Fixation for Resurfacing Arthroplasty Endoprostheses. *Biomimetics* **2024**, *9*, 99. https://doi.org/10.3390/biomimetics9020099.
3. Popkov, A.; Kononovich, N.; Dubinenko, G.; Gorbach, E.; Shastov, A.; Tverdokhlebov, S.; Popkov, D. Long Bone Defect Filling with Bioactive Degradable 3D-Implant: Experimental Study. *Biomimetics* **2023**, *8*, 138. https://doi.org/10.3390/biomimetics8020138.
4. Ensoylu, M.; Deliormanlı, A.M.; Atmaca, H. Preparation, Characterization, and Drug Delivery of Hexagonal Boron Nitride-Borate Bioactive Glass Biomimetic Scaffolds for Bone Tissue Engineering. *Biomimetics* **2023**, *8*, 10. https://doi.org/10.3390/biomimetics8010010.
5. Matos, R.J.R.; Silva, J.C.; Soares, P.I.P.; Borges, J.P. Polyvinylpyrrolidone Nanofibers Incorporating Mesoporous Bioactive Glass for Bone Tissue Engineering. *Biomimetics* **2023**, *8*, 206. https://doi.org/10.3390/biomimetics8020206.
6. Ferro, V.M.; Silva, B.C.; Macedo, D.F.; Fernandes, N.F.; Silva, A.P. TCP Doped with Metal Ions Reinforced with Tetragonal and Cubic Zirconia. *Biomimetics* **2023**, *8*, 599. https://doi.org/10.3390/biomimetics8080599.
7. Tavares, C.; Vieira, T.; Silva, J.C.; Borges, J.P.M.R.; Lança, M.C. Bioactive Hydroxyapatite Aerogels with Piezoelectric Particles. *Biomimetics* **2024**, *9*, 143. https://doi.org/10.3390/biomimetics9030143.

Part II. Biomimetic Scaffolds for Maxillofacial Surgery and Oral Surgery

8. Ramezanzade, S.; Aeinehvand, M.; Ziaei, H.; Khurshid, Z.; Keyhan, S.O.; Fallahi, H.R.; Melville, J.C.; Saeinasab, M.; Sefat, F. Reconstruction of Critical Sized Maxillofacial Defects Using Composite Allogeneic Tissue Engineering: Systematic Review of Current Literature. *Biomimetics* **2023**, *8*, 142. https://doi.org/10.3390/biomimetics8020142.

9. Gauthier, R.; Attik, N.; Chevalier, C.; Salles, V.; Grosgogeat, B.; Gritsch, K.; Trunfio-Sfarghiu, A.-M. 3D Electrospun Polycaprolactone Scaffolds to Assess Human Periodontal Ligament Cells Mechanobiological Behaviour. *Biomimetics* **2023**, *8*, 108. https://doi.org/10.3390/biomimetics8010108.

Part III. Biomimetic Scaffolds for General Surgery

10. Johnston, A.; Callanan, A. Recent Methods for Modifying Mechanical Properties of Tissue-Engineered Scaffolds for Clinical Applications. *Biomimetics* **2023**, *8*, 205. https://doi.org/10.3390/biomimetics8020205.
11. Dey, K.; Sandrini, E.; Gobetti, A.; Ramorino, G.; Lopomo, N.F.; Tonello, S.; Sardini, E.; Sartore, L. Designing Biomimetic Conductive Gelatin-Chitosan–Carbon Black Nanocomposite Hydrogels for Tissue Engineering. *Biomimetics* **2023**, *8*, 473. https://doi.org/10.3390/biomimetics8060473.

References

1. Wang, W.; Yeung, K.W.K. Bone grafts and biomaterials substitutes for bone defect repair: A review. *Bioact. Mater.* **2017**, *2*, 224–247. [CrossRef] [PubMed]
2. Vidal, L.; Kampleitner, C.; Brennan, M.Á.; Hoornaert, A.; Layrolle, P. Reconstruction of Large Skeletal Defects: Current Clinical Therapeutic Strategies and Future Directions Using 3D Printing. *Front. Bioeng. Biotechnol.* **2020**, *8*, 61. [CrossRef] [PubMed]
3. Tsang, S.J.; Ferreira, N.; Simpson, A.H.R.W. The reconstruction of critical bone loss: The holy grail of orthopaedics. *Bone Joint Res.* **2022**, *11*, 409–412. [CrossRef] [PubMed]
4. Huang, E.E.; Zhang, N.; Shen, H.; Li, X.; Maruyama, M.; Utsunomiya, T.; Gao, Q.; Guzman, R.A.; Goodman, S.B. Novel Techniques and Future Perspective for Investigating Critical-Size Bone Defects. *Bioengineering* **2022**, *9*, 171. [CrossRef] [PubMed]
5. Feng, Y.; Zhu, S.; Mei, D.; Li, J.; Zhang, J.; Yang, S.; Guan, S. Application of 3D Printing Technology in Bone Tissue Engineering: A Review. *Curr. Drug Deliv.* **2021**, *18*, 847–861. [CrossRef] [PubMed]
6. Kämmerer, P.W.; Al-Nawas, B. Bone reconstruction of extensive maxillomandibular defects in adults. *Periodontol. 2000* **2023**, *93*, 340–357. [CrossRef] [PubMed]
7. Cojocaru, F.D.; Balan, V.; Verestiuc, L. Advanced 3D Magnetic Scaffolds for Tumor-Related Bone Defects. *Int. J. Mol. Sci.* **2022**, *23*, 16190. [CrossRef] [PubMed]
8. Nikolova, M.P.; Chavali, M.S. Recent advances in biomaterials for 3D scaffolds: A review. *Bioact. Mater.* **2019**, *4*, 271–292. [CrossRef]
9. Rahimnejad, M.; Rezvaninejad, R.; Rezvaninejad, R.; França, R. Biomaterials in bone and mineralized tissue engineering using 3D printing and bioprinting technologies. *Biomed. Phys. Eng. Express* **2021**, *7*, 062001. [CrossRef]
10. Yazdanpanah, Z.; Johnston, J.D.; Cooper, D.M.L.; Chen, X. 3D Bioprinted Scaffolds for Bone Tissue Engineering: State-of-the-Art and Emerging Technologies. *Front. Bioeng. Biotechnol.* **2022**, *10*, 824156. [CrossRef]
11. Pitacco, P.; Sadowska, J.M.; O'Brien, F.J.; Kelly, D.J. 3D bioprinting of cartilaginous templates for large bone defect healing. *Acta Biomater.* **2023**, *156*, 61–74. [CrossRef]
12. Huang, L.; Chen, L.; Chen, H.; Wang, M.; Jin, L.; Zhou, S.; Gao, L.; Li, R.; Li, Q.; Wang, H.; et al. Biomimetic Scaffolds for Tendon Tissue Regeneration. *Biomimetics* **2023**, *8*, 246. [CrossRef]
13. Mulbauer, G.; Matthew, H.W.T. Biomimetic Scaffolds in Skeletal Muscle Regeneration. *Discoveries* **2019**, *7*, e90. [CrossRef]
14. Park, J.Y.; Park, S.H.; Kim, M.G.; Park, S.H.; Yoo, T.H.; Kim, M.S. Biomimetic Scaffolds for Bone Tissue Engineering. *Adv. Exp. Med. Biol.* **2018**, *1064*, 109–121. [CrossRef] [PubMed]
15. Jiang, S.; Wang, M.; He, J. A review of biomimetic scaffolds for bone regeneration: Toward a cell-free strategy. *Bioeng. Transl. Med.* **2020**, *6*, e10206. [CrossRef]
16. Badekila, A.K.; Kini, S.; Jaiswal, A.K. Fabrication techniques of biomimetic scaffolds in three-dimensional cell culture: A review. *J. Cell Physiol.* **2021**, *236*, 741–762. [CrossRef] [PubMed]
17. Bahraminasab, M. Challenges on optimization of 3D-printed bone scaffolds. *BioMed Eng. OnLine* **2020**, *19*, 69. [CrossRef] [PubMed]
18. Hollister, S.J.; Maddox, R.D.; Taboas, J.M. Optimal design and fabrication of scaffolds to mimic tissue properties and satisfy biological constraints. *Biomaterials* **2002**, *23*, 4095–4103. [CrossRef] [PubMed]
19. Egan, P.F. Integrated Design Approaches for 3D Printed Tissue Scaffolds: Review and Outlook. *Materials* **2019**, *12*, 2355. [CrossRef]
20. Marew, T.; Birhanu, G. Three dimensional printed nanostructure biomaterials for bone tissue engineering. *Regen. Ther.* **2021**, *18*, 102–111. [CrossRef]
21. Li, J.; Yuan, H.; Chandrakar, A.; Moroni, L.; Habibovic, P. 3D porous Ti6Al4V-beta-tricalcium phosphate scaffolds directly fabricated by additive manufacturing. *Acta Biomater.* **2021**, *126*, 496–510. [CrossRef] [PubMed]
22. Bertsch, C.; Maréchal, H.; Gribova, V.; Lévy, B.; Debry, C.; Lavalle, P.; Fath, L. Biomimetic Bilayered Scaffolds for Tissue Engineering: From Current Design Strategies to Medical Applications. *Adv. Healthc. Mater.* **2023**, *12*, e2203115. [CrossRef] [PubMed]

23. Pugliese, R.; Graziosi, S. Biomimetic scaffolds using triply periodic minimal surface-based porous structures for biomedical applications. *SLAS Technol.* **2023**, *28*, 165–182. [CrossRef] [PubMed]
24. Zielińska, A.; Karczewski, J.; Eder, P.; Kolanowski, T.; Szalata, M.; Wielgus, K.; Szalata, M.; Kim, D.; Shin, S.R.; Słomski, R.; et al. Scaffolds for drug delivery and tissue engineering: The role of genetics. *J. Control Release* **2023**, *359*, 207–223. [CrossRef]
25. Peng, Y.; Zhuang, Y.; Liu, Y.; Le, H.; Li, D.; Zhang, M.; Liu, K.; Zhang, Y.; Zuo, J.; Ding, J. Bioinspired gradient scaffolds for osteochondral tissue engineering. *Exploration* **2023**, *3*, 20210043. [CrossRef]
26. Pattnaik, A.; Sanket, A.S.; Pradhan, S.; Sahoo, R.; Das, S.; Pany, S.; Douglas, T.E.L.; Dandela, R.; Liu, Q.; Rajadas, J.; et al. Designing of gradient scaffolds and their applications in tissue regeneration. *Biomaterials* **2023**, *296*, 122078. [CrossRef]
27. Grelewski, P.G.; Kwaśnicka, M.; Bar, J.K. Properties of scaffolds as carriers of mesenchymal stem cells for use in bone engineering. *Polim. Med.* **2023**, *53*, 129–139. [CrossRef] [PubMed]
28. Re, F.; Borsani, E.; Rezzani, R.; Sartore, L.; Russo, D. Bone Regeneration Using Mesenchymal Stromal Cells and Biocompatible Scaffolds: A Concise Review of the Current Clinical Trials. *Gels* **2023**, *9*, 389. [CrossRef]
29. Dubey, A.; Vahabi, H.; Kumaravel, V. Antimicrobial and Biodegradable 3D Printed Scaffolds for Orthopedic Infections. *ACS Biomater. Sci. Eng.* **2023**, *9*, 4020–4044. [CrossRef]
30. Liu, Y.; He, L.; Cheng, L.; Li, X.; Gao, M.; Li, Q.; Gao, J.; Ramalingam, M. Enhancing Bone Grafting Outcomes: A Comprehensive Review of Antibacterial Artificial Composite Bone Scaffolds. *Med. Sci. Monit.* **2023**, *29*, e939972. [CrossRef]
31. Pádua, A.S.; Figueiredo, L.; Silva, J.C.; Borges, J.P. Chitosan scaffolds with mesoporous hydroxyapatite and mesoporous bioactive glass. *Prog. Biomater.* **2023**, *12*, 137–153. [CrossRef] [PubMed]
32. Tavares, F.J.T.M.; Soares, P.I.P.; Silva, J.C.; Borges, J.P. Preparation and In Vitro Characterization of Magnetic CS/PVA/HA/pSPIONs Scaffolds for Magnetic Hyperthermia and Bone Regeneration. *Int. J. Mol. Sci.* **2023**, *24*, 1128. [CrossRef] [PubMed]
33. De Sá, K.D.; Figueira, D.R.; Miguel, S.P.; Correia, T.R.; Silva, A.P.; Correia, I.J. 3D Scaffolds Coated with Nanofibers Displaying Bactericidal Activity for Bone Tissue Applications. *Int. J. Polym. Mater. Polym. Biomater.* **2017**, *66*, 432–442. [CrossRef]
34. Sartore, L.; Pasini, C.; Pandini, S.; Dey, K.; Ferrari, M.; Taboni, S.; Chan, H.H.L.; Townson, J.; Viswanathan, S.; Mathews, S.; et al. Hybrid Core-Shell Polymer Scaffold for Bone Tissue Regeneration. *Int. J. Mol. Sci.* **2022**, *23*, 4533. [CrossRef] [PubMed]
35. Re, F.; Sartore, L.; Borsani, E.; Ferroni, M.; Baratto, C.; Mahajneh, A.; Smith, A.; Dey, K.; Almici, C.; Guizzi, P.; et al. Mineralization of 3D Osteogenic Model Based on Gelatin-Dextran Hybrid Hydrogel Scaffold Bioengineered with Mesenchymal Stromal Cells: A Multiparametric Evaluation. *Materials* **2021**, *14*, 3852. [CrossRef] [PubMed]
36. Popkov, A.; Tverdokhlebov, S.; Muradisinov, S.; Popkov, D. First Clinical Case of Ilizarov Femur Lengthening over a Bioactive and Degradable Intramedullary Implant. *Case Rep. Orthop.* **2023**, *2023*, 7547590. [CrossRef] [PubMed]
37. Popkov, A.; Kononovich, N.; Gorbach, E.; Popkov, D. Osteointegration technology in long bone defect reconstruction: Experimental study. *Acta Bioeng. Biomech.* **2020**, *22*, 85–91. [CrossRef] [PubMed]
38. Farano, V.; Cresswell, M.; Gritsch, K.; Jackson, P.; Attik, N.; Grosgogeat, B.; Maurin, J.C. Bioactivity evaluation of collagen-based scaffolds containing a series of Sr-doped melt-quench derived phosphate-based glasses. *J. Mater. Sci. Mater. Med.* **2018**, *29*, 101. [CrossRef] [PubMed]
39. Miranda, D.G.; Malmonge, S.M.; Campos, D.M.; Attik, N.G.; Grosgogeat, B.; Gritsch, K. A chitosan-hyaluronic acid hydrogel scaffold for periodontal tissue engineering. *J. Biomed. Mater. Res. B Appl. Biomater.* **2016**, *104*, 1691–1702. [CrossRef]
40. Ensoylu, M.; Deliormanlı, A.M.; Atmaca, H. Hexagonal Boron Nitride/PCL/PLG Coatings on Borate Bioactive Glass Scaffolds for Bone Regeneration. *J. Inorg. Organomet. Polym.* **2022**, *32*, 1551–1566. [CrossRef]
41. Ensoylu, M.; Atmaca, H.; Deliormanlı, A.M. Fabrication and in vitro characterization of macroporous WS2/bioactive glass scaffolds for biomedical applications. *J. Aust. Ceram Soc.* **2022**, *58*, 397–409. [CrossRef]
42. Slavkovic, V.; Palic, N.; Milenkovic, S.; Živić, F.; Grujovic, N. Thermo-Mechanical Characterization of 4D-Printed Biodegradable Shape-Memory Scaffolds Using Four-Axis 3D-Printing System. *Materials* **2023**, *16*, 5186. [CrossRef] [PubMed]
43. Milenkovic, S.; Slavkovic, V.; Fragassa, C.; Grujovic, N.; Palić, N.; Živić, F. Effect of the raster orientation on strength of the continuous fiber reinforced PVDF/PLA composites, fabricated by hand-layup and fused deposition modeling. *Compos. Struct.* **2021**, *270*, 114063. [CrossRef]
44. Uklejewski, R.; Rogala, P.; Winiecki, M. *Prototype of a Biomimetic Multi-Spiked Connecting Scaffold for a New Generation of Resurfacing Endoprostheses*, 1st ed.; CRC Press/Taylor & Francis Group: Boca Raton, FL, USA, 2024. [CrossRef]
45. Uklejewski, R.; Winiecki, M.; Krawczyk, P.; Tokłowicz, R. Native Osseous CaP Biomineral Coating on a Biomimetic Multi-Spiked Connecting Scaffold Prototype for Cementless Resurfacing Arthroplasty Achieved by Combined Electrochemical Deposition. *Materials* **2019**, *12*, 3994. [CrossRef] [PubMed]
46. Gao, Y.; Bate, T.S.R.; Callanan, A. A Unification of Nanotopography and Extracellular Matrix in Electrospun Scaffolds for Bioengineered Hepatic Models. *ACS Appl. Bio Mater.* **2023**, *6*, 2158–2171. [CrossRef]
47. Conde-González, A.; Dutta, D.; Wallace, R.; Callanan, A.; Bradley, M. Rapid fabrication and screening of tailored functional 3D biomaterials. *Mater. Sci. Eng. C Mater. Biol Appl.* **2020**, *108*, 110489. [CrossRef] [PubMed]
48. Conde-González, A.; Glinka, M.; Dutta, D.; Wallace, R.; Callanan, A.; Oreffo, R.O.C.; Bradley, M. Rapid fabrication and screening of tailored functional 3D biomaterials: Validation in bone tissue repair—Part II. *Biomater. Adv.* **2023**, *145*, 213250. [CrossRef] [PubMed]

49. Aghmiuni, A.I.; Keshel, S.H.; Rezaei-tavirani, M.; Sefat, F.; Khojasteh, A.; Soleimani, M.; Pakdel, F. Effect of PEG Molecular Weight and Volume Ratio of Chitosan/PEG and Silk Fibroin on Physicomechanical Properties of Chitosan/PEG-SF Scaffold as a Bio-mimetic Substrate in Skin-tissue Engineering Applications. *Fibers. Polym.* **2022**, *23*, 3358–3368. [CrossRef]
50. Bazgir, M.; Saeinasab, M.; Zhang, W.; Zhang, X.; Min Tsui, K.; Maasoumi Sarvestani, A.; Nawaz, S.; Coates, P.; Youseffi, M.; Elies, J.; et al. Investigation of Cell Adhesion and Cell Viability of the Endothelial and Fibroblast Cells on Electrospun PCL, PLGA and Coaxial Scaffolds for Production of Tissue Engineered Blood Vessel. *J. Funct. Biomater.* **2022**, *13*, 282. [CrossRef]

biomimetics

MDPI

Review

Review of Spider Silk Applications in Biomedical and Tissue Engineering

Marija Branković [1,2], Fatima Zivic [2,*], Nenad Grujovic [2], Ivan Stojadinovic [3,4], Strahinja Milenkovic [2] and Nikola Kotorcevic [2]

1 Institute for Information Technologies, University of Kragujevac, Jovana Cvijića bb, 34000 Kragujevac, Serbia; marija.brankovic99@gmail.com
2 Faculty of Engineering, University of Kragujevac, Liceja Knezevine Srbije 1A, 34000 Kragujevac, Serbia; gruja@kg.ac.rs (N.G.); strahinja.milenkovic@fink.rs (S.M.); nidzakotorcevic@gmail.com (N.K.)
3 Clinic for Orthopaedics and Traumatology, University Clinical Center, Zmaj Jovina 30, 34000 Kragujevac, Serbia; stojadinovic.ivan.78@gmail.com
4 Faculty of Medical Sciences, University of Kragujevac, Svetozara Markovića 69, 34000 Kragujevac, Serbia
* Correspondence: zivic@kg.ac.rs

Abstract: This review will present the latest research related to the production and application of spider silk and silk-based materials in reconstructive and regenerative medicine and tissue engineering, with a focus on musculoskeletal tissues, and including skin regeneration and tissue repair of bone and cartilage, ligaments, muscle tissue, peripheral nerves, and artificial blood vessels. Natural spider silk synthesis is reviewed, and the further recombinant production of spider silk proteins. Research insights into possible spider silk structures, like fibers (1D), coatings (2D), and 3D constructs, including porous structures, hydrogels, and organ-on-chip designs, have been reviewed considering a design of bioactive materials for smart medical implants and drug delivery systems. Silk is one of the toughest natural materials, with high strain at failure and mechanical strength. Novel biomaterials with silk fibroin can mimic the tissue structure and promote regeneration and new tissue growth. Silk proteins are important in designing tissue-on-chip or organ-on-chip technologies and micro devices for the precise engineering of artificial tissues and organs, disease modeling, and the further selection of adequate medical treatments. Recent research indicates that silk (films, hydrogels, capsules, or liposomes coated with silk proteins) has the potential to provide controlled drug release at the target destination. However, even with clear advantages, there are still challenges that need further research, including clinical trials.

Keywords: spider silk; recombinant spider silk proteins; tissue engineering; drug delivery; bone and cartilage; ligament and muscle repair; repair of peripheral nerves; tissue-on-chip; organ-on-chip

Citation: Branković, M.; Zivic, F.; Grujovic, N.; Stojadinovic, I.; Milenkovic, S.; Kotorcevic, N. Review of Spider Silk Applications in Biomedical and Tissue Engineering. *Biomimetics* **2024**, *9*, 169. https://doi.org/10.3390/biomimetics9030169

Academic Editor: Rosalyn Abbott

Received: 15 January 2024
Revised: 1 March 2024
Accepted: 4 March 2024
Published: 11 March 2024

1. Introduction

The development of biomaterials aims for sustainable and biobased materials [1], including natural sustainable materials that can mimic tissue structures, such as silk-based biomaterials [2]. Silk-based biomaterials have been intensively studied for diverse applications in biomedical and tissue engineering [3–7]. The application of scaffolds with silk fibroin has been studied for bone scaffolding and ligament, cartilage, and tendon scaffolds, as well as for wound dressings and skin scaffolds [8–12]. Spider silk was selected and analyzed in this article as a natural sustainable biocompatible material, with excellent mechanical properties in comparison to other organic based materials, and a material which has been used for centuries in different applications. Biomedical applications of spider silk still account for a smaller part of its applications, whereas other industries have widely used it, even though there is a clear potential for its application in tissue engineering, including in musculoskeletal tissues, due to its favorable mechanical properties and other material properties that can further be customized in composite structures.

Spider silk has been used in many different applications, including tissue engineering [13,14]. In ancient times, spider silk was used to stop bleeding, where it served as an astringent [15,16]. The first clinical use of spider silk was in the 18th century, when it was used for suturing, whereas nowadays the application of silk as a biomaterial has been widely studied due to its excellent biocompatibility, high toughness, and ability to support tissue growth, especially for bone and ligament tissues [17,18]. Silk is one of the toughest natural materials and its fibers exhibit high strain at failure and very high mechanical strength [19]. Since its natural collection is limited, fabrication of the recombinant spider silk protein has begun [20]. Consequently, this technology has enabled modern biomedical applications [21]. All its excellent properties (very good mechanical characteristics, excellent biocompatibility, low density, and biodegradation) have shown high potential in tissue engineering [13]. The major clinical application of silk is in silk sutures, with a rather limited number of clinical uses in other medical applications, such as in cosmetics, wound dressing, breast reconstruction, and the treatment of gynecological conditions [22].

For musculoskeletal tissue regeneration (bone and cartilage tissue engineering), silk-based biomaterials offer a unique combination of properties and possibilities of molecular-level modifications and tailoring to the specific tissue scaffold [9,12,23,24]. Damage to the cartilage or degenerative conditions have influenced millions of patients, and in many cases joint replacements are the only possible treatments. However, these traditional treatments do not regenerate cartilage, just relieve patients of pain, unlike novel biomaterials with silk fibroin that can mimic the tissue structure and promote cartilage regeneration and new tissue growth [10]. Ligament tissue regeneration is complex and still under research, whereas silk-based materials show great promise [25]. For the repair of damaged intervertebral discs by using silk (hydrogels with silk and silk-based scaffolds), preclinical studies have shown excellent results [26], but clinical studies are still missing.

Research has shown that composites and polymer blends that contain silk have promising properties for hard tissue engineering, but also for soft tissues [8,27]. Smart and bioactive materials [28–30] have properties that enable them to interact with the surrounding tissue. With adequate stimuli, a controlled healing process can be started which results in repaired, new tissue. This is especially important for complex defects of the bones. Another advantage of silk-based biomaterials is their antibacterial and antimicrobial properties, but the possible tailoring of those effects in tissue scaffolds is still under study [31]. Bioactivity of the biomaterials used for tissue regeneration is complex [12], whereas biomineralization, or the formation of biominerals, is especially important for hard bone tissues. It has been shown that silk proteins can meditate that biomineralization process [32]. Self-healing biomaterial has been produced from spider silk [33], promising smart biomaterials. Further advancements can be expected with the development of new organ printing and biofabrication technologies [34,35], such as bioprinting with bio-inks containing silk fibroin [36,37], electrospun scaffolds containing silk fibroin [5,38,39], or 4D printing for biomedical applications [40].

This review will present the latest research related to the production and application of spider silk in reconstructive and regenerative medicine and tissue engineering, including nanomedicine and drug delivery systems.

2. Structure and Properties of Spider Silk

Spidroin proteins, composed of an N-terminal domain, repeated motifs, and a C-terminal domain, determine the hierarchical structure of spider silk [41]. Although spidroin I and II are believed to be the main silk proteins [42], the identification of more than 20 silk genes suggests that the number of spidroins in silk glands is higher than anticipated [43,44].

The primary amino acids in spider silk are glycine, alanine, and serine [16]. The method of forced spinning has revealed that the microstructure and tensile behavior of spider silk fibers are influenced by the silking force exerted on the dope [45]. Furthermore, the strength of spider silk is highly dependent on the size and orientation of the nanocrystals [46]. Rheological properties of natural silk are complex and shear thinning governs the behavior of the fiber during the spinning process [47]. On the other hand, ion electrodiffusion governs silk electrogelation—the formation of a gel structure from an aqueous silk fibroin solution with the presence of electricity [48].

The microstructure of silk fibers is semi-crystalline, as shown in Figure 1, due to the presence of two phases: crystalline and amorphous (non-crystalline), as shown in [49]. The nanocrystalline phase is a result of the specific polypeptide secondary structure. Namely, in places with a high concentration of the amino acid alanine (polyalanine regions), several antiparallel β-sheets will form and group. These sheets are networked in an amorphous phase rich in glycine [50]. Weak hydrogen bonds are responsible for the superiority of this biopolymer in terms of its mechanical properties [51]. SEM and AFM imaging revealed that the silk thread (with a diameter of 4–5 μm) consists of many silk fibers with diameters in a range of 40–80 nm [46].

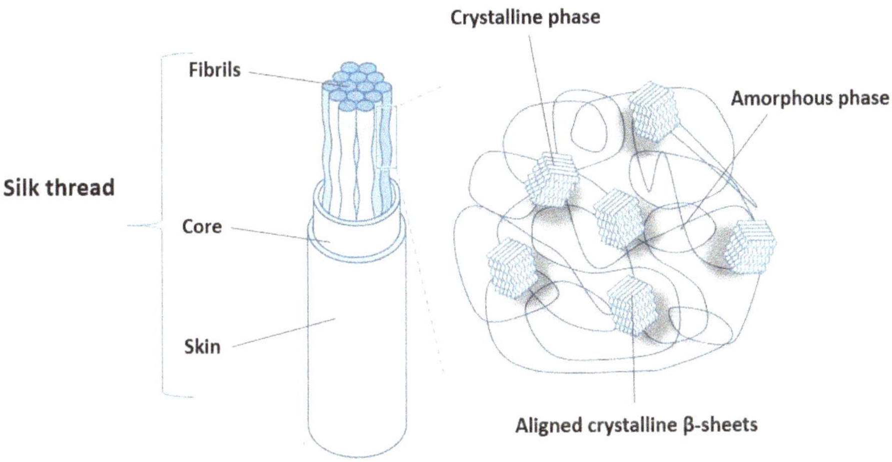

Figure 1. Structural organization of spider silk thread.

When cobweb fibers are exposed to high humidity or water, their length shortens significantly [52]. This phenomenon is called supercontraction, and the silk fibers are in the ground state [53]. The discovery of supercontraction made it possible to define different types of silk threads, that is, to establish a correlation between the mechanical properties of silk materials, the ecological niche, and the evolution of a given species [54]. In addition, this behavior of spider silk has been used to develop smart materials sensitive to temperature [55] and humidity. In this regard, the overall behavior of any silk fiber can be defined by the parameter α* that represents experimental magnitude that is determined from the reference Major Ampullate Silk (MAS) fibers [56].

3. Natural Synthesis of Spider Fibers

Spiders are cold-blooded organisms (poikilotherms) that can spin silk fibers with very high mechanical strength and toughness [57]. The temperature in their environment affects the speed of spinning of the web, and thus its mechanical and structural properties [58]. Likewise, silk varies widely in composition, depending on the specific source (spiders produce silk using seven different types of glands) [59].

Depending on life needs, spiders produce seven silk types (Tables 1 and 2) [60], thanks to the various silk glands located at the rear end of the abdomen [61]. These types have different properties depending on whether they serve as a shelter, a means of catching prey, part of the love game, or as a particular thread the spider uses to escape in case of danger. Among them, "dragline" silk has been investigated in the most detail [13].

Major ampullate fibers have very high tensile strength and toughness (Table 3). Consequently, they form the frame and radially distributed supports of the spider web [13]. Their structure is layered, and each of the layers has a different role, e.g., the transmission of pheromones and recognition of gender and species, protection against microorganisms, and protection against damage by physicochemical agents or mechanical support. Generally, spider silk consists of at least two types of proteins: proline-free spidroins (MaSp1) and proline-rich spidroins (MaSp2) [62]. In addition to differences in the content of this amino acid, the spidroins also differ in hydropathicity. The MaSp1 class is hydrophobic, and the MaSp2 is predominantly hydrophilic [13].

Within the Major Ampullate (MA) gland, we distinguish four regions:

- The "tail" zone, which is responsible for the synthesis and secretion of spider web proteins;
- Lumen (bag) used for protein accumulation;
- Fiber alignment channel;
- Output for final fiber production.

As proteins travel from the lumen along the channel, they undergo elongation, promoting hydrophobic and hydrogen-bonding interactions. This is followed by the alignment of the proteins in the solution, resulting in stiffer and stronger fibers. Finally, the cobweb, excreted in liquid form, hardens very quickly in contact with air.

However, the collection of natural spider silk does not yield significant quantities. According to the Animal Welfare Act and European directives, animal stress is reduced to an absolute minimum, with no consequences during the collection procedure. For this purpose, spiders were fixed with gauze and needles on Styrofoam without anesthesia [63]. MA silk was isolated from the major ampullary glands using forceps. In this method, a mechanical stimulus is sufficient to initiate spider silk production, and the collection continues until the animal becomes distressed [63]. In each orb-weaver spider, the silk composition varies depending on the MaSp1 and MaSp2 spidroin content. For instance, Nephila claviceps has 81% MaSp1 and 19% MaSp2, while Argiope aurantia has 41% MaSp1 and 59% MaSp2 [7].

Table 1. Different types of spider silk (fibers).

Major Ampullate Silk	Major ampullate silk is produced in the main ampullary glands. These fibers serve to allow escape from predators. Also, they are used for the web's outer rim and spokes. In this way, the other threads can be attached to them. They have a strength five times greater than steel and three times greater than Kevlar [59].
Minor Ampullate Silk	Minor ampullate silk is produced in the secondary ampullary glands. It has a role in the spiral formation of the network. Unlike MA fibers, it does not contain proline. Also, it has a reduced content of glutamate [59].
Flagelliform Silk	Capture-spiral (flagelliform) silk is produced in the flagelliform glands. It is used for catching prey [59].
Tubiliform Silk	Tubiliform (cylindriform) silk is produced in the tubiliform (cylindriform) glands. It is used for protective egg sacs [59].
Aciniform Silk	Aciniform silk is produced in the aciniform glands. It is a wrapping silk usedfor the immobilization of prey [64].
Pyriform Silk	Pyriform silk is produced in the pyriform glands. It functions like a glue, and connects the web to different materials [65].
Aggregate Silk	Aggregate silk is made in the aggregate glands. It produces aqueous gluey substances, making the capture threads sticky [66].

Table 2. Function and composition of different types of spider silk obtained from [16] under a Creative Commons License Type, CC BY 4.0.

Glands	Type of Spider Silk	Composition
Aggregate	Aqueous cement	ASG1, ASG2
Pyriform	Core fiber of capture spiral	PySp1, PySp2
Tubuliform	Egg-case silk	TuSp1, ECP-1, ECP-2
Flagelliform	Spiral silk	Flag
Aciniform	Capture silk	AcSp1
Minor ampullate	Dragline silk, framework silk	MiSp1, MiSp2
Major ampullate	Dragline silk, framework silk, radial silk	MaSp1, MaSp2

Table 3. Comparative presentation of the mechanical properties of spider silk and other fibers.

Material	Tensile Strength (Mpa)	Elongation (%)	Toughness (kJ/kg)
Dragline (MA) silk	4000	35	400
Silkworm silk	600	20	60
Kevlar 49	3600	5	30
Ligament	150	5	5
Bone	160	3	3

Spiders produce silk naturally in the form of fibers. However, silk produced through recombinant techniques involves the extraction of spider silk proteins in the form of powder. This technique allows for the combination of spider silk proteins with various materials to create fibers with different mechanical and structural properties. Although using expression systems makes recombinant production cost-effective, the process of purifying spider silk protein powder is both time-consuming and expensive [67].

4. Recombinant Production of Spider Silk

Biotechnological production has opened new approaches to produce spider silk proteins from other sources like bacteria, plants, yeasts, cells, or animals, to provide cost-efficient and stable fabrication [68]. The most commonly used proteins are derived from sequences isolated from the species *Nephila clavipes* and *Araneus diadematus* [13].

Recombinant protein production involves the following steps [69]:

- Determining the sequence of nucleotides in natural DNA (isolation of the desired sequence that encodes the target protein);
- Designing recombinant DNA;
- Selection of the vector that will enable the transmission of the desired sequence;
- Transmission of the vector into the host's organism (bacteria, yeast, plants, insect cells, mammalian cells, and transgenic animals);
- Cultivation/production of proteins in the host organism;
- Isolation of the obtained proteins.

A schematic representation of the recombinant spider silk proteins' production, sources, and possible biomaterials is shown in Figure 2.

Organisms used as hosts for silk protein fabrication can be different and the most commonly used hosts, as shown in Figure 2, are bacteria like *Escherichia coli (E. coli)*, yeasts like *Pichia pastoris*, and mammalian cells like hamster kidney cells, but insect cells can also be used, like *Spodoptera frugiperda*. *E. coli* proved to be the most suitable host due to a high density of cells that grow fast and can be easily transformed. Biotechnological manipulation and production allow the modification and improvement of silk characteristics [70].

Significant progress has been made in recombinant spider silk production, but further improvements are necessary to overcome the main challenges of high investments and small product yields [67]. Additionally, the commercial use of recombinant spider silk has been limited due to the inability to produce spidroins at their natural size. For example, transgenic mammals and insects have the potential to produce larger proteins, which are easier to purify. However, growing such organisms is costly, and the yields are typically low. On the other hand, *E. coli* is relatively affordable to cultivate, but it is not efficient in expressing larger spidroins [71].

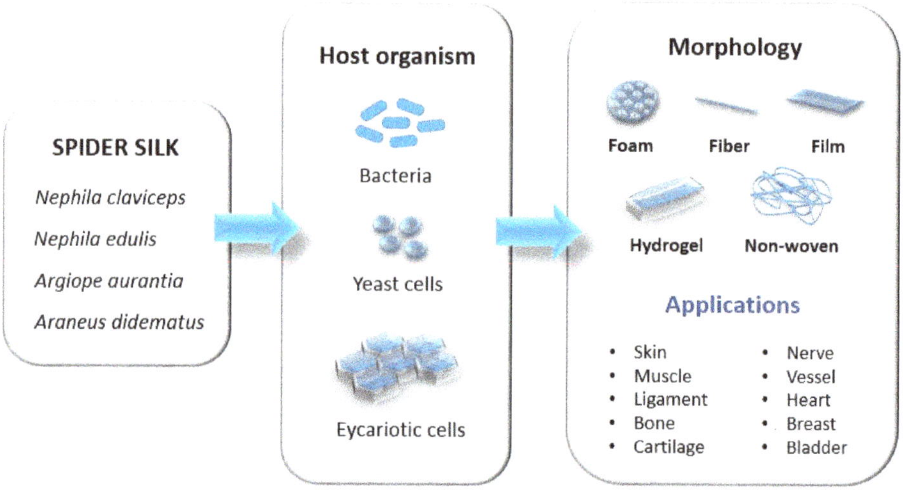

Figure 2. Production of recombinant spider silk proteins, sources, and possible biomaterials. Adapted from "Spider Silk for Tissue Engineering Applications" under Creative Common CC BY 4.0 license; the original work can be found at [13].

It is a common misunderstanding that the choice of host organism for producing recombinant spider silk only depends on the intended application. Advancements in bioinformatics have facilitated a more thorough examination of the distinct repetitive segments that significantly impact the structure, properties, and function of spidroin. These insights can enhance the efficiency of current host platforms and stimulate the development of novel production techniques to cater to specific production requirements [71].

Producing spider silk proteins using bacteria is a challenging task due to the high molecular weight of the proteins and their long repetitive regions, containing high levels of glycine and alanine. However, recent studies using *E. coli* have tried to overcome this issue by using alternative methods that involve the intein system to assemble protein subunits. This has led to the creation of chimeric fibers with impressive mechanical properties [72,73]. Additionally, other bacteria like *Corynebacterium glutamicum* [74], *Salmonella typhimurium* [75], and *Rhodovulum sulfidophilum* [76] have also been studied for their potential benefits in recombinant production. Among yeasts, *Pichia pastoris* [77] and *Saccharomyces cerevisiae* [78] successfully expressed these proteins.

It is not practical to use cell cultures from mammals or insects to produce structural proteins. However, scientists have had some success in expressing spider silk protein in certain types of cells, such as bovine mammary epithelial cells, hamster kidney cells [79], African green monkey kidney cells [80], and silkworm neuronal cells [81].

The use of transgenic plants represents an economical approach to produce large recombinant proteins, but in low yields [82]. In this context, various plants have been studied for the production of recombinant silk proteins, such as the species *Nicotiana tabacum* [83,84], *Solanum tuberosum* [85], *Medicago sativa* [86], and representatives of the genus Arabidopsis [87]. Also, transgenic animals have played a significant role in the production of recombinant spider silk through genetic engineering. Among them, transgenic silkworms have proven to be the most effective due to their ability to spin fibers [88]. Transgenic goats produced milk containing recombinant spider silk proteins, up to 0.5 g/L. Purified protein powder was obtained from such milk using special filtration techniques [89]. However, attempts to express recombinant spider silk proteins in sheep hair follicles have been unsuccessful [90].

5. Spider Silk Structures

Tissue regeneration is the process of tissue renewal and regrowth. Biomaterials, together with cells and bioactive factors, are the "building blocks" for making structures that resemble living tissues, through their combination in specific conditions. Appropriate substrates and scaffolds for tissue engineering must support the activity of cells in terms of their adhesion, migration, proliferation, and differentiation. Today, three-dimensional porous systems are the most adequate for cellular nutrition, respiration, and metabolism [13]. The level of porosity and the size of the pores has critical influence on the formation of the bone and bone regeneration [91]. The degradation of scaffolds over time allows space for the growth of new tissue, while adequate scaffold strength supports already-existing tissues [13].

5.1. Spider Silk in Fiber form (1D)

In Kuhbier et al.'s (2010) research, the process for obtaining cobweb fibers is detailed [92]. Once collected, the fibers are combined into bundles of 60–120 individual fibers and stored on large polypropylene tubes. These tubes are kept in an environment suitable for spiders for 6 months. After this period, spider silk sutures are produced using a miniature knitting machine and stored until biomechanical testing is conducted [93]. Sutures are commonly used to repair musculoskeletal tissue and therefore need to be able to withstand continuous mechanical stress. To test this, spider silk sutures and commercial Prolene®6-0 sutures (ProNorth Medical, online store, Canada) were subjected to 1000 stretching cycles. The study results revealed that spider silk was not affected by continuous use, while the failure load of Prolene®6-0 was significantly reduced. Furthermore, Prolene®showed a 24% ± 1.9 increase in strain, while spider silk only showed an increase of 7.2% ± 0.48 [94].

Spider silk fibers are attractive due to their mechanical strength and stability, biocompatibility, and good surface-to-volume ratio. On that note, MA silk has been investigated as a potential treatment agent for tendon ruptures [12,23,94]. Using ordinary sutures, the rate of successful regeneration of tendons was low, and an infection or reaction of the organism was often caused, so there is an incompatibility with the mechanical properties of the tendon tissue [13].

In another study, parallel silk fibers were studied as a substrate for developing human neurons. The neuronal cell bodies came into contact with the spider silk fibers, and over four weeks, ganglion-like structures formed [95].

Also, the woven spider silk was used in skin reconstruction. After sterilization, two weeks of fibroblast cultivation, and the addition of keratinocytes, a two-layer skin model was formed [96]. Furthermore, these fibers have been tested in preclinical models related to the reconstruction of the bladder [97].

5.2. Spider Silk Coatings (2D)

Spider silk coatings have been studied for different biomaterial applications, especially biological response, physicochemical characterization, and parameters that determine the final coating properties [98]. These coatings can be customized from aspects of different properties aiming to support better scaffolds in tissue engineering and natural-based materials as coatings on implants, but also for the development of biosensors [99] and to serve in surface functionalization for bioactive materials [100,101]. The design of new thin films based on spider silk showed possibilities of tailoring morphologies and hydrophobicity, as very important properties of the biomaterial surface, thus opening wide application areas for silk-based coatings and thin films [102]. Drug delivery systems can also utilize the possibility to customize spider-silk-coating properties [103].

The use of a coating is commonly considered to be a surface modification technique in the case of issues with surface responses. For example, the surface of a silicone breast implant can be coated with a thin film of spider silk to prevent fibrous tissue formation, which is a common issue [104]. Furthermore, significant improvement in the biocompatibility of the implant was observed, as well as reduced postoperative inflammation [104]. Films are made by dipping the test samples into a silk protein solution three times for 120 s. After each dip, the samples are left to dry for 300 s at room temperature. The formation of the β-sheet is induced by treating the films with KH2PO4 (1M solution) for 120 s, followed by air drying for 120 s. Finally, all test samples are rinsed with a 0.9% w/v NaCl solution [104]. During the initial stage of inflammation, certain types of cells such as CD4+, CD8+, CD68+, and TGFß1+ cells, along with pro-inflammatory cytokines IL-6 and TNF-a, are likely to appear. Compared to uncoated implants, silicone implants coated with spider silk showed a significantly lower expression of all the mentioned factors [104]. Spider silk coatings on different polymer catheters showed low cell adhesion, and almost no response from the surrounding tissue, and with good biocompatibility this can be a good coating material for catheters [105]. Silk structures have inspired the design of thin films to serve as bioelectronic interfaces (interfaces between tissue and electronics), as a very significant element for further development of flexible bioelectronics, including shape-adaptive biomaterials [106].

5.3. Three-Dimensional Constructs

Porous biomaterials have been proven to be the best candidates for guided cell growth and the better acceptance of implants by the body, or for tissue engineering and bone regeneration [91]. A porous foam made from recombinant protein spider silk pNSR-16, with a pore size of 250~350 nm, showed that fibroblasts form cell-rich zones on the surface and inside the structure [107]. To create the foams, a mixture of spider silk solution and granular NaCl (used as a porogen) was prepared and placed in a container. The mixture was heated in an oven at a temperature range of 55~60 °C for 30 min. As a result, NaCl-silk protein blocks were obtained. These blocks were first soaked in ethanol to induce a β-sheet structural transition and then in distilled water to extract the salt [107]. The cytotoxicity of the scaffold was tested in vitro on NIH-3T3 cells. Within six days, the entire surface of the scaffold was completely covered with cells, with very little space between them. Each subsequent day, the number of cells increased, clearly confirming that NIH-3T3 cells can easily attach, grow unhindered, and secrete the extracellular matrix on the pNSR-16 recombinant spider silk protein scaffold [107]. Likewise, the foam made of recombinant protein 4RepCT enabled the human stem cells to integrate and deploy [13]. The differentiation of human mesenchymal cells into adipocyte lineage was also studied using porous foams. This research has shown positive results because of lipid droplets commonly found in adipocytes (fat cells) [108]. Resorbable membranes have been studied for controlled bone regeneration [109].

In the tissue engineering of soft tissues, scaffolds need to support the surrounding tissue with adequate mechanical strength to enable cell growth and proliferation, and in cases of biodegradable implants, the rate of biodegradation needs to be controlled in accordance with the surrounding tissues. Porosity and other properties of the silk-based foams can be tailored to desired properties according to the surrounding tissues, even up to highly porous scaffolds [110]. Porous materials with a high toughness of incorporated fibers can absorb a high quantity of mechanical energy [18], which recommends them for the construction of bone tissue scaffolds.

Hydrogels are polymer networks with a water content above 95% and a high swelling rate [111]. Spider silk hydrogels can be used in a biofabrication combined with living cells to create a hierarchical tissue-like structure [112]. Cytocompatible bioinks suitable for cells and 3D printing are currently challenging to develop [113]. In this regard, hydrogels from recombinant spider silk proteins have shown promising properties [13].

The protein eADF4(C16) is an engineered version of the spider silk protein that mimics the repetitive part of the dragline silk fibroin ADF4 found in the garden cross spider

(*Araneus diadematus*). The optimal method for creating hydrogels from eADF4(C16) involves the dialysis of low-concentration protein solutions out of 6 M guanidinium thiocyanate (GdmSCN) into 10 mM Tris/HCl (pH of 7.5), followed by dialysis against a PEG solution. Hydrogels obtained in this way are often combined with living cells, such as fibroblasts, to create tissue-like structures [13].

Antimicrobial hydrogels based on hyaluronic acid and spider silk (HA/Ss) are one more interesting example. These hydrogels are made by dissolving hyaluronic acid in 0.1 mol/L MES (4-(N-morpholino)ethanesulfonic acid) and mixing it with spider silk dissolved in trifluoroacetic acid (TFA). The resulting HA/Ss mixture is then polymerized using 100 mM NHS (N-Hydroxysuccinimide) and 100 mM EDC [1-ethyl-3-(3-dimethylaminopropyl) carbodiimide hydrochloride]) Exceptional antimicrobial activity against both Gram-positive (*Micrococcus sulfuricum*) and Gram-negative *E. coli* bacteria is due to the presence of hyaluronic acid, which inhibits protein synthesis, and spider silk, which prevents bacterial adhesion [114]. Antimicrobial properties of spider silk are shown in Figure 3.

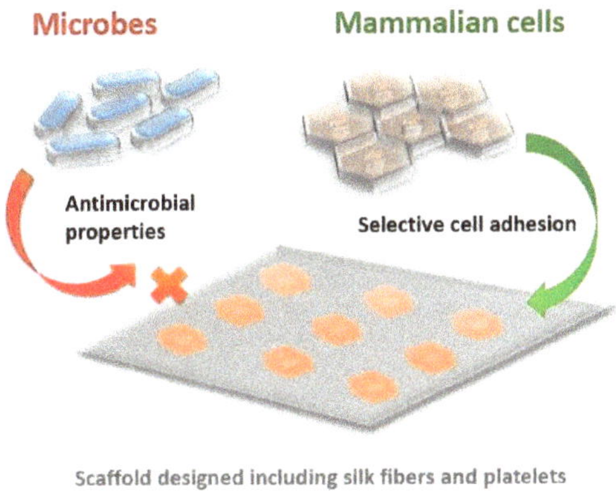

Figure 3. Antimicrobial properties of spider silk.

Hydrogel scaffolds can be customized to enable the tailoring and control of cell migration, proliferation, and differentiation by controlling the transport of nutrients and oxygen and by mimicking the surrounding natural tissue [115]. A comprehensive review of hydrogels with silk fibroin aimed for tissue engineering and drug delivery systems is given in [116,117]. Hydrogels with a combination of silk fibroin with other bioactive compounds like alginate have shown promising properties for wound healing [118]. Hydrogel-based embolic agents for transcatheter arterial embolization showed very good results in endovascular embolization [119]. The sincorporation of smart properties has provoked research interest in making hydrogels that can sense external stimulus, and hence can be used in medical diagnostics [120] or drug delivery systems [121].

Considering the development of advanced technologies, such as the 3D printing of gels [122], possibilities for customized solutions in which silk fibroin can be modified and combined with other materials to form printable hydrogels [123] and functional silk protein hydrogels [124] are realistically wide-ranging.

Hydrogels with silk can be utilized for drug delivery in osteoarthritis treatment [125,126] or can act as self-healing scaffolds for bone regeneration, through the filling of bone defects [127]. Self-healing silk hydrogel can also be used as a flexible strain sensor with self-adhesive property [128] that can be used for the development of wearables, with immense significance, as well, for medical diagnostics. In cartilage tissue engineering, biomimetic hydrogels have shown excellent properties for constructing biomimetic scaffolds that support regeneration, but they still have challenges regarding mechanical properties [129]. Recent research has studied possibilities of injectable composite hydrogels with silk fibroin in spinal treatments [130,131].

Recent advances in additive manufacturing and the availability of stem cells have opened up exciting possibilities for the development of bionic organs, including tissue-on-chip and organ-on-chip designs [132]. Organ-on-chip, as shown in Figure 4, contains microdevices with cells, microfluidics, and extracellular matrix scaffolds and can replicate microtissue and associated physiological processes, making it an ideal substitute for animal models in preclinical trials [133]. Several such chips have already been designed, including the 3D-printed ACL-ON-CHIP [134], which is used for the precise engineering of ligaments and their surrounding environment. Designing these micro-devices is important to understand the in vivo responses of silk proteins [135].

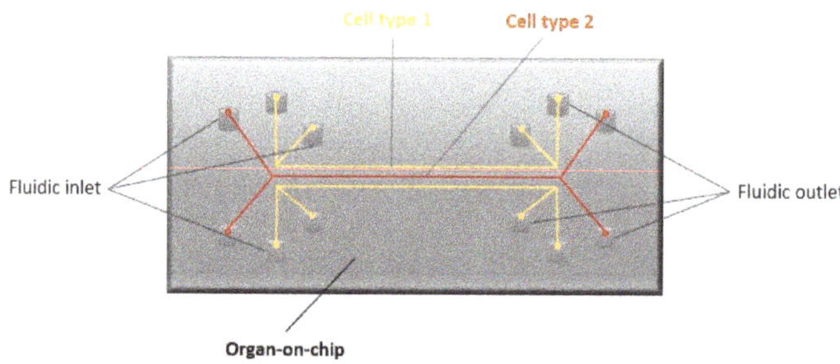

Figure 4. Organ-on-chip.

6. Applications in Tissue Engineering

6.1. Skin Regeneration

Engineered scaffolds of connecting tissues and skin have greatly benefited from all the excellent properties of silk-based materials, including the development of wound dressings with silk compounds [3,5,11]. One study presented the application of materials from recombinant spider silk for wound dressings on models of rats. The experiment used silk from pNSR-16 and pNSR-32 protein in second-degree burns. Interestingly, the materials from recombinant proteins of spider silk showed much better results than the control group (collagen), related to the skin regeneration [136]. Compared to the positive and negative control groups, wound healing was significantly faster in the group treated with recombinant spider silk proteins ($p < 0.01$). On the 14th day after treatment, in the sample with pNSR-16 and pNSR-32, the wound tissue was regenerated by newly formed, densely distributed epidermal cells, with a thickening of the subcutaneous tissue and a decrease in inflammatory cells. On the other hand, control groups did not show complete recovery even after the 21st day [136]. To improve cell adhesion, scientists modified silk with fibronectin [13]. In the case of third-degree burns in the animal model, the microporous dressing showed accelerated healing compared to a commercially available DuoDERM patch. Furthermore, histological analyses confirmed that wound healing in animals was accompanied by vascularization [13].

6.2. Bone and Cartilage Tissue Repair

The bone naturally consists of inorganic (predominantly calcium-phosphate) and organic substances (predominantly collagen). Accordingly, materials made from recombinant spider silk proteins can be biomineralized and used in bone regeneration [32]. Silk fibroin can be applied for bone tissue growth, proved by both in vitro and in vivo tests [8]. The MaSp1 class of proteins combined with the BSP fusion protein induce calcium-phosphate deposition but also a good adhesion of mesenchymal stem cells (it significantly contributed to their differentiation) and the noticeable synthesis of type 2 collagen in cartilage cells (Figure 5) [137].

Currently used bone grafts are commonly positioned to support bone healing in the case of large fractures, such as the one shown in Figure 6, or to provide the additional stabilization of spinal disks in case of spine fractures, as shown in Figure 7. For multifragmented fractures (Figure 6), bone grafts must be applied to support the bone healing process and the use of advanced silk-based scaffolds should enable active bone grafting, even though it is not part of clinical practice yet. Better fusion provided by the bone grafts for complex spinal fractures (Figure 7) also needs bioactive materials, and this is yet another example where silk-based scaffolding would significantly assist in medical treatments. However, clinical applications of silk-based scaffolding are still underway and need clinical trials before official approval from regulatory bodies.

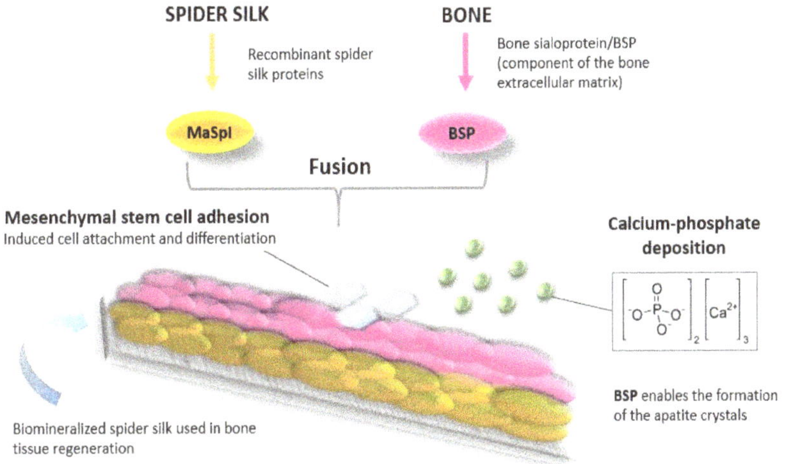

Figure 5. Use of spider silk protein (major ampullate spidroin protein) in the fabrication of scaffolds for bone tissue regeneration.

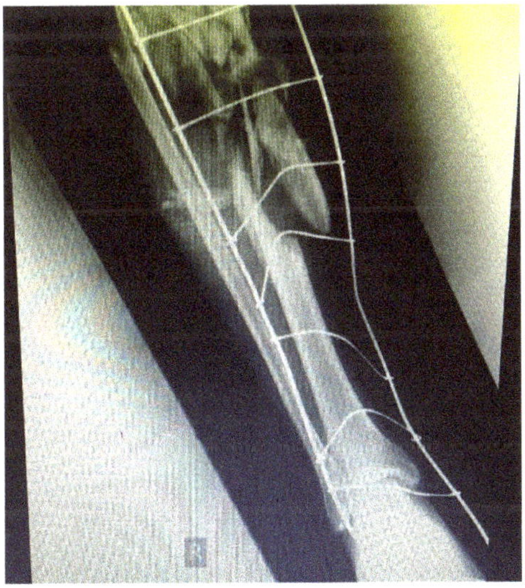

(a)

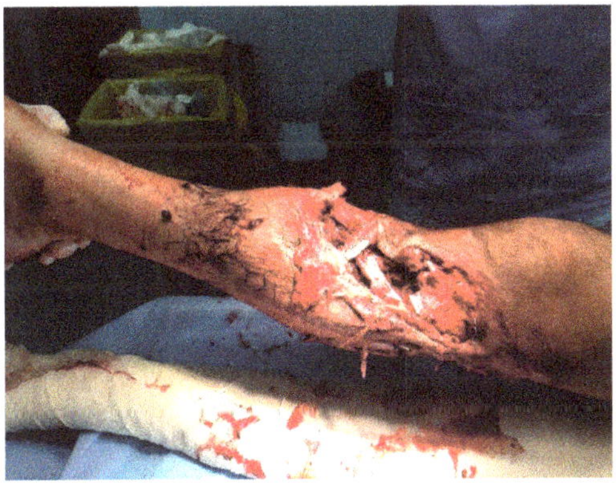

(b)

Figure 6. Open multifragmented fracture of the lower leg. Bone graft was applied. (**a**) radiography image of the fracture; (**b**) real image of the complex fracture before the surgical procedure.

The results showed that mineralized silk still has a performance comparable to many natural and artificial fibers [138]. Moreover, the hybrid composite exhibited mechanical properties similar to bone in strength and modulus of elasticity. This makes silk superior to many other biomaterials used for tissue engineering in bones. Fine-tuning of the biomineralization parameters led to the controlled incorporation of hydroxyapatite onto native spider silk, maintaining good mechanical properties [138]. Natural bio-based nanomaterials can provide outstanding scaffold properties for bone tissue engineering [139] and cartilage regeneration [9,10].

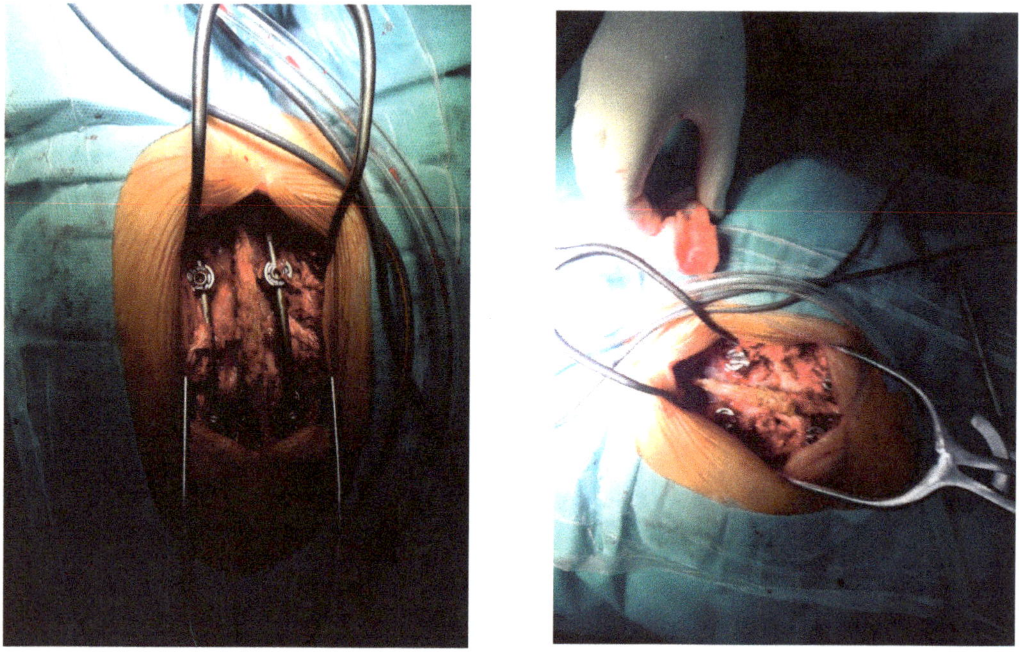

(**a**) (**b**)

Figure 7. Intraoperative image after stabilization of spinal fracture. After positioning of the transpedicular screws, bone grafts are applied for better final fusion. (**a**,**b**) two views of the stabilization method during the surgical procedure, with prepared bone graft shown in the right image.

6.3. *Vascularization*

The effectiveness of recombinant scaffolds and spider silk foams was even proved in the vascularization process, when a grouping of endothelial cells formed millimeter-branched beginnings of future blood vessels [108]. Considering this, the spider silk matrix has been used to develop artificial blood vessels [140]. The biomimetic design of structures with silk nanofibers has enabled constructs of vascular grafts with porous structures for blood vessel regeneration [141,142] or for small vascular grafts [143,144].

6.4. *Ligament Repair*

Advances in bioengineering research and orthopedic surgery have provided alternative solutions for the regeneration of musculoskeletal tissues of the human body [12,23,24]. Research shows that spider silk can withstand stresses similar to the ligament itself and even higher, which makes this biomaterial a good alternative for anterior cruciate ligament (ACL) of the knee replacement [134,145]. A biomechanical comparison was performed between an ACL composed of collagen fibers and an artificial spider silk construct. The ANSIS software was used to simulate the forces that occur during the anatomical movements of the ACL. The analysis showed that spider silk tissue outperforms ACL, thanks to its exceptional mechanical resistance, as it can withstand stresses of up to 2.5 MPa [145].

However, only experimental research still exists for the use of fibroin in ACL grafts, without clinical trials [25], even though research results are very promising [146]. Scaffolds made of silk that exhibit osteogenic function, interference screws, and tunnel fillers have been prototyped, with further research related to the stimulation of signaling pathways [25].

6.5. Muscle Tissue Repair

Musculoskeletal tissue engineering has benefitted from silk-based materials [12,23]. Nanofibrous scaffolds made from a mixture of silk, PLA, and collagen induce enhanced the adhesion, proliferation, and maturation of myoblasts. Recombinant spider silk proteins have been constructed in film form to examine their effect on rat cardiomyocytes. Cardiomyocytes cultured on eADF4 (k16) films successfully responded to extracellular stimuli and properly propagated electrical impulses. Apart from the fact that these films enable the adhesion of the most important types of cells in the heart tissue, they are non-cytotoxic and do not cause pharmacological or hypertrophic effects [147]. In addition, cardiomyocytes exhibited a higher expression of connexin 43, the protein responsible for the propagation of electrical impulses between cells [147]. It is challenging to design fully biomimetic muscle tissue together with the alignment of cells and adequate tissue responses. Nanofibers with silk fibroin have been studied for simple biomimetic skeletal muscle structures [148]. Conductive biomaterials can more efficiently promote muscle tissue growth [128,149,150]. Composite structures and blends, such as combinations of silk fibroin and gelatine, have been studied for skeletal muscle tissue engineering, including the design of structures that can be further used in flexible electronics and medical diagnostics [151].

6.6. Repair of Peripheral Nerves

Spider silk has a proven capacity to guide cell proliferation and migration and enhance peripheral nerve regeneration [152]. Axon regeneration in peripheral nerve damage can be promoted by implanting specific biodegradable guidance channels which are able to guide the cells while present. Spider silk fibers were suitable for the human neuron culture and a study showed their very good adhesion, cell body migration, differentiation, and neurite (axon) extension, resembling ganglion structures [63]. After ten months, axons were regenerated with the presence of myelination, thus indicating that Schwann cells migrated through the constructs (Figure 8). Moreover, spider silk impregnated with collagen fibers is successful in cell differentiation and neural network formation. Neuronal cells were fully capable of activating action potentials, and in them, there was an increased expression of SNAP-25 protein, which is an indicator of the existence of functional synapses [153].

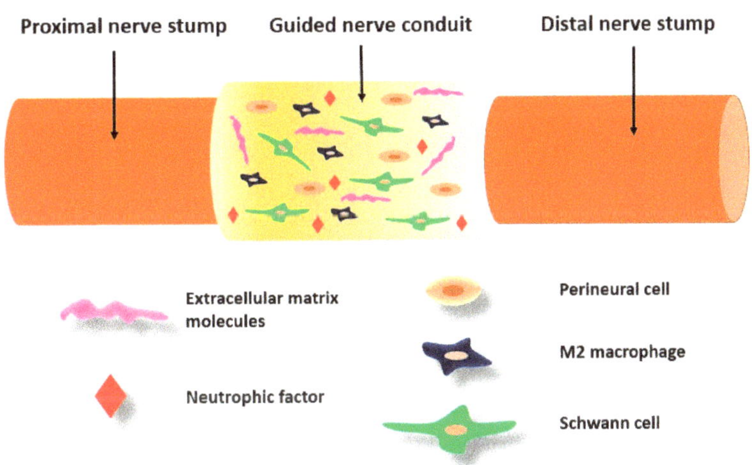

Figure 8. Peripheral nerve tissue engineering.

In addition, cultured neural progenitor cells, which have the potential to transform into neurons, astrocytes, and oligodendrocytes, showed successful differentiation on recombinant spider silk [154]. The most significant limiting factor for regeneration is the length of the gap. However, these nerve channels have been successfully applied in the peripheral nervous systems of animals [63].

7. Spider Silk in Nanomedicine

Unlike silkworms, the breeding of spiders is quite limited due to their cannibalistic and territorial nature. Although considerable efforts have been made to produce artificial recombinant spider silk, insufficient understanding of its nanostructure has prevented commercial success [155]. Nanomaterials are natural, accidentally obtained, or manufactured materials ranging in size from 1 nm to 100 nm, and the smallest changes at the nano level result in dramatic changes in the macro world [156].

Nanomedicine has emerged, offering potential solutions to the treatment of different conditions in relation to using silk-based materials, such as osteoarthritis treatment [129], or Achilles tendinopathy [157]. Research with gold nanoparticles in silk hydrogel used as a media for the laser treatment of subcutaneous bacterial abscesses has shown promising results for this complex medical condition [158].

7.1. Drug Delivery Systems

Soft porous natural materials represent very suitable materials for drug delivery systems, and silk-based biomaterials have shown different possibilities for tailoring such systems [3,91,103,109,125], including in the scope of bone tissue engineering [8]. Silk fibroin has shown good adjustability to suit different drug delivery systems, including hydrogels [116,117,121] or coatings and thin films [103]. Bioactivity and in vivo responses of silk proteins are important properties of the silk-based materials, utilized in this sense for tissue growth and remodeling [29,135]. It is also important that silk fibers can be used in the design of nanomaterials for drug delivery systems [159].

As is known, the biggest obstacle for an anticancer drug is controlled release at the target destination. Considering the characteristics of this material, the use of silk products (such as films, hydrogels, capsules, or liposomes coated with silk proteins) can potentially overcome that problem (Figure 9) [7].

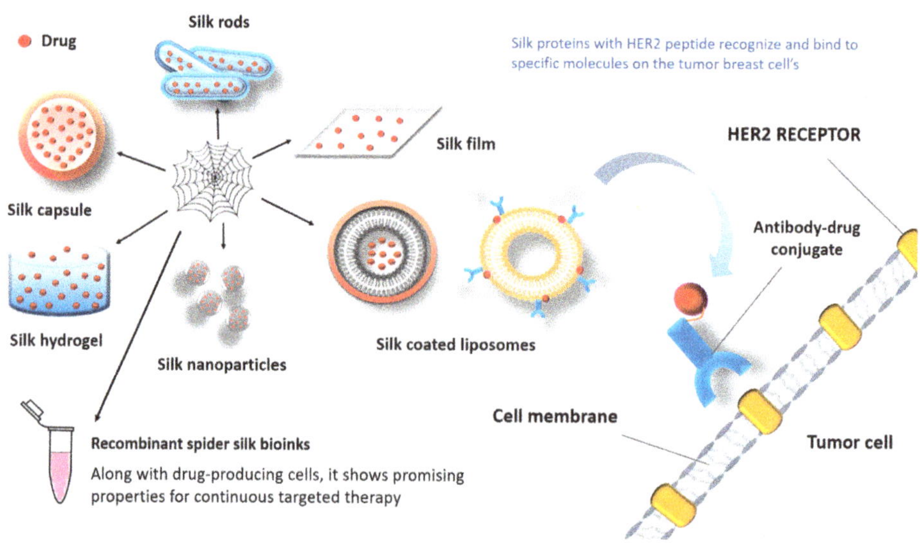

Figure 9. Silk-based biomaterials for chemotherapeutic delivery.

Although surgery, radiotherapy, and chemotherapy have achieved substantial progress in cancer therapy, recent research has been focused on systems for the targeted delivery of drugs (therapeutics, hormones, inorganic nanoparticles, etc.) [160]. Silk represents an excellent tool in cancer therapy due to numerous advantages, including biocompatibility, biodegradability, the possibility of varied shaping, storing in a dried state [161], and the absence of systemic toxicity [162]. The basis of all systems for targeted drug delivery are nanomaterials constructed in the form of liposomes, micelles, dendrimers, and nanoparticles [163]. This type of treatment ensures a remarkably higher concentration of active substances in tumor tissues [164]. Moreover, these systems are naturally removed from the body due to enzymatic degradation (without harmful byproducts) [165]. The smaller the proportion of the crystalline phase (β-sheet), the faster the degradation occurs [166].

Increased selectivity is achieved by combining silk proteins with peptides (e.g., F3, Lyp1, CGKRK, or HER2 peptide) that recognize and bind to specific molecules on the surface of tumor cells (Figure 9) [167–169]. Since silk is hydrophobic, hydrophobic drugs give better results in combination with this biomaterial [170]. Also, negatively charged molecules are released faster than positively charged molecules [171].

Silk-based drug delivery systems can be applied locally or systemically [172]. For the needs of local delivery, various two-dimensional and three-dimensional systems are used. Two-dimensional structures include thin films, and coatings, made of silk fibers, while three-dimensional implants include hydrogels, foams, and porous scaffolds. Most of these structures have demonstrated the ability of sustained drug release over four weeks [172], inhibiting tumor growth in vivo [173]. Local delivery also involves transdermal methods that include microneedles for non-invasive and painless drug release.

This approach allows drug release by swelling and dissolution after passing through the skin [174]. Silk fibroin microneedles have been developed as electro-responsive material for specific drug delivery (insulin) [175]. Hydrogels that can have bioactive responses to external stimulus have become of great significance for the development of drug delivery systems and medical diagnostics [121].

Systemic drug delivery involves capsules [176], spheres [177], and particles containing active substances, which are released by diffusion or degradation of the material. Nanosized systems can reach the smallest capillaries and then be incorporated into cells through physiological barriers, which is crucial in cancer therapy. With this in mind, silk-based nanoparticles have been designed to deliver chemotherapeutics in tumor tissues [178].

Some of the techniques for producing silk nanoparticles include:

- Desolvation [179];
- Electrospraying [180];
- Ionic liquids [181];
- Laminar jet break-up [182];
- Microemulsion [183];
- Microfluidics [184];
- Milling technologies [185];
- Salting out [186];
- Self-aggregation [187];
- Sol-gel techniques [188];
- Supercritical fluids [189].

One way of controlling drug release using spider silk is thanks to pH-dependent carriers. Specifically, silk in combination with iron oxide nanoparticles [190] results in a limited release of the drug into the blood, that is, an increased release of the drug in the tumor tissue, which represents a more acidic environment (due to increased metabolic activity, more glucose brakes down and a large amount of lactic acid is produced) [191]. Moreover, tumor tissues are not homogeneous (they consist of different types of cells), so drug carriers can be modified to target the cells of the tumor microenvironment [172].

Septic arthritis is a medical condition that causes inflammation in the joints, bones, and cartilage. It is caused by a type of bacteria called *Staphylococcus aureus*, and currently available antibiotics are becoming less effective due to bacterial resistance [192]. However, scientists have successfully created a conjugate using spider silk and a thrombin-sensitive peptide (TSP) to deliver the antibiotic vancomycin directly to the affected area. Spider silk was dissolved in a MES (4-(N-morpholino)ethanesulfonic acid) reaction buffer. Then, 100 mM EDC [1-ethyl-3-(3-dimethylaminopropyl) carbodiimide hydrochloride] and 200 mM NHS (N-Hydroxysuccinimide) were added to activate it, which enabled the chemical binding of TSP. Finally, the encapsulation of vancomycin was carried out using the method described in the paper available at [193].

This conjugate has the potential to trigger drug release in the presence of specific enzymes produced by *Staphylococcus aureus* bacteria. Drug release from the conjugate particles amounted to 84.4% after 24 h of incubation, while plain spider silk particles only released 15.6% of the drug. Additionally, bacterial cultures were obtained from rat knee joint synovial fluid. The bacterial culture treated with the conjugate particles had an average of 40 CFU/mL, whereas the culture treated with plain silk had an average of 810 CFU/mL. This confirms the effectiveness of the conjugate particles in triggering drug release in the presence of infection [194].

7.2. Nanocomposites and Biomimetics

With the advancement of technology, the production of nanocomposites and nanomaterials inspired by spider silk or using silk-based materials has been investigated through different approaches [106,195,196]. For instance, to create a material with high stiffness, strength, and toughness, scientists combined silk with nanocellulose. The results showed that this composite could replace plastic (ecological importance) and serve as a basis for fabric production, even in medical implants [197]. In addition, spider silk was reinforced with graphene microparticles and carbon nanotubes, resulting in the strongest known fiber [198]. However, not only the structure of individual spider silk fibers was considered. Here, the design of the entire spider's web served as a template for different devices and applications, which is the subject of research in biomimetics. Silk-based microspheres can be added to the cell cultures to enhance cell growth and adhesion [199]. The biomimetic approach in material design has opened up many new directions in material structures, as well as final applications [129,141] and including adjustments of existing fabrication technologies for silk fibers [30,200].

In bone tissue regeneration, the development of new nanocomposites, and especially those derived from nature, has opened new research directions [148,201]. Combination with different nanoparticles, such as silver nanoparticles, can enhance antibacterial effects [202]. The combination of titanium dioxide nanotubes as drug carriers with zeolite-based compounds and silk fibroin has shown promising properties for drug delivery systems [203]. Silk fibroin in composite structures can enable multifunctional drug delivery microcarriers [159,204].

Combinations of silk fibroin with hydroxyapatite and/or graphene oxide as nanocomposites have enabled the construction of porous scaffolds with very good mechanical properties and improved capabilities for bone tissue regeneration [205]. Nanocomposite structures have also been studied to enable the design of in vitro models for cancer treatments in bone tissues [206].

Spider silk, as one of the silk variations, has been used in many applications, including biomedical ones. It has been studied and analyzed, with clear potential proven in experimental lab studies, towards smart biomaterials and novel composite structures that can mimic natural tissues. However, clinical applications are rather limited to the specific areas of cosmetics, wound dressing, breast reconstruction, or certain other treatments, and it is almost without use in musculoskeletal tissue engineering. Common biomaterials considered for bone tissue scaffolds are focused on those that can primarily provide mechanical strength. New development directions towards smart or biodegradable structures should encounter less-acknowledged materials like silk-based ones for load bearing applications, especially considering that the tailoring of properties within composite structures can be achieved. In such applications, spider silk and silk fibroin can provide additional properties that are not fully exhibited with currently used biomaterials, beside a very important sustainability. However, production costs are still very high and challenging and further research focus should be directed toward discovering or upgrading fabrication procedures to enable the wider availability of spider silk.

8. Conclusions

This review presented the latest research related to the potential applications of spider silk in reconstructive and regenerative medicine and tissue engineering, including nanomedicine and drug delivery systems, with a focus on musculoskeletal tissues. Spider silk is a natural material with long-established use in different applications. Silk-based materials have emerged as sustainable, natural materials that can provide significant benefits in the development of advanced biomaterials for medical implants and tissue engineering. Bone and cartilage, muscle, and tendon tissue engineering, as well as the advanced design of skin and vascular tissues, can greatly benefit from bioactive and smart biomaterials, incorporating silk-based biomaterials. Silk proteins in bioactive materials that mimic the tissue structure can enhance tissue regeneration and growth. Silk proteins have shown good adjustability to suit different drug delivery systems.

Natural spider silk synthesis and the further recombinant production of spider silk proteins have been reviewed. Silk-based biomaterials used for tissue engineering applications (scaffolds, hydrogels, films, fibers, or nanoparticles in drug delivery) have been successfully designed and developed for various tissues such as bones, tendons, ligaments, skin, muscles, and nerves. However, the use of natural spider silk remains limited due to low production yields and difficulties in cultivation. Hence, the only practical solution is the production of recombinant spider silk proteins.

Preliminary research results and tissue engineering examples with silk-based materials showed no inflammatory reactions, but in vivo studies of spider silk-based materials remain limited. The degradation of fibers by macrophages confirmed the property of biodegradability, and consequently the newly grown tissue can replace the silk material. Furthermore, the by-products of silk degradation are non-toxic, which means they can be recognized and neutralized by the immune system, unlike many synthetic polymers. In this regard, another important characteristic is the rate of degradation. Spider silk materials remain mechanically stable for a significant time, without increase in brittleness or susceptibility to tearing under physiological conditions, which is a desired material property in neurological-related applications. Bioactive bone grafts with conductive properties for signaling pathways are the latest research direction promising great advancements.

Silk proteins are important in designing tissue-on-chip or organ-on-chip technologies and micro devices that have started to be used for the precise engineering of artificial tissues and organs, disease modeling, and the further selection of adequate medical treatments. Recent research indicates that silk (films, hydrogels, capsules, or liposomes coated with silk proteins) has the potential to provide controlled drug release at the target destination.

The specific properties of spider silk material show some clear advantages and disadvantages. However, a comprehensive understanding of its nanostructure and associated mechanisms in bioactive material systems is needed, as well as further clinical trials to gain approval for its use in tissue engineering.

Author Contributions: Conceptualization, M.B., F.Z. and N.G.; methodology, M.B., F.Z. and I.S.; validation, F.Z. and N.G.; formal analysis, M.B., F.Z. and N.G.; investigation, M.B., F.Z., N.G., I.S., S.M. and N.K.; resources, F.Z. and N.G.; data curation, M.B., F.Z., N.G., I.S., S.M. and N.K.; writing—original draft preparation, M.B. and F.Z.; writing—review and editing, M.B., F.Z., N.G., I.S., S.M. and N.K.; visualization, M.B., F.Z. and I.S.; supervision, F.Z. and N.G.; funding acquisition, F.Z. and N.G. All authors have read and agreed to the published version of the manuscript.

Funding: This paper is funded through the EIT's HEI Initiative DEEPTECH-2M project, supported by EIT Digital and coordinated by EIT RawMaterials, funded by the European Union.

Institutional Review Board Statement: Not applicable.

Data Availability Statement: No new data were created or analyzed in this study. Data sharing is not applicable to this article.

Conflicts of Interest: The authors declare no conflicts of interest. The funders had no role in the design of the study; in the collection, analyses, or interpretation of data; in the writing of the manuscript; or in the decision to publish the results.

References

1. Miserez, A.; Yu, J.; Mohammadi, P. Protein-Based Biological Materials: Molecular Design and Artificial Production. *Chem. Rev.* **2023**, *123*, 2049–2111. [CrossRef]
2. Beppu, M.M.; Lopes, L.M.; Queiroz, C.D.S.; Moraes, M.A.D. Silk-Based Natural Biomaterials: Fundamentals and Biomedical Applications. In *Handbook of Natural Polymers*; Elsevier: Amsterdam, The Netherlands, 2023; Volume 1, pp. 421–440. ISBN 978-0-323-99853-6.
3. Babu, P.J.; Suamte, L. Applications of Silk-Based Biomaterials in Biomedicine and Biotechnology. *Eng. Regen.* **2024**, *5*, 56–69. [CrossRef]
4. Bhatt, A.S.; Santhosh, G. Biopolymer Matrices: Current Trends in Tissue Engineering. In *Tailored Functional Materials for Clean and Sustainable Development*; Apple Academic Press: New York, NY, USA, 2023; pp. 139–155. ISBN 978-1-00-339476-1.
5. Dos Santos, F.V.; Siqueira, R.L.; De Morais Ramos, L.; Yoshioka, S.A.; Branciforti, M.C.; Correa, D.S. Silk Fibroin-Derived Electrospun Materials for Biomedical Applications: A Review. *Int. J. Biol. Macromol.* **2024**, *254*, 127641. [CrossRef]
6. Holland, C.; Numata, K.; Rnjak-Kovacina, J.; Seib, F.P. The Biomedical Use of Silk: Past, Present, Future. *Adv. Healthc. Mater.* **2019**, *8*, 1800465. [CrossRef]
7. Bandyopadhyay, A.; Chowdhury, S.K.; Dey, S.; Moses, J.C.; Mandal, B.B. Silk: A Promising Biomaterial Opening New Vistas towards Affordable Healthcare Solutions. *J. Indian Inst. Sci.* **2019**, *99*, 445–487. [CrossRef]
8. Melke, J.; Midha, S.; Ghosh, S.; Ito, K.; Hofmann, S. Silk Fibroin as Biomaterial for Bone Tissue Engineering. *Acta Biomater.* **2016**, *31*, 1–16. [CrossRef] [PubMed]
9. Zhou, Z.; Cui, J.; Wu, S.; Geng, Z.; Su, J. Silk Fibroin-Based Biomaterials for Cartilage/Osteochondral Repair. *Theranostics* **2022**, *12*, 5103–5124. [CrossRef] [PubMed]
10. Su, X.; Wei, L.; Xu, Z.; Qin, L.; Yang, J.; Zou, Y.; Zhao, C.; Chen, L.; Hu, N. Evaluation and Application of Silk Fibroin Based Biomaterials to Promote Cartilage Regeneration in Osteoarthritis Therapy. *Biomedicines* **2023**, *11*, 2244. [CrossRef]
11. Sun, W.; Gregory, D.A.; Tomeh, M.A.; Zhao, X. Silk Fibroin as a Functional Biomaterial for Tissue Engineering. *Int. J. Mol. Sci.* **2021**, *22*, 1499. [CrossRef] [PubMed]
12. Wu, R.; Li, H.; Yang, Y.; Zheng, Q.; Li, S.; Chen, Y. Bioactive Silk Fibroin-Based Hybrid Biomaterials for Musculoskeletal Engineering: Recent Progress and Perspectives. *ACS Appl. Bio Mater.* **2021**, *4*, 6630–6646. [CrossRef]
13. Salehi, S.; Koeck, K.; Scheibel, T. Spider Silk for Tissue Engineering Applications. *Molecules* **2020**, *25*, 737. [CrossRef]
14. Wang, Y.; Kim, H.-J.; Vunjak-Novakovic, G.; Kaplan, D.L. Stem Cell-Based Tissue Engineering with Silk Biomaterials. *Biomaterials* **2006**, *27*, 6064–6082. [CrossRef]
15. Goel, A. Surgical Sutures—A Review. *Off. Sci. J. Delhi Ophthalmol. Soc.* **2015**, *26*, 159–162. [CrossRef]
16. Gu, Y.; Yu, L.; Mou, J.; Wu, D.; Zhou, P.; Xu, M. Mechanical Properties and Application Analysis of Spider Silk Bionic Material. *e-Polymers* **2020**, *20*, 443–457. [CrossRef]
17. Altman, G.H.; Diaz, F.; Jakuba, C.; Calabro, T.; Horan, R.L.; Chen, J.; Lu, H.; Richmond, J.; Kaplan, D.L. Silk-Based Biomaterials. *Biomaterials* **2003**, *24*, 401–416. [CrossRef] [PubMed]
18. Hsia, Y.; Gnesa, E.; Tang, S.; Jeffery, F.; Geurts, P.; Zhao, L.; Franz, A.; Vierra, C. Identification and Synthesis of Novel Biomaterials Based on Spider Structural Silk Fibers. *Appl. Phys. A* **2011**, *105*, 301–309. [CrossRef]
19. Xu, G.; Toh, G.W.; Du, N.; Liu, X.Y. Spider Silk: The Toughest Natural Polymer. In *Natural Polymers. 1: Composites*; RSC Green Chemistry; RSC Publishing: Cambridge, UK, 2012; ISBN 978-1-84973-402-8.
20. DeFrancesco, L. Hanging on a Thread. *Nat. Biotechnol.* **2017**, *35*, 496–499. [CrossRef] [PubMed]
21. Aigner, T.; DeSimone, E.; Scheibel, T. Biomedical applications of recombinant silk-based materials. *Adv. Mater.* **2018**, *30*, e1704636. [CrossRef]
22. Foppiani, J.A.; Weidman, A.; Alvarez, A.H.; Valentine, L.; Devi, K.; Kaplan, D.L.; Lin, S.J. Clinical Use of Non-Suture Silk-Containing Products: A Systematic Review. *Biomimetics* **2023**, *8*, 45. [CrossRef]
23. Nie, K.; Zhou, S.; Li, H.; Tian, J.; Shen, W.; Huang, W. Advanced Silk Materials for Musculoskeletal Tissue Regeneration. *Front. Bioeng. Biotechnol.* **2023**, *11*, 1199507. [CrossRef]
24. Zhang, L.; Zhang, W.; Hu, Y.; Fei, Y.; Liu, H.; Huang, Z.; Wang, C.; Ruan, D.; Heng, B.C.; Chen, W.; et al. Systematic Review of Silk Scaffolds in Musculoskeletal Tissue Engineering Applications in the Recent Decade. *ACS Biomater. Sci. Eng.* **2021**, *7*, 817–840. [CrossRef]
25. Hahn, J.; Gögele, C.; Schulze-Tanzil, G. Could an Anterior Cruciate Ligament Be Tissue-Engineered from Silk? *Cells* **2023**, *12*, 2350. [CrossRef]

26. Croft, A.S.; Spessot, E.; Bhattacharjee, P.; Yang, Y.; Motta, A.; Wöltje, M.; Gantenbein, B. Biomedical Applications of Silk and Its Role for Intervertebral Disc Repair. *JOR Spine* **2022**, *5*, e1225. [CrossRef]
27. Dey, S.; Jaiswal, C.; Shome, S.; Bhar, B.; Bandyopadhyay, A.; Manikumar, K.; Dadheech, R.; Mandal, B.B. Photocrosslinkable Silk-Based Biomaterials for Regenerative Medicine and Healthcare Applications. *Regen. Eng. Transl. Med.* **2023**, *9*, 181–201. [CrossRef]
28. Intravaia, J.T.; Graham, T.; Kim, H.S.; Nanda, H.S.; Kumbar, S.G.; Nukavarapu, S.P. Smart Orthopedic Biomaterials and Implants. *Curr. Opin. Biomed. Eng.* **2023**, *25*, 100439. [CrossRef] [PubMed]
29. Stratton, S.; Shelke, N.B.; Hoshino, K.; Rudraiah, S.; Kumbar, S.G. Bioactive Polymeric Scaffolds for Tissue Engineering. *Bioact. Mater.* **2016**, *1*, 93–108. [CrossRef] [PubMed]
30. Khan, A.Q.; Shafiq, M.; Li, J.; Yu, K.; Liu, Z.; Zhou, X.; Zhu, M. Recent Developments in Artificial Spider Silk and Functional Gel Fibers. *SmartMat* **2023**, *4*, e1189. [CrossRef]
31. Schäfer, S.; Aavani, F.; Köpf, M.; Drinic, A.; Stürmer, E.K.; Fuest, S.; Grust, A.L.C.; Gosau, M.; Smeets, R. Silk Proteins in Reconstructive Surgery: Do They Possess an Inherent Antibacterial Activity? A Systematic Review. *Wound Repair Regen.* **2023**, *31*, 99–110. [CrossRef] [PubMed]
32. Ma, L.; Sun, Y.; Cheng, Q.; Yang, Z.; Wang, J.; Xu, Z.; Yang, M.; Shuai, Y. Silk Protein-Mediated Biomineralization: From Bioinspired Strategies and Advanced Functions to Biomedical Applications. *ACS Appl. Mater. Interfaces* **2023**, *15*, 33191–33206. [CrossRef] [PubMed]
33. Chen, W.; Wang, R.; Yu, S.; Chen, J.; Kao, Y.; Wang, T.; Chang, P.; Sheu, H.; Chen, S.; Liu, W.; et al. Self-Healable Spider Dragline Silk Materials. *Adv. Funct. Mater.* **2023**, *33*, 2303571. [CrossRef]
34. Jang, J.; Chae, S.; Yoon, J.; Kim, H.; Park, W. *Organ Printing*, 2nd ed.; IOP Publishing: Bristol, UK, 2023; ISBN 978-0-7503-5122-5.
35. Pedde, R.D.; Mirani, B.; Navaei, A.; Styan, T.; Wong, S.; Mehrali, M.; Thakur, A.; Mohtaram, N.K.; Bayati, A.; Dolatshahi-Pirouz, A.; et al. Emerging Biofabrication Strategies for Engineering Complex Tissue Constructs. *Adv. Mater.* **2017**, *29*, 1606061. [CrossRef] [PubMed]
36. Sultan, M.T.; Lee, O.J.; Lee, J.S.; Park, C.H. Three-Dimensional Digital Light-Processing Bioprinting Using Silk Fibroin-Based Bio-Ink: Recent Advancements in Biomedical Applications. *Biomedicines* **2022**, *10*, 3224. [CrossRef] [PubMed]
37. Wang, H.-Y.; Zhao, S.-X.; Li, J.-X.; Zhang, Y.-Q. Silk Fibroin Improves the Biological Properties of Egg White-Based Bioink for the Bioprinting of Tissue Engineering Materials. *ACS Omega* **2023**, *8*, 46685–46696. [CrossRef]
38. Ye, L.; Liu, X.; Li, K.; Li, X.; Zhu, J.; Yang, S.; Xu, L.; Yang, M.; Yan, Y.; Yan, J. A Bioinspired Synthetic Fused Protein Adhesive from Barnacle Cement and Spider Dragline for Potential Biomedical Materials. *Int. J. Biol. Macromol.* **2023**, *253*, 127125. [CrossRef] [PubMed]
39. Zhang, X.; Reagan, M.R.; Kaplan, D.L. Electrospun Silk Biomaterial Scaffolds for Regenerative Medicine. *Adv. Drug Deliv. Rev.* **2009**, *61*, 988–1006. [CrossRef]
40. Antezana, P.E.; Municoy, S.; Ostapchuk, G.; Catalano, P.N.; Hardy, J.G.; Evelson, P.A.; Orive, G.; Desimone, M.F. 4D Printing: The Development of Responsive Materials Using 3D-Printing Technology. *Pharmaceutics* **2023**, *15*, 2743. [CrossRef] [PubMed]
41. Xu, M.; Lewis, R. Structure of a protein superfiber: Spider dragline silk. *Proc. Natl. Acad. Sci. USA* **1990**, *87*, 7120–7124. [CrossRef] [PubMed]
42. Vollrath, F.; Knight, D. Liquid crystalline spinning of spider silk. *Nature* **2001**, *410*, 541–548. [CrossRef]
43. Jorge, I.; Ruiz, V.; Lavado-García, J.; Vázquez, J.; Hayashi, C.; Rojo, F.J.; Atienza, J.M.; Elices, M.; Guinea, G.V.; Pérez-Rigueiro, J. Expression of Spidroin Proteins in the Silk Glands of Golden Orb-weaver Spiders. *J. Exp. Zoolog. B Mol. Dev. Evol.* **2022**, *338*, 241–253. [CrossRef]
44. Babb, P.L.; Lahens, N.F.; Correa-Garhwal, S.M.; Nicholson, D.N.; Kim, E.J.; Hogenesch, J.B.; Kuntner, M.; Higgins, L.; Hayashi, C.Y.; Agnarsson, I.; et al. The Nephila Clavipes Genome Highlights the Diversity of Spider Silk Genes and Their Complex Expression. *Nat. Genet.* **2017**, *49*, 895–903. [CrossRef]
45. Elices, M.; Guinea, G.V.; Plaza, G.R.; Real, J.I.; Pérez-Rigueiro, J. Example of Microprocessing in a Natural Polymeric Fiber: Role of Reeling Stress in Spider Silk. *J. Mater. Res.* **2006**, *21*, 1931–1938. [CrossRef]
46. Du, N.; Liu, X.Y.; Narayanan, J.; Li, L.; Lim, M.L.; Li, D. Design of Superior Spider Silk: From Nanostructure to Mechanical Properties. *Biophys. J.* **2006**, *91*, 4528–4535. [CrossRef] [PubMed]
47. Kojic, N.; Bico, J.; Clasen, C.; McKinley, G.H. Ex Vivo Rheology of Spider Silk. *J. Exp. Biol.* **2006**, *209*, 4355–4362. [CrossRef] [PubMed]
48. Kojic, N.; Panzer, M.J.; Leisk, G.G.; Raja, W.K.; Kojic, M.; Kaplan, D.L. Ion Electrodiffusion Governs Silk Electrogelation. *Soft Matter* **2012**, *8*, 6897. [CrossRef]
49. Pérez-Rigueiro, J.; Elices, M.; Plaza, G.R.; Guinea, G.V. Basic Principles in the Design of Spider Silk Fibers. *Molecules* **2021**, *26*, 1794. [CrossRef]
50. Simmons, A.; Michal, C.; Jelinski, L. Molecular Orientation and Two-Component Nature of the Crystalline Fraction of Spider Dragline Silk. *Science* **1996**, *271*, 84–87. [CrossRef]
51. Keten, S.; Buehler, M. Nanostructure and molecular mechanics of spider dragline silk protein assemblies. *J. R. Soc. Interface* **2010**, *7*, 1709–1721. [CrossRef]
52. Work, R. Dimensions, Birefringences, and Force-Elongation Behavior of Major and Minor Ampullate Silk Fibers from Orb-Web-Spinning Spiders—The Effects of Wetting on These Properties. *Text. Res. J.* **1977**, *47*, 650–662. [CrossRef]

53. Elices, M.; Pérez-Rigueiro, J.; Plaza, G.; Guinea, G. Recovery in Spider Silk Fibers. *J. Appl. Polym. Sci.* **2004**, *92*, 3537–3541. [CrossRef]
54. Madurga, R.; Plaza, G.; Blackledge, T.; Guinea, G.; Elices, M.; Pérez-Rigueiro, J. Material Properties of Evolutionary Diverse Spider Silks Described by Variation in a Single Structural Parameter. *Sci. Rep.* **2016**, *6*, 18991. [CrossRef]
55. Gu, L.; Jiang, Y.; Hu, J. Structure Design and Property of Spider Silk-Inspired Shape Memory Materials. *Mater. Today Proc.* **2019**, *16*, 1491–1496. [CrossRef]
56. Blamires, S.; Lozano-Picazo, P.; Bruno, A.L.; Arnedo, M.; Ruiz-León, Y.; González-Nieto, D.; Rojo, F.J.; Elices, M.; Guinea, G.V.; Pérez-Rigueiro, J. The Spider Silk Standardization Initiative (S3I): A Powerful Tool to Harness Biological Variability and to Systematize the Characterization of Major Ampullate Silk Fibers Spun by Spiders from Suburban Sydney, Australia. *J. Mech. Behav. Biomed. Mater.* **2023**, *140*, 105729. [CrossRef] [PubMed]
57. Jin, H.-J.; Kaplan, D.L. Mechanism of Silk Processing in Insects and Spiders. *Nature* **2003**, *424*, 1057–1061. [CrossRef] [PubMed]
58. Schmuck, B.; Greco, G.; Bäcklund, F.G.; Pugno, N.M.; Johansson, J.; Rising, A. Impact of Physio-Chemical Spinning Conditions on the Mechanical Properties of Biomimetic Spider Silk Fibers. *Commun. Mater.* **2022**, *3*, 83. [CrossRef]
59. Gatwiri, J.; Kamweru, P.K. Spider Silk: A Natural Marvel of Mechanical and Structural Strength. *Afr. J. Biol. Sci.* **2021**, *3*, 1–16.
60. Altman, G.H.; Chen, J.; Horan, R.L.; Horan, D.J. Method of Forming an Implantable Knitted Fabric Comprising Silk Fibroin Fibers. U.S. Patent No. 8,628,791, 24 March 2011.
61. Gosline, J.; DeMont, M.; Denny, M. The Structure and Properties of Spider Silk. *Endeavour* **1986**, *10*, 37–43. [CrossRef]
62. Tokareva, O.; Jacobsen, M.; Buehler, M.; Wong, J.; Kaplan, D. Structure–Function– Property–Design Interplay in Biopolymers: Spider Silk. *Acta Biomater.* **2014**, *10*, 1612–1626. [CrossRef]
63. Roloff, F.; Strauß, S.; Vogt, P.M.; Bicker, G.; Radtke, C. Spider Silk as Guiding Biomaterial for Human Model Neurons. *BioMed. Res. Int.* **2014**, *2014*, 906819. [CrossRef]
64. Tremblay, M.-L.; Xu, L.; Lefèvre, T.; Sarker, M.; Orrell, K.E.; Leclerc, J.; Rainey, J.K. Spider Wrapping Silk Fibre Architecture Arising from Its Modular Soluble Protein Precursor. *Sci. Rep.* **2015**, *5*, 11502. [CrossRef]
65. Simmons, J.R.; Xu, L.; Rainey, J.K. Recombinant Pyriform Silk Fiber Mechanics Are Modulated by Wet-Spinning Conditions. *ACS Biomater. Sci. Eng.* **2019**, *5*, 4985–4993. [CrossRef]
66. Moon, M.-J. Fine Structure of the Aggregate Silk Nodules in the Orb-Web Spider Nephila Clavata. *Anim. Cells Syst.* **2018**, *22*, 421–428. [CrossRef]
67. Bittencourt, D.M.D.C.; Oliveira, P.; Michalczechen-Lacerda, V.A.; Rosinha, G.M.S.; Jones, J.A.; Rech, E.L. Bioengineering of Spider Silks for the Production of Biomedical Materials. *Front. Bioeng. Biotechnol.* **2022**, *10*, 958486. [CrossRef]
68. Ramezaniaghdam, M.; Nahdi, N.D.; Reski, R. Recombinant Spider Silk: Promises and Bottlenecks. *Front. Bioeng. Biotechnol.* **2022**, *10*, 835637. [CrossRef]
69. Heidebrecht, A.; Scheibel, T. Recombinant Production of Spider Silk Proteins. *Adv. Appl. Microbiol.* **2013**, *82*, 115–153.
70. Humenik, M.; Smith, A.; Scheibel, T. Recombinant Spider Silks—Biopolymers with Potential for Future Applications. *Polymers* **2011**, *3*, 640–661. [CrossRef]
71. Whittall, D.R.; Baker, K.V.; Breitling, R.; Takano, E. Host Systems for the Production of Recombinant Spider Silk. *Trends Biotechnol.* **2021**, *39*, 560–573. [CrossRef]
72. Lin, S.; Chen, G.; Liu, X.; Meng, Q. Chimeric Spider Silk Proteins Mediated by Intein Result in Artificial Hybrid Silks. *Biopolymers* **2016**, *105*, 385–392. [CrossRef]
73. Bowen, C.H.; Dai, B.; Sargent, C.J.; Bai, W.; Ladiwala, P.; Feng, H.; Huang, W.; Kaplan, D.L.; Galazka, J.M.; Zhang, F. Recombinant Spidroins Fully Replicate Primary Mechanical Properties of Natural Spider Silk. *Biomacromolecules* **2018**, *19*, 3853–3860. [CrossRef] [PubMed]
74. Jin, Q.; Pan, F.; Hu, C.-F.; Lee, S.Y.; Xia, X.-X.; Qian, Z.-G. Secretory Production of Spider Silk Proteins in Metabolically Engineered Corynebacterium Glutamicum for Spinning into Tough Fibers. *Metab. Eng.* **2022**, *70*, 102–114. [CrossRef] [PubMed]
75. Widmaier, D.M.; Tullman-Ercek, D.; Mirsky, E.A.; Hill, R.; Govindarajan, S.; Minshull, J.; Voigt, C.A. Engineering the *Salmonella* Type III Secretion System to Export Spider Silk Monomers. *Mol. Syst. Biol.* **2009**, *5*, 309. [CrossRef] [PubMed]
76. Foong, C.P.; Higuchi-Takeuchi, M.; Malay, A.D.; Oktaviani, N.A.; Thagun, C.; Numata, K. A Marine Photosynthetic Microbial Cell Factory as a Platform for Spider Silk Production. *Commun. Biol.* **2020**, *3*, 357. [CrossRef]
77. Jansson, R.; Lau, C.H.; Ishida, T.; Ramström, M.; Sandgren, M.; Hedhammar, M. Functionalized Silk Assembled from a Recombinant Spider Silk Fusion Protein (Z-4RepCT) Produced in the Methylotrophic Yeast *Pichia pastoris*. *Biotechnol. J.* **2016**, *11*, 687–699. [CrossRef]
78. Sidoruk, K.V.; Davydova, L.I.; Kozlov, D.G.; Gubaidullin, D.G.; Glazunov, A.V.; Bogush, V.G.; Debabov, V.G. Fermentation Optimization of a Saccharomyces Cerevisiae Strain Producing 1F9 Recombinant Spidroin. *Appl. Biochem. Microbiol.* **2015**, *51*, 766–773. [CrossRef]
79. Lazaris, A.; Arcidiacono, S.; Huang, Y.; Zhou, J.-F.; Duguay, F.; Chretien, N.; Welsh, E.A.; Soares, J.W.; Karatzas, C.N. Spider Silk Fibers Spun from Soluble Recombinant Silk Produced in Mammalian Cells. *Science* **2002**, *295*, 472–476. [CrossRef]
80. Grip, S.; Rising, A.; Nimmervoll, H.; Storckenfeldt, E.; McQueen-Mason, S.J.; Pouchkina-Stantcheva, N.; Vollrath, F.; Engström, W.; Fernandez-Arias, A. Transient Expression of a Major Ampullate Spidroin 1 Gene Fragment from *Euprosthenops* sp. in Mammalian Cells. *Cancer Genom. Proteom.* **2006**, *3*, 83–87.

81. Zhang, Y.; Hu, J.; Miao, Y.; Zhao, A.; Zhao, T.; Wu, D.; Liang, L.; Miikura, A.; Shiomi, K.; Kajiura, Z.; et al. Expression of EGFP-Spider Dragline Silk Fusion Protein in BmN Cells and Larvae of Silkworm Showed the Solubility Is Primary Limit for Dragline Proteins Yield. *Mol. Biol. Rep.* **2008**, *35*, 329–335. [CrossRef] [PubMed]
82. Schillberg, S.; Raven, N.; Spiegel, H.; Rasche, S.; Buntru, M. Critical Analysis of the Commercial Potential of Plants for the Production of Recombinant Proteins. *Front. Plant Sci.* **2019**, *10*, 720. [CrossRef] [PubMed]
83. Weichert, N.; Hauptmann, V.; Helmold, C.; Conrad, U. Seed-Specific Expression of Spider Silk Protein Multimers Causes Long-Term Stability. *Front. Plant Sci.* **2016**, *7*, 6. [CrossRef] [PubMed]
84. Peng, C.A.; Russo, J.; Gravgaard, C.; McCartney, H.; Gaines, W.; Marcotte, W.R. Spider Silk-like Proteins Derived from Transgenic Nicotiana Tabacum. *Transgenic Res.* **2016**, *25*, 517–526. [CrossRef] [PubMed]
85. Scheller, J.; Gührs, K.-H.; Grosse, F.; Conrad, U. Production of Spider Silk Proteins in Tobacco and Potato. *Nat. Biotechnol.* **2001**, *19*, 573–577. [CrossRef] [PubMed]
86. Hugie, M.R. Expression Systems for Synthetic Spider Silk Protein Production. Ph.D. Thesis, Utah State University, Logan, UT, USA, 2019. [CrossRef]
87. Yang, J.; Barr, L.A.; Fahnestock, S.R.; Liu, Z.-B. High Yield Recombinant Silk-like Protein Production in Transgenic Plants through Protein Targeting. *Transgenic Res.* **2005**, *14*, 313–324. [CrossRef]
88. Zhang, X.; Xia, L.; Day, B.A.; Harris, T.I.; Oliveira, P.; Knittel, C.; Licon, A.L.; Gong, C.; Dion, G.; Lewis, R.V.; et al. CRISPR/Cas9 Initiated Transgenic Silkworms as a Natural Spinner of Spider Silk. *Biomacromolecules* **2019**, *20*, 2252–2264. [CrossRef] [PubMed]
89. Tucker, C.L.; Jones, J.A.; Bringhurst, H.N.; Copeland, C.G.; Addison, J.B.; Weber, W.S.; Mou, Q.; Yarger, J.L.; Lewis, R.V. Mechanical and Physical Properties of Recombinant Spider Silk Films Using Organic and Aqueous Solvents. *Biomacromolecules* **2014**, *15*, 3158–3170. [CrossRef] [PubMed]
90. Li, H.; Chen, S.; Piao, S.; An, T.; Wang, C. Production of Artificial Synthetic Spidroin Gene 4S-Transgenic Cloned Sheep Embryos Using Somatic Cell Nuclear Transfer. *Anim. Biotechnol.* **2021**, *32*, 616–626. [CrossRef] [PubMed]
91. Karageorgiou, V.; Kaplan, D. Porosity of 3D Biomaterial Scaffolds and Osteogenesis. *Biomaterials* **2005**, *26*, 5474–5491. [CrossRef]
92. Kuhbier, J.W.; Allmeling, C.; Reimers, K.; Hillmer, A.; Kasper, C.; Menger, B.; Brandes, G.; Guggenheim, M.; Vogt, P.M. Interactions between Spider Silk and Cells—NIH/3T3 Fibroblasts Seeded on Miniature Weaving Frames. *PLoS ONE* **2010**, *5*, e12032. [CrossRef] [PubMed]
93. Kuhbier, J.W.; Reimers, K.; Kasper, C.; Allmeling, C.; Hillmer, A.; Menger, B.; Vogt, P.M.; Radtke, C. First Investigation of Spider Silk as a Braided Microsurgical Suture. *J. Biomed. Mater. Res. B Appl. Biomater.* **2011**, *97*, 381–387. [CrossRef] [PubMed]
94. Hennecke, K.; Redeker, J.; Kuhbier, J.; Strauss, S.; Allmeling, C.; Kasper, C.; Vogt, P. Bundles of Spider Silk, Braided into Sutures, Resist Basic Cyclic Tests: Potential Use for Flexor Tendon Repair. *PLoS ONE* **2013**, *8*, e61100. [CrossRef]
95. Allmeling, C.; Jokuszies, A.; Reimers, K.; Kall, S.; Vogt, P. Use of Spider Silk Fibres as an Innovative Material in a Biocompatible Artificial Nerve Conduit. *J. Cell. Mol. Med.* **2006**, *10*, 770–777. [CrossRef]
96. Wendt, H.; Hillmer, A.; Reimers, K.; Kuhbier, J.; Schäfer-Nolte, F.; Allmeling, C.; Vogt, P. Artificial Skin—Culturing of Different Skin Cell Lines for Generating an Artificial Skin Substitute on Cross-Weaved Spider Silk Fibres. *PLoS ONE* **2011**, *6*, e21833. [CrossRef]
97. Steins, A.; Dik, P.; Müller, W.; Vervoort, S.; Reimers, K.; Kuhbier, J.; Schepers, K. In Vitro Evaluation of Spider Silk Meshes as a Potential Biomaterial for Bladder Reconstruction. *PLoS ONE* **2015**, *10*, e0145240. [CrossRef]
98. Trossmann, V.T.; Lentz, S.; Scheibel, T. Factors Influencing Properties of Spider Silk Coatings and Their Interactions within a Biological Environment. *J. Funct. Biomater.* **2023**, *14*, 434. [CrossRef]
99. Borkner, C.B.; Elsner, M.B.; Scheibel, T. Coatings and Films Made of Silk Proteins. *ACS Appl. Mater. Interfaces* **2014**, *6*, 15611–15625. [CrossRef]
100. Nilebäck, L.; Hedin, J.; Widhe, M.; Floderus, L.S.; Krona, A.; Bysell, H.; Hedhammar, M. Self-Assembly of Recombinant Silk as a Strategy for Chemical-Free Formation of Bioactive Coatings: A Real-Time Study. *Biomacromolecules* **2017**, *18*, 846–854. [CrossRef]
101. Zhou, S.; Wang, Q.; Yang, W.; Wang, L.; Wang, J.; You, R.; Luo, Z.; Zhang, Q.; Yan, S. Development of a Bioactive Silk Fibroin Bilayer Scaffold for Wound Healing and Scar Inhibition. *Int. J. Biol. Macromol.* **2024**, *255*, 128350. [CrossRef] [PubMed]
102. Krishnaji, S.T.; Huang, W.; Rabotyagova, O.; Kharlampieva, E.; Choi, I.; Tsukruk, V.V.; Naik, R.; Cebe, P.; Kaplan, D.L. Thin Film Assembly of Spider Silk-like Block Copolymers. *Langmuir* **2011**, *27*, 1000–1008. [CrossRef] [PubMed]
103. Lamprogiannis, L.; Karamitsos, A.; Karagkiozaki, V.; Tsinopoulos, I.; Gioti, M.; Fatouros, D.G.; Dimitrakos, S.; Logothetidis, S. Design and Fabrication of Drug-eluting Polymeric Thin Films for Applications in Ophthalmology. *IET Nanobiotechnol.* **2018**, *12*, 1074–1079. [CrossRef] [PubMed]
104. Zeplin, P.; Maksimovikj, N.; Jordan, M.; Nickel, J.; Lang, G.; Leimer, A.; Scheibel, T. Spider Silk Coatings as a Bioshield to Reduce Periprosthetic Fibrous Capsule Formation. *Adv. Funct. Mater.* **2014**, *24*, 2658–2666. [CrossRef]
105. Borkner, C.B.; Wohlrab, S.; Möller, E.; Lang, G.; Scheibel, T. Surface Modification of Polymeric Biomaterials Using Recombinant Spider Silk Proteins. *ACS Biomater. Sci. Eng.* **2017**, *3*, 767–775. [CrossRef] [PubMed]
106. Yi, J.; Zou, G.; Huang, J.; Ren, X.; Tian, Q.; Yu, Q.; Wang, P.; Yuan, Y.; Tang, W.; Wang, C.; et al. Water-Responsive Supercontractile Polymer Films for Bioelectronic Interfaces. *Nature* **2023**, *624*, 295–302. [CrossRef]
107. Wang, H.; Xue, Z.; Wei, M.; Chen, D.; Li, M. A Novel Scaffold from Recombinant Spider Silk Protein in Tissue Engineering. In *Advanced Materials Research*; Trans Tech Publications Ltd.: Stafa-Zurich, Switzerland, 2010; Volume 152.

108. Johansson, U.; Widhe, M.; Shalaly, N.; Arregui, I.; Nilebäck, L.; Tasiopoulos, C.; Hedhammar, M. Assembly of Functionalized Silk Together with Cells to Obtain Proliferative 3D Cultures Integrated in a Network of ECM-like Microfibers. *Sci. Rep.* **2019**, *9*, 6291. [CrossRef] [PubMed]
109. Abtahi, S.; Chen, X.; Shahabi, S.; Nasiri, N. Resorbable Membranes for Guided Bone Regeneration: Critical Features, Potentials, and Limitations. *ACS Mater. Au* **2023**, *3*, 394–417. [CrossRef]
110. Schacht, K.; Vogt, J.; Scheibel, T. Foams Made of Engineered Recombinant Spider Silk Proteins as 3D Scaffolds for Cell Growth. *ACS Biomater. Sci. Eng.* **2016**, *2*, 517–525. [CrossRef] [PubMed]
111. Rammensee, S.; Huemmerich, D.; Hermanson, K.; Scheibel, T.; Bausch, A. Rheological Characterization of Hydrogels Formed by Recombinantly Produced Spider Silk. *Appl. Phys.* **2006**, *82*, 261–264. [CrossRef]
112. Jungst, T.; Smolan, W.; Schacht, K.; Scheibel, T.; Groll, J. Strategies and Molecular Design Criteria for 3D Printable Hydrogels. *Chem. Rev.* **2016**, *116*, 1496–1539. [CrossRef]
113. Nie, L.; Sun, Y.; Okoro, O.V.; Deng, Y.; Jiang, G.; Shavandi, A. Click Chemistry for 3D Bioprinting. *Mater. Horiz.* **2023**, *10*, 2727–2763. [CrossRef]
114. Withanage, S.; Savin, A.; Nikolaeva, V.; Kiseleva, A.; Dukhinova, M.; Krivoshapkin, P.; Krivoshapkina, E. Native Spider Silk-Based Antimicrobial Hydrogels for Biomedical Applications. *Polymers* **2021**, *13*, 1796. [CrossRef]
115. Spicer, C.D. Hydrogel Scaffolds for Tissue Engineering: The Importance of Polymer Choice. *Polym. Chem.* **2020**, *11*, 184–219. [CrossRef]
116. Kapoor, S.; Kundu, S.C. Silk Protein-Based Hydrogels: Promising Advanced Materials for Biomedical Applications. *Acta Biomater.* **2016**, *31*, 17–32. [CrossRef]
117. Ilić-Stojanović, S.; Nikolić, L.; Cakić, S. A Review of Patents and Innovative Biopolymer-Based Hydrogels. *Gels* **2023**, *9*, 556. [CrossRef]
118. Zahra, D.; Shokat, Z.; Ahmad, A.; Javaid, A.; Khurshid, M.; Ashfaq, U.A.; Nashwan, A.J. Exploring the Recent Developments of Alginate Silk Fibroin Material for Hydrogel Wound Dressing: A Review. *Int. J. Biol. Macromol.* **2023**, *248*, 125989. [CrossRef]
119. Li, X.; Ullah, M.W.; Li, B.; Chen, H. Recent Progress in Advanced Hydrogel-Based Embolic Agents: From Rational Design Strategies to Improved Endovascular Embolization. *Adv. Healthc. Mater.* **2023**, *12*, 2202787. [CrossRef]
120. Wang, Z.; Wei, H.; Huang, Y.; Wei, Y.; Chen, J. Naturally Sourced Hydrogels: Emerging Fundamental Materials for next-Generation Healthcare Sensing. *Chem. Soc. Rev.* **2023**, *52*, 2992–3034. [CrossRef]
121. Vázquez-González, M.; Willner, I. Stimuli-Responsive Biomolecule-Based Hydrogels and Their Applications. *Angew. Chem. Int. Ed.* **2020**, *59*, 15342–15377. [CrossRef]
122. Li, J.; Wu, C.; Chu, P.K.; Gelinsky, M. 3D Printing of Hydrogels: Rational Design Strategies and Emerging Biomedical Applications. *Mater. Sci. Eng. R Rep.* **2020**, *140*, 100543. [CrossRef]
123. Hasturk, O.; Jordan, K.E.; Choi, J.; Kaplan, D.L. Enzymatically Crosslinked Silk and Silk-Gelatin Hydrogels with Tunable Gelation Kinetics, Mechanical Properties and Bioactivity for Cell Culture and Encapsulation. *Biomaterials* **2020**, *232*, 119720. [CrossRef]
124. Zheng, H.; Zuo, B. Functional Silk Fibroin Hydrogels: Preparation, Properties and Applications. *J. Mater. Chem. B* **2021**, *9*, 1238–1258. [CrossRef]
125. Liu, L.; Tang, H.; Wang, Y. Polymeric Biomaterials: Advanced Drug Delivery Systems in Osteoarthritis Treatment. *Heliyon* **2023**, *9*, e21544. [CrossRef]
126. Li, S.; Xiong, Y.; Zhu, H.; Ma, T.; Sun, X.; Xiao, J. Microenvironment-Responsive Nanosystems for Osteoarthritis Therapy. *Eng. Regen.* **2024**, *5*, 92–110. [CrossRef]
127. Shi, L.; Wang, F.; Zhu, W.; Xu, Z.; Fuchs, S.; Hilborn, J.; Zhu, L.; Ma, Q.; Wang, Y.; Weng, X.; et al. Self-Healing Silk Fibroin-Based Hydrogel for Bone Regeneration: Dynamic Metal-Ligand Self-Assembly Approach. *Adv. Funct. Mater.* **2017**, *27*, 1700591. [CrossRef]
128. Zheng, H.; Lin, N.; He, Y.; Zuo, B. Self-Healing, Self-Adhesive Silk Fibroin Conductive Hydrogel as a Flexible Strain Sensor. *ACS Appl. Mater. Interfaces* **2021**, *13*, 40013–40031. [CrossRef]
129. Ngadimin, K.D.; Stokes, A.; Gentile, P.; Ferreira, A.M. Biomimetic Hydrogels Designed for Cartilage Tissue Engineering. *Biomater. Sci.* **2021**, *9*, 4246–4259. [CrossRef]
130. Hu, J.; Chen, B.; Guo, F.; Du, J.; Gu, P.; Lin, X.; Yang, W.; Zhang, H.; Lu, M.; Huang, Y.; et al. Injectable Silk Fibroin/Polyurethane Composite Hydrogel for Nucleus Pulposus Replacement. *J. Mater. Sci. Mater. Med.* **2012**, *23*, 711–722. [CrossRef]
131. Liu, Y.; Zhao, Z.; Guo, C.; Huang, Z.; Zhang, W.; Ma, F.; Wang, Z.; Kong, Q.; Wang, Y. Application and Development of Hydrogel Biomaterials for the Treatment of Intervertebral Disc Degeneration: A Literature Review. *Front. Cell Dev. Biol.* **2023**, *11*, 1286223. [CrossRef]
132. Shanmugam, D.K.; Anitha, S.C.; Souresh, V.; Madhavan, Y.; Sampath, S.; Catakapatri Venugopal, D.; Saravanan, M. Current Advancements in the Development of Bionic Organs Using Regenerative Medicine and 3D Tissue Engineering. *Mater. Technol.* **2023**, *38*, 2242732. [CrossRef]
133. Sontheimer-Phelps, A.; Hassell, B.; Ingber, D. Modelling Cancer in Microfluidic Human Organs-on Chips. *Nat. Rev. Cancer* **2019**, *19*, 65–81. [CrossRef] [PubMed]
134. Bakirci, E.; Guenat, O.; Ahmad, S.; Gantenbein, B. Tissue Engineering Approaches for the Repair and Regeneration of the Anterior Cruciate Ligament: Towards 3D Bioprinted ACL-on-Chip. *Eur. Cells Mater.* **2022**, *44*, 21–42. [CrossRef] [PubMed]

135. Thurber, A.E.; Omenetto, F.G.; Kaplan, D.L. In Vivo Bioresponses to Silk Proteins. *Biomaterials* **2015**, *71*, 145–157. [CrossRef] [PubMed]
136. Baoyong, L.; Jian, Z.; Denglong, C.; Min, L. Evaluation of a New Type of Wound Dressing Made from Recombinant Spider Silk Protein Using Rat Models. *Burns* **2010**, *36*, 891–896. [CrossRef] [PubMed]
137. Gomes, S.; Leonor, I.; Mano, J.; Reis, R.; Kaplan, D. Spider silk-bone sialoprotein fusion proteins for bone tissue engineering. *Soft Matter* **2011**, *7*, 4964–4973. [CrossRef]
138. Dellaquila, A.; Greco, G.; Campodoni, E.; Mazzocchi, M.; Mazzolai, B.; Tampieri, A.; Sandri, M. Optimized Production of a High-Performance Hybrid Biomaterial: Biomineralized Spider Silk for Bone Tissue Engineering. *J. Appl. Polym. Sci.* **2019**, *137*, 48739. [CrossRef]
139. Wang, F.; Gu, Z.; Yin, Z.; Zhang, W.; Bai, L.; Su, J. Cell Unit-Inspired Natural Nano-Based Biomaterials as Versatile Building Blocks for Bone/Cartilage Regeneration. *J. Nanobiotechnology* **2023**, *21*, 293. [CrossRef]
140. Dastagir, K.; Dastagir, N.; Limbourg, A.; Reimers, K.; Straub, S.; Vogt, P. In Vitro Construction of Artificial Blood Vessels Using Spider Silk as a Supporting Matrix. *J. Mech. Behav. Biomed. Mater.* **2020**, *101*, 103436. [CrossRef]
141. Ding, Z.; Wang, Y.; Chen, F.; Hu, X.; Cheng, W.; Lu, Q.; Kaplan, D.L. Biomimetic Vascular Grafts with Circumferentially and Axially Oriented Microporous Structures for Native Blood Vessel Regeneration. *Adv. Funct. Mater.* **2024**, *34*, 2308888. [CrossRef]
142. Cattaneo, I.; Figliuzzi, M.; Azzollini, N.; Catto, V.; Farè, S.; Tanzi, M.C.; Alessandrino, A.; Freddi, G.; Remuzzi, A. In Vivo Regeneration of Elastic Lamina on Fibroin Biodegradable Vascular Scaffold. *Int. J. Artif. Organs* **2013**, *36*, 166–174. [CrossRef]
143. Soffer, L.; Wang, X.; Zhang, X.; Kluge, J.; Dorfmann, L.; Kaplan, D.L.; Leisk, G. Silk-Based Electrospun Tubular Scaffolds for Tissue-Engineered Vascular Grafts. *J. Biomater. Sci. Polym. Ed.* **2008**, *19*, 653–664. [CrossRef] [PubMed]
144. Catto, V.; Farè, S.; Cattaneo, I.; Figliuzzi, M.; Alessandrino, A.; Freddi, G.; Remuzzi, A.; Tanzi, M.C. Small Diameter Electrospun Silk Fibroin Vascular Grafts: Mechanical Properties, in Vitro Biodegradability, and in Vivo Biocompatibility. *Mater. Sci. Eng. C* **2015**, *54*, 101–111. [CrossRef]
145. Cardenas, L.; Sandoval, O.L.; Hurtado, D. Biomechanical Analysis of Spider Silk for Anterior Cruciate Ligament Reconstruction. *ARPN J. Eng. Appl. Sci.* **2016**, *11*, 52063642.
146. Farè, S.; Torricelli, P.; Giavaresi, G.; Bertoldi, S.; Alessandrino, A.; Villa, T.; Fini, M.; Tanzi, M.C.; Freddi, G. In Vitro Study on Silk Fibroin Textile Structure for Anterior Cruciate Ligament Regeneration. *Mater. Sci. Eng. C* **2013**, *33*, 3601–3608. [CrossRef] [PubMed]
147. Petzold, J.; Aigner, T.; Touska, F.; Zimmermann, K.; Scheibel, T.; Engel, F. Surface Features of Recombinant Spider Silk Protein eADF4(K16)-Made Materials Are Well-Suited for Cardiac Tissue Engineering. *Adv. Funct. Mater.* **2017**, *27*, 1701427. [CrossRef]
148. Wang, L.; Wu, Y.; Guo, B.; Ma, P.X. Nanofiber Yarn/Hydrogel Core–Shell Scaffolds Mimicking Native Skeletal Muscle Tissue for Guiding 3D Myoblast Alignment, Elongation, and Differentiation. *ACS Nano* **2015**, *9*, 9167–9179. [CrossRef]
149. Wu, Y.; Wang, L.; Guo, B.; Ma, P.X. Interwoven Aligned Conductive Nanofiber Yarn/Hydrogel Composite Scaffolds for Engineered 3D Cardiac Anisotropy. *ACS Nano* **2017**, *11*, 5646–5659. [CrossRef]
150. Dong, R.; Ma, P.X.; Guo, B. Conductive Biomaterials for Muscle Tissue Engineering. *Biomaterials* **2020**, *229*, 119584. [CrossRef]
151. Brooks, A.K.; Ramsey, R.G.; Zhang, N.; Yadavalli, V.K. Tunable Light-Actuated Interpenetrating Networks of Silk Fibroin and Gelatin for Tissue Engineering and Flexible Biodevices. *ACS Biomater. Sci. Eng.* **2023**, *9*, 5793–5803. [CrossRef]
152. Allmeling, C.; Jokuszies, A.; Reimers, K.; Kall, S.; Choi, C.Y.; Brandes, G.; Kasper, C.; Scheper, T.; Guggenheim, M.; Vogt, P.M. Spider Silk Fibres in Artificial Nerve Constructs Promote Peripheral Nerve Regeneration. *Cell Prolif.* **2008**, *41*, 408–420. [CrossRef] [PubMed]
153. Pawar, K.; Welzel, G.; Haynl, C.; Schuster, S.; Scheibel, T. Recombinant Spider Silk and Collagen-Based Nerve Guidance Conduits Support Neuronal Cell Differentiation and Functionality In Vitro. *ACS Appl. Bio Mater.* **2019**, *2*, 4872–4880. [CrossRef] [PubMed]
154. Lewicka, M.; Rebellato, P.; Lewicki, J.; Uhlén, P.; Rising, A.; Hermanson, O. Recombinant Spider Silk Protein Matrices Facilitate Multi-Analysis of Calcium-Signaling in Neural Stem Cell-Derived AMPA-Responsive Neuron. *BioRxiv* **2019**, 579292. [CrossRef]
155. Hu, J. Spider Silks and Their Nanostructures. In Proceedings of the Global Summit on Nanomaterials: Applications and Properties Nanomaterials NANOMAT22, Dubai, United Arab Emirates, 23–25 March 2022.
156. Zivic, F.; Grujovic, N.; Ul Ahad, I.; Brabazon, D. Introduction—The Current Status and Momentum. In *Commercialization of Nanotechnologies—A Case Study Approach*; Springer: Cham, Switzerland, 2018.
157. Sayin, D.; Gundogdu, G.; Kilic-Erkek, O.; Gundogdu, K.; Coban, H.S.; Abban-Mete, G. Silk Protein Sericin: A Promising Therapy for Achilles Tendinopathy—Evidence from an Experimental Rat Model. *Clin. Rheumatol.* **2023**, *42*, 3361–3373. [CrossRef] [PubMed]
158. Kojic, N.; Pritchard, E.M.; Tao, H.; Brenckle, M.A.; Mondia, J.P.; Panilaitis, B.; Omenetto, F.; Kaplan, D.L. Focal Infection Treatment Using Laser-Mediated Heating of Injectable Silk Hydrogels with Gold Nanoparticles. *Adv. Funct. Mater.* **2012**, *22*, 3793–3798. [CrossRef] [PubMed]
159. Jacob, J.; Haponiuk, J.T.; Thomas, S.; Gopi, S. Biopolymer Based Nanomaterials in Drug Delivery Systems: A Review. *Mater. Today Chem.* **2018**, *9*, 43–55. [CrossRef]
160. Rasekh, M.; Pisapia, F.; Nokhodchi, A. Innovative Applications of Electrospun Nanofibers in Cancer Research. *J. Drug Deliv. Sci. Technol.* **2024**, *91*, 105255. [CrossRef]
161. Helfricht, N.; Klug, M.; Mark, A.; Kuznetsov, V.; Blum, C.; Scheibel, T.; Papastavrou, G. Surface Properties of Spider Silk Particles in Solution. *Biomater. Sci.* **2013**, *1*, 1166–1171. [CrossRef]

162. Jastrzębska, K.; Kucharczyk, K.; Florczak, A.; Dondajewska, E.; Mackiewicz, A.; Dams-Kozlowska, H. Silk as an Innovative Biomaterial for Cancer Therapy. *Rep. Pract. Oncol. Radiother.* **2015**, *20*, 87–98. [CrossRef] [PubMed]
163. Mohamed, N.; Hamad, M.; Hafez, M.; Wooley, K.; Elsabahy, M. Nanomedicine in Management of Hepatocellular Carcinoma: Challenges and Opportunities. *Int. J. Cancer* **2017**, *140*, 1475–1484. [CrossRef]
164. Aquib, M.; Juthi, A.; Farooq, M.; Ali, M.; Janabi, A.; Bavi, S.; Wang, B. Advances in Local and Systemic Drug Delivery Systems for Post-Surgical Cancer Treatment. *J. Mater. Chem. B* **2020**, *8*, 8507–8518. [CrossRef] [PubMed]
165. Doppalapudi, S.; Jain, A.; Domb, A.; Khan, W. Biodegradable Polymers for Targeted Delivery of Anti-Cancer Drugs. *Expert Opin. Drug Deliv.* **2016**, *13*, 891–909. [CrossRef] [PubMed]
166. Jin, H.; Park, J.; Karageorgiou, V.; Kim, U.; Valluzzi, R.; Cebe, P.; Kaplan, D. Water-Stable Silk Films with Reduced β-Sheet Content. *Adv. Funct. Mater.* **2015**, *15*, 1241–1247. [CrossRef]
167. Laakkonen, P.; Vuorinen, K. Homing Peptides as Targeted Delivery Vehicles. *Integr. Biol.* **2010**, *2*, 326–337. [CrossRef] [PubMed]
168. Numata, K.; Mieszawska-Czajkowska, A.; Kvenvold, L.; Kaplan, D. Silk-Based Nanocomplexes with Tumor-Homing Peptides for Tumor-Specific Gene Delivery. *Macromol. Biosci.* **2012**, *12*, 75–82. [CrossRef]
169. Florczak, A.; Jastrzebska, K.; Mackiewicz, A.; Dams-Kozlowska, H. Blending Two Bioengineered Spider Silks to Develop Cancer Targeting Spheres. *Mater. Chem. B* **2017**, *5*, 3000–3011. [CrossRef]
170. Hines, D.; Kaplan, D. Characterization of Small Molecule Controlled Release from Silk Films. *Macromol. Chem. Phys.* **2012**, *214*, 280–294. [CrossRef]
171. Jastrzebska, K.; Florczak, A.; Kucharczyk, K.; Lin, Y.; Wang, Q.; Mackiewicz, A.; Dams-Kozlowska, H. Delivery of Chemotherapeutics Using Spheres Made of Bioengineered Spider Silks Derived from MaSp1 and MaSp2 Proteins. *Nanomedicine* **2018**, *13*, 439–454. [CrossRef] [PubMed]
172. Florczak, A.; Deptuch, T.; Kucharczyk, K.; Dams-Kozlowska, H. Systemic and Local Silk-Based Drug Delivery Systems for Cancer Therapy. *Cancers* **2021**, *13*, 5389. [CrossRef] [PubMed]
173. Xie, M.; Fan, D.; Chen, Y.; Zhao, Z.; He, X.; Li, G.; Lan, P. An Implantable and Controlled Drug-Release Silk Fibroin Nanofibrous Matrix to Advance the Treatment of Solid Tumour Cancers. *Biomaterials* **2016**, *103*, 33–43. [CrossRef] [PubMed]
174. Gao, Y.; Hou, M.; Yang, R.; Zhang, L.; Xu, Z.; Kang, Y.; Xue, P. Highly Porous Silk Fibroin Scaffold Packed in PEGDA/Sucrose Microneedles for Controllable Transdermal Drug Delivery. *Biomacromolecules* **2019**, *20*, 1334–1345. [CrossRef] [PubMed]
175. Qi, Z.; Tao, X.; Tan, G.; Tian, B.; Zhang, L.; Kundu, S.C.; Lu, S. Electro-Responsive Silk Fibroin Microneedles for Controlled Release of Insulin. *Int. J. Biol. Macromol.* **2023**, *242*, 124684. [CrossRef] [PubMed]
176. Shchepelina, O.; Drachuk, I.; Gupta, M.; Lin, J.; Tsukruk, V. Silk-on-Silk Layer-by-Layer Microcapsules. *Adv. Mater.* **2011**, *23*, 4655–4660. [CrossRef] [PubMed]
177. Wang, X.; Yucel, T.; Lu, Q.; Hu, X.; Kaplan, D. Silk Nanospheres and Microspheres from Silk/Pva Blend Films for Drug Delivery. *Biomaterials* **2010**, *31*, 1025–1035. [CrossRef]
178. Florczak, A.; Grzechowiak, I.; Deptuch, T.; Kucharczyk, K.; Kaminska, A.; Dams-Kozlowska, H. Silk Particles as Carriers of Therapeutic Molecules for Cancer Treatment. *Materials* **2020**, *13*, 4946. [CrossRef]
179. Wongpinyochit, T.; Johnston, B.; Seib, F. Manufacture and Drug Delivery Applications of Silk Nanoparticles. *J. Vis. Exp.* **2016**, *116*, e54669.
180. Cao, Y.; Liu, F.; Chen, Y.; Yu, T.; Lou, D.; Guo, Y.; Ran, H. Drug Release from Core-Shell PVA/Silk Fibroin Nanoparticles Fabricated by One-Step Electrospraying. *Sci. Rep.* **2017**, *7*, 11913. [CrossRef] [PubMed]
181. Lozano-Perez, A.; Montalban, M.; Aznar-Cervantes, S.; Cragnolini, F.; Cenis, J.; Villora, G. Production of Silk Fibroin Nanoparticles Using Ionic Liquids and High-Power Ultrasounds. *J. Appl. Polym. Sci.* **2014**, *132*, 41702–41709. [CrossRef]
182. Wenk, E.; Wandrey, A.; Merkle, H.; Meinel, L. Silk Fibroin Spheres as a Platform for Controlled Drug Delivery. *J. Controll. Release* **2008**, *132*, 26–34. [CrossRef]
183. Myung, S.; Kim, H.; Kim, Y.; Chen, P.; Jin, H. Fluorescent Silk Fibroin Nanoparticles Prepared Using a Reverse Microemulsion. *Macromol. Res.* **2008**, *16*, 604–608. [CrossRef]
184. Solomun, J.; Totten, J.; Wongpinyochit, T.; Florence, A.; Seib, F. Manual Versus Microfluidic Assisted Nanoparticle Manufacture: Impact of Silk Fibroin Stock on Nanoparticle Characteristics. *ACS Biomater. Sci. Eng.* **2020**, *6*, 2796–2804. [CrossRef]
185. Kazemimostaghim, M.; Rajkhowa, R.; Tsuzuki, T.; Wang, X. Ultrafine Silk Powder from Biocompatible Surfactant-Assisted Milling. *Powder Technol.* **2013**, *185*, 87–95. [CrossRef]
186. Sun, N.; Lei, R.; Xu, J.; Kundu, S.; Cai, Y.; Yao, J.; Ni, Q. Fabricated Porous Silk Fibroin Particles for pHresponsive Drug Delivery and Targeting of Tumor Cells. *J. Mater. Sci.* **2019**, *54*, 3319–3330. [CrossRef]
187. Subia, B.; Kundu, S. Drug Loading and Release on Tumor Cells Using Silk Fibroin-Albumin Nanoparticles as Carriers. *Nanotechnology* **2013**, *24*, 035103. [CrossRef]
188. Yu, J.; Zhang, Z.; Ni, Y.; Xiong, Y.; Xu, W. Preparation and Characterization of a Novel Composite Based on Hyperbranched Polysilane and Fullerene. *J. Appl. Polym. Sci.* **2007**, *105*, 821–826. [CrossRef]
189. Zhao, Z.; Xie, M.; Li, Y.; Chen, A.; Li, G.; Zhang, J.; Li, S. Formation of Curcumin Nanoparticles via Solution-Enhanced Dispersion by Supercritical CO_2. *Int. J. Nanomed.* **2015**, *10*, 3171–3181. [CrossRef]
190. Kucharczyk, K.; Rybka, J.; Hilgendorff, M.; Krupinski, M.; Slachcinski, M.; Mackiewicz, A.; Dams-Kozlowska, H. Composite Spheres Made of Bioengineered Spider Silk and Iron Oxide Nanoparticles for Theranostics Applications. *PLoS ONE* **2019**, *14*, e0219790. [CrossRef]

191. Herold, H.; Dobl, A.; Wohlrab, S.; Humenik, M.; Scheibel, T. Designed Spider Silk-Based Drug Carrier for Redox- or pH-Triggered Drug Release. *Biomacromolecules* **2020**, *21*, 4904–4912. [CrossRef]
192. Colavite, P.; Sartori, A. Septic Arthritis: Immunopathogenesis, Experimental Models and Therapy. *J. Venom. Anim. Toxins Trop. Dis.* **2014**, *20*, 19. [CrossRef]
193. Hasan, R.; Wohlers, A.; Shreffler, J.; Mulinti, P.; Ostlie, H.; Schaper, C.; Brooks, B.; Brooks, A. An Antibiotic-Releasing Bone Void Filling (ABVF) Putty for the Treatment of Osteomyelitis. *Materials* **2020**, *13*, 5080. [CrossRef]
194. Mulinti, P.; Shreffler, J.; Hasan, R.; Dea, M.; Brooks, A.E. Infection Responsive Smart Delivery of Antibiotics Using Recombinant Spider Silk Nanospheres. *Pharmaceutics* **2021**, *13*, 1358. [CrossRef] [PubMed]
195. Zuluaga-Vélez, A.; Quintero-Martinez, A.; Orozco, L.M.; Sepúlveda-Arias, J.C. Silk Fibroin Nanocomposites as Tissue Engineering Scaffolds—A Systematic Review. *Biomed. Pharmacother.* **2021**, *141*, 111924. [CrossRef] [PubMed]
196. Mehrotra, S.; Chouhan, D.; Konwarh, R.; Kumar, M.; Jadi, P.K.; Mandal, B.B. Comprehensive Review on Silk at Nanoscale for Regenerative Medicine and Allied Applications. *ACS Biomater. Sci. Eng.* **2019**, *5*, 2054–2078. [CrossRef]
197. Mohammadi, P.; Aranko, S.; Landowski, C.; Ikkala, O.; Jaudzems, K.; Wagermaier, W.; Linder, M. Biomimetic Composites with Enhanced Toughening Using Silk-Inspired Triblock Proteins and Aligned Nanocellulose Reiforcements. *Sci. Adv.* **2019**, *5*, 9–13. [CrossRef] [PubMed]
198. Luca, V.; Nicola, P. *Nanotube Superfiber Materials*, 2nd ed.; Elsevier: Amsterdam, The Netherlands, 2019.
199. Ornithopoulou, E.; Åstrand, C.; Gustafsson, L.; Crouzier, T.; Hedhammar, M. Self-Assembly of RGD-Functionalized Recombinant Spider Silk Protein into Microspheres in Physiological Buffer and in the Presence of Hyaluronic Acid. *ACS Appl. Bio Mater.* **2023**, *6*, 3696–3705. [CrossRef]
200. Wang, X.; Ye, X.; Guo, J.; Dai, X.; Yu, S.; Zhong, B. Modeling the 3-Dimensional Structure of the Silkworm's Spinning Apparatus in Silk Production. *Acta Biomater.* **2024**, *174*, 217–227. [CrossRef]
201. Bharadwaz, A.; Jayasuriya, A.C. Recent Trends in the Application of Widely Used Natural and Synthetic Polymer Nanocomposites in Bone Tissue Regeneration. *Mater. Sci. Eng. C* **2020**, *110*, 110698. [CrossRef] [PubMed]
202. Zhang, J.; Zhao, J.; Chen, S.; Jiao, X.; Li, X.; Chen, W. Synthesis, in Vitro Biocompatibility and Antibacterial Property of Novel Silk fibroin@Ag Spheres. *Mater. Lett.* **2024**, *357*, 135681. [CrossRef]
203. Liu, Y.; Chen, S. ZIF-8 Nanoparticles Combined with Fibroin Protein Co-Modified TiO_2 Nanotube Arrays to Construct a Drug Sustained-Release Platform. *Mater. Lett.* **2024**, *356*, 135541. [CrossRef]
204. Zhang, H.; Chen, H.; Lu, L.; Wang, H.; Zhao, Y.; Chai, R. Natural Multifunctional Silk Microcarriers for Noise-Induced Hearing Loss Therapy. *Adv. Sci.* **2024**, *11*, 2305215. [CrossRef]
205. Li, X.; Hajinur Hirad, A.; Alarfaj, A.A.; Li, H.; Santhanam, R. A Convergent Fabrication of Graphene Oxide/Silk Fibroin/Hydroxyapatite Nanocomposites Delivery Improved Early Osteoblast Cell Adhesion and Bone Regeneration. *Arab. J. Chem.* **2024**, *17*, 105468. [CrossRef]
206. Katti, K.S.; Jasuja, H.; Kar, S.; Katti, D.R. Nanostructured Biomaterials for in Vitro Models of Bone Metastasis Cancer. *Curr. Opin. Biomed. Eng.* **2021**, *17*, 100254. [CrossRef]

biomimetics

MDPI

Review

Towards the First Generation of Biomimetic Fixation for Resurfacing Arthroplasty Endoprostheses

Ryszard Uklejewski [1,*], Mariusz Winiecki [1,*], Mikołaj Dąbrowski [2] and Piotr Rogala [3]

1 Department of Constructional Materials and Biomaterials, Faculty of Materials Engineering, Kazimierz Wielki University, Jan Karol Chodkiewicz Street 30, 85-064 Bydgoszcz, Poland
2 Adult Spine Orthopaedics Department, Wiktor Dega Orthopaedic and Rehabilitation Clinical Hospital, Poznan University of Medical Sciences, 28 Czerwca 1956 Street 135/147, 61-545 Poznan, Poland; mdabrowski@ump.edu.pl
3 Institute of Health Sciences, Hipolit Cegielski State College of Higher Education, Card. Stefan Wyszyński Street 38, 62-200 Gniezno, Poland; gabinet.rogala@gmail.com
* Correspondence: uklejew@ukw.edu.pl (R.U.); winiecki@ukw.edu.pl (M.W.)

Abstract: This paper presents advances in designs of resurfacing arthroplasty endoprostheses that occurred through their historical generations. The critical characteristics of contemporary generation hip resurfacing arthroplasty endoprostheses are given and the failures resulting from the specific generation cemented and short stem fixation of the femoral component are reviewed. On the background of these failures, the critical need arises for an alternative approach to the fixation of components of resurfacing arthroplasty leading towards the first generation of biomimetic fixation for resurfacing arthroplasty endoprostheses. The state of the art of the completed bioengineering research on the first biomimetic fixation for resurfacing arthroplasty endoprostheses is presented. This new design type of completely cementless and stemless resurfacing arthroplasty endoprostheses of the hip joint (and other joints), where endoprosthesis components are embedded in the surrounding bone via the prototype biomimetic multi-spiked connecting scaffold (MSC-Scaffold), initiates the first at all generations of biomimetic endoprostheses of diarthrodial joints.

Keywords: resurfacing arthroplasty; resurfacing endoprostheses; biomimetic multi-spiked connecting scaffold (MSC-Scaffold); biomimetic fixation

Citation: Uklejewski, R.; Winiecki, M.; Dąbrowski, M.; Rogala, P. Towards the First Generation of Biomimetic Fixation for Resurfacing Arthroplasty Endoprostheses. *Biomimetics* **2024**, *9*, 99. https://doi.org/10.3390/biomimetics9020099

Academic Editor: Maria Grazia Cascone

Received: 21 December 2023
Revised: 21 January 2024
Accepted: 6 February 2024
Published: 8 February 2024

1. Introduction

Hip resurfacing arthroplasty (HRA) involves replacing the femoral head acetabular articular cartilage, and subchondral bone with prosthetic components designed to replace the removed articular cartilage and subchondral periarticular bone to minimize the change in overall joint kinematics [1]. The important benefit of advocating for HRA over the long-stem total hip arthroplasty (THA) is the possibility of preserving at the initial operation a bone stock. Replacing diseased tissue with near anatomic-sized femoral component retains the potential for revision since the femoral canal was not violated [2,3]. Along with the bone tissue preservation, the HRA design solutions and applied fixation techniques are assumed, contrary to the long-stem THA systems, to allow near-physiological load transfer in periarticular bone allowing the recreation of closely native hip kinematics and bone biomechanics [4]. With its excellent functional outcome [5–7], HRA remains a reasonable alternative to THA in the appropriate patient cohort [8], and by many surgeons, it is considered an excellent option for hip reconstruction in young patients and/or high activity level patients diagnosed with osteonecrosis of the femoral head or acetabular dysplasia [9].

The contemporary generation of HRA endoprostheses has been utilized for over 20 years, while resurfacing arthroplasty has a hundred-year-long history [10–13]. From its beginning, the concept of resurfacing arthroplasty evolved through the variety of designs of endoprosthesis components, different material choices used and through the changes in fixation methods. Through the decades, its success has varied widely and was always limited by specific technological limitations of particular eras. The registered failures of resurfacing arthroplasty endoprostheses coupled with new technological capabilities are the foundations of occurring innovations and indications for the new designs. To achieve successful long-term results, apart from design features of resurfacing arthroplasty endoprostheses and applied manufacturing processes, the key importance also are patient selection and surgeon experience [9].

This review aims to present chronologically the background of the chronological milestones through the past consecutive design generations of resurfacing arthroplasty endoprostheses and critical insight into contemporary HRA endoprostheses, the state-of-art bioengineering research on the first generation of biomimetic fixation for resurfacing arthroplasty endoprostheses.

2. Milestones of Early Materials and Designs of Resurfacing Arthroplasty Endoprostheses

The first design considered to be the start basis for later-known HRA was devised in 1923 by Smith-Petersen and applied to regenerate worn and damaged articular cartilage [14]. The thin, ball-shaped, hollow hemisphere manufactured of glass, which fits over the ball of the hip joint, was placed between the femoral head and the acetabulum and intended to stimulate cartilage regeneration on both sides of the moulded glass joint. The glass shell was supposed to be removed after the restoration of the cartilage. Even though the glass was biocompatible and provided a smooth surface for motion, the system failed immediately because it could not withstand physiological weight-bearing stress. Different types of glass with improved properties were used in further developed versions of this design, leading to attempts with other materials, such as Viscaloid (a derivative of celluloid) in 1925, Pyrex in 1933, and Bakelite in 1937 [15]. The performance of these solutions was far from satisfactory, causing severe inflammatory reactions due to material wear debris in the joint. The glass mould arthroplasty could not withstand weight-bearing pressure and failed shortly after surgery [14].

Around the same time, Groves introduced one of the first ivory hip arthroplasties [16]. In 1927, he replaced the femoral head in an ankylosed hip with a stemmed ivory head replacement. The design was similar to contemporary HRA endoprostheses [17].

In the 1930s, Vitallium®, a new alloy of cobalt, chrome, and molybdenum (CoCrMo), was developed, and in 1938, Smith-Petersen incorporated it in his mould arthroplasty [14,18,19]. This allowed movement between the cup and the bone surfaces of the acetabulum and the femoral head and neck and since then Vitallium® mould arthroplasty has been widely performed. Its survival rate and functional outcome were inferior to total hip arthroplasty, and it was found to provide long-term function of the hip joint [15,18,20], but the general material performance and implant survival turned out to be unpredictable and poor [20–23]. Although Vitallium® was found to be an inert and durable material for this type of surgery, its surface characteristics were less than adequate [24].

Further efforts to explore new materials for joint arthroplasty led the Judet brothers to introduce an acrylic femoral resurfacing implant in 1946 [25]. In this technique, also called hemiarthroplasty, the artificial femoral stem was inserted into the cavity of femoral marrow with or without any kind of cementing. The Judet hip yielded favorable early results; this implant was poorly tolerated, and acrylic wear debris elicited an osteolytic reaction within the hip joint and surrounding tissue, leading to acetabular erosion and implant failure [25]. The acrylic material was soon discarded in favor of CoCrMo.

Sir John Charnley, in the 1950s, introduced hip resurfacing of the low friction material, Teflon (polytetrafluoroethylene, PTFE), to produce thin shells, which were used to "resurface" the femur and acetabulum. In this design, the acetabulum was lined with a thin shell of the material, while the head of the femur was similarly covered with a hollow sphere. It was hoped that the motion would take place preferentially between the two slippery

PTFE surfaces rather than between one of the PTFE surfaces and the bone to which it was attached. Charnley assumed that if the PTFE implant remained stationary in relation to the subjacent bone, there seemed some hope that mechanical bonding might improve if bone grew into irregularities provided for this purpose [26]. This design ended shortly, causing postoperative osteolysis at a very early stage due to PTFE wear debris, leading to catastrophic failure in many cases [27].

Further developments were undertaken to improve on the early failures. In 1964, Townley, based on Judet and Judet's design, developed the CoCrMo femoral component to articulate with a polyurethane acetabular cup [28]. Due to the ultimately demonstrated poor wear characteristics of the polyurethane cup and significant osteolysis, it was substituted with polyethylene (PE). Since it still demonstrated a high failure rate, the PE cup was replaced by a CoCrMo acetabular component. Although there are known earlier examples of the use of a metal-on-metal bearing couple applied in hip replacement, see Wiles in 1938 [29] or Haboush in 1951 [30], Müller and Boltzy were the first who published their results and are recognized as the metal-on-metal resurfacing arthroplasty pioneers [31].

Their initial concept featured a femoral component made from CoCrMo articulating with small Teflon or PE pads, known as "sliding bearings", fixed to the acetabular cup's inner surface. In 1968, Müller [32] shifted away from metal-on-metal articulation, opting instead for a metal-on-PE prosthesis with a curved stem. However, a high early re-operation rate of 50% prompted a revision of this approach [33]. In a similar vein, Gérard developed an HRA system in 1970, which utilized both a CoCrMo femoral head and acetabulum, allowing movement between the components as well as between the components and the surrounding bone [34]. In 1972, PE was implemented as a bearing material in the acetabular cup of this system. Nonetheless, this was discontinued in 1975 following reports about enhanced wear and osteolysis [34,35].

Also, in 1972, Nishio combined a Urist cementless acetabular cup [36] with his own femoral head of CoCrMo, which due to the growing trend for metal-on-PE designs in 1975, was substituted with the acetabular component with a PE-lined cementless socket [37,38]. In 1976, Salzer used the first ceramic-on-ceramic resurfacing endoprosthesis, which had both components made of alumina (Al_2O_3) [39]. Its acetabular component had three pegs for primary stability, and the femoral cup was twisted on to the prepared head. This solution was soon abandoned due to the high rates of loosening resulting from the lack of secondary fixation.

The high failure rates of these early-generation devices were the consequence of unproven material selection, the limitations in materials properties, and manufacturing of the time, crude implant design, lack of solid implant fixation at the bone–implant interface, and unrefined surgical techniques and instrumentation. These resurfacing arthroplasties had an unacceptably high wear rate resulting in failure secondary to osteolysis (i.e., aseptic loosening or inflammatory bone resorption) [40–42].

In the development of resurfacing implant generations of the 1970s and 1980s, particular attention has been paid to improved materials and enhanced fixation strategies to improve survivorship. New bearing materials were incorporated into hip resurfacing prostheses, and the concept of cementing components to ensure initial rigid fixation of endoprosthesis components within the bone became a way of eliminating issues associated with current cementless designs.

In 1971, Trentani [43] in Italy and Furuya [44] in Japan independently carried out the first cemented double-cup arthroplasty. Trentani and Paltrinieri developed hip resurfacing with a cemented stainless steel femoral head articulating with an acetabular ultra-high molecular weight polyethylene (UHMWPE) cup. Furuya implanted resurfacings using a stainless steel acetabular component articulating with a high-density polyethylene (HDPE) femoral component fixed with cement. In 1972, Freeman [45,46] first implanted a cemented double cup arthroplasty consisting of an HDPE femoral head coupled with a CoCrMo acetabular cup, but in 1974, the design was modified to consist of a CoCrMo femoral head and an HDPE cup. In 1973, Eicher and Capello [47] developed a cemented hip resurfacing using a metal femoral and a polyethylene acetabular component.

Similar solutions in which cemented resurfacing endoprostheses included metal or ceramic femoral heads articulating against polyethylene acetabular cups and a reversed design where a polyethylene head articulated against a metal cup were introduced independently by Wagner [48], Amstutz [2,49,50], Eicher and Capello [47], and Tanaka [51]. These designs also failed, and their poor performance and outcomes were explained by stress shielding, significant loss of blood supply, and consequential compromise to the femoral head [52,53].

In the early 1980s, the development of cementless fixation for resurfacing endoprosthesis components derived from the belief that cement was the main cause of implant failure [54].

In 1982, Amstutz introduced a plasma-sprayed (PS) metal-backed polyethylene acetabular component for use with cement, and in 1983, he implanted the first cementless resurfacing arthroplasty with a Ti-6Al-4V femoral component articulating with modular UHMWPE acetabular liners and a porous backing of pure titanium (Ti) mesh [49,50].

In 1987, Amstutz et al. presented another cementless, porous surface replacement. In this design, porous-coated or sintered beads coated with a metallic acetabular shell was used for bone fixation on both the acetabular shell with the UHMWPE insert and the CoCrMo porous-coated femoral component [55]. Although these designs achieved adequate initial cementless fixation, there were more failures with the femoral than the acetabular component due to the high wear rate of the thin polyethylene line [56,57].

In 1989, Buechel and Pappas introduced a Ti-6Al-4V modular acetabular component mated with a Ti-6Al-4V femoral head coated with a titanium nitride ceramic material. Although their laboratory simulations on the use of titanium nitride ceramic film against polyethylene in a joint couple were encouraging [58], the case report at the 11-year follow-up showed severe metallosis with catastrophic wear on the polyethylene liner [59].

A chronological review of early generations of resurfacing arthroplasty endoprostheses is presented in Table 1.

Table 1. Early resurfacing arthroplasty endoprostheses (chronological review).

Surgeon/Designer	Introduced	Materials	Fixation	Reference
Smith-Petersen	1923	Glass		[14]
Smith-Petersen	1925	Viscaloid		
Groves	1927	Ivory		[16]
Smith-Petersen	1933	Pyrex		
Smith-Petersen	1937	Bakelite		
Smith-Petersen	1938	Vitallium® (CoCrMo)		[14,18,19]
Judet and Judet	1946	Acrylic		[25]
Charnley	1951	Teflon		[26,27]
Townley	1964	CoCrMo femoral head/polyurethane acetabular cup		[28]
Müller and Boltzy	1968	CoCrMo femoral head/PE acetabular cup CoCrMo femoral head/CoCrMo acetabular cup		[31]
Gérard	1970	CoCrMo femoral head/CoCrMo acetabular cup CoCrMo femoral head/PE acetabular cup		[34,35]
Trentani & Paltrinieri	1971	stainless steel femoral head/UHMWPE acetabular cup	Cemented	[42]
Furuya	1971	HDPE femoral head/stainless steel acetabular cup	Cemented	[43]
Nishio	1972	CoCrMo femoral head/CoCrMo acetabular cup	Cementless	[37]
Nishio	1975	CoCrMo femoral head/polyethylene acetabular cup	Cementless	[38]

Table 1. *Cont.*

Surgeon/Designer	Introduced	Materials	Fixation	Reference
Freeman	1972 1974	HDPE femoral head/CoCrMo acetabular cup CoCrMo femoral head/UHMWPE acetabular cup	Cemented Cemented	[45,46]
Eicher and Capello	1973	CoCrMo femoral head/UHMWPE acetabular cup	Cemented	[47]
Wagner	1974	CoCrMo femoral head/UHMWPE acetabular cup Al_2O_3 femoral head/UHMWPE acetabular cup	Cemented	[48]
Amstutz	1975	CoCrMo femoral head/UHMWPE acetabular cup	Cemented	[49,50]
Salzer	1976	Al_2O_3 femoral head/Al_2O_3 acetabular cup	Cementless	[39]
Tanaka	1978	CoCrMo femoral head/UHMWPE acetabular cup	Cemented	[50]
Amstutz	1982	CoCrMo femoral head/UHMWPE PS metal-backed acetabular cup	Cementless (Press fit)	[49,50]
Amstutz	1983	Ti-6Al-4V femoral head with sintered titanium fiber mesh/UHMWPE liner with porous backing of pure titanium mesh	Cementless	[49,50]
Amstutz	1987	Porous-coated CoCrMo femoral head UHMWPE insert Porous-coated or sintered bead-coated metallic acetabular cup	Cementless	[55]
Buechel and Pappas	1989	Titanium nitride-coated Ti-6Al-4V femoral head, UHMWPE liner, Ti-6Al-4V acetabular cup	Cementless	[57,59]

The results of hip resurfacing in the 1970s and 1980s were disappointing. Failures were the result of poor materials (conventional polyethylene was susceptible to wear [60,61]), poor implant design, inadequate instrumentation, and an imprecise surgical technique [62]. The extensive damage to the acetabulum was partially due to the significant bone removal needed for the acetabular component and its cement mantle, largely attributed to periprosthetic osteolysis. Moreover, the combination of a large articulation diameter with thin polyethylene cups or liners resulted in rapid deterioration and the generation of a considerable amount of biologically active particles, causing bone loss and loosening of the implant. Failures of hip resurfacing in the 1970s and 1980s were also attributed to avascular necrosis of the femoral head and acetabular component loosening due to high frictional torque, but with present knowledge, it is clear that the production of large volumes of biologically active particulate wear debris induced osteolysis that led to bone loss and implant loosening and caused a high incidence of fractures of the femoral neck [63]. The consistency of peri-implant bone destruction with wear particle-induced osteolysis, not avascular necrosis, was confirmed by the retrieval studies carried out by Howie et al. [64] and Campbell et al. [65]. It led to these solutions of hip resurfacing being largely abandoned by the mid-1980s [62].

3. Characteristics of Contemporary Hip Resurfacing Arthroplasty Endoprostheses

Resurfacing arthroplasty has been experiencing a renaissance since the early 1990s. The first two designs to appear were introduced in the early 1990s by Wagner [66] and McMinn [67]. From this time on, all further hip resurfacing devices used exclusively CoCrMo metal-on-metal bearings. Both of these first systems were cementless. Wagner's endoprosthesis had a threaded internal geometry of the femoral component and a grit-blasted Ti surface coating at the bone interfaces, while McMinn's endoprosthesis had anti-rotation ridges and a short epiphyseal stem to assist with femoral component alignment and stability; the first was coated with hydroxyapatite (HA) and later press fit, while the acetabular component had HA coating and peripheral fins for rotational stability. The experience gained with these solutions has shown the enduring fixation of the acetabular components [68] and improved results in terms of loosening at the early stage [69] but has demonstrated poor outcomes in longer-term follow-up [54,70,71]. The Wagner system was discontinued, while subsequent modifications to the McMinn design involved cement fixation of both components, and then the hybrid configuration evolved into the development of the Cormet™ (Corin Group, Cirencester, UK) in 1997 and current Cormet 2000 (Corin Medical Ltd., Cirencester, UK) in 2007, as well as Birmingham Hip Resurfacing (BHR)™ (Smith & Nephew, Memphis, TN, USA) endoprostheses in 1997. Meanwhile, Amstutz began a series of innovations that culminated in the Conserve Plus™ (Wright Medical Technology Inc., Arlington, TN, USA) in 1993 and began implanting it at the end of 1996 [71]. It was a hybrid (i.e., porous fixation with sintered beads on the acetabular side and cemented on the femoral side) [54].

At the turn of the 21st century, most hip resurfacing systems were hybrid with a thin-walled one-piece cementless acetabular component and a cemented femoral component. The femoral components featured a short epiphyseal stem designed for alignment during insertion.

These systems offered several benefits, such as enhanced durability in fixation, reduced wear, improved bone tissue protection, and a decreased rate of complications, particularly fractures and sprains. Numerous clinical studies and joint registry reports provided extensive evidence indicating positive outcomes and survival of surface implants. They proved to be a success, with 96%, 92%, and 88.5% survivorship at ten years for Cormet, BHR, and Converse, respectively [72–76].

Following these successes, in the early 2000s, numerous hybrid hip resurfacing systems emerged, e.g., the Durom™ (Zimmer Inc., Warsaw, IN, USA) introduced in 2001, the Articular Surface Replacement (ASR™) (DePuy Orthopaedics Inc., Warsaw, IN, USA), introduced in 2003, the Icon™ (IO International Orthopaedics Holding, Geisingen, Germany) and the ReCap™ (Biomet Inc., Warsaw, IN, USA), both introduced in 2004, the ADEPT™ hip resurfacing system (Finsbury Orthopaedics Ltd., Leatherhead, UK), introduced in 2005, the MITCH hip resurfacing system (Stryker, Kalamazoo, MI, USA), introduced in 2006, the ROMAX® Resurfacing System (Medacta, Castel San Pietro, Switzerland), introduced in 2008, the DynaMoM hip resurfacing prosthesis (Tornier, Saint-Ismier, France), introduced in 2008, and the Minimally Invasive Hip Resurfacing (MIHR) International® metal-on-metal (MoM) hip system (Comis Orthopaedics Ltd., Birmingham, UK), introduced in 2009 [77]. A list of contemporary generations of HRA endoprostheses is presented in Table 2.

Table 2. Contemporary HRA endoprostheses.

System	Introduced	Femoral Component Material and Fixation	Acetabular Component Bearing Material/Bone-Contacting Material	References
Wagner's	1991	CoCrMo cementless, press-fit	CoCrMo/grit-blasted Ti coating, cementless	[66]
McMinn's	1992	CoCrMo cementless, initially HA coating, then cemented	CoCrMo/HA coating, cementless	[67]
Conserve Plus™ (Wright Medical Technology Inc., Arlington, TN, USA)	1996	CoCrMo cemented	CoCrMo/CrCrMo beads + HA coating, cementless	[78–80]
BHR™ (Smith & Nephew, Memphis, TN, USA)	1997	CoCrMo cemented	CoCrMo/CrCrMo beads + HA coating, cementless	[81,82]
Cormet™ (Corin Group, Cirencester, UK)	1997	CoCrMo cemented (cementless option with PS Ti, HA coating)	CoCrMo/PS Ti + HA coating, cementless	[83,84]
Durom™ (Zimmer Inc., Warsaw, IN, USA)	2001	CoCrMo cemented	CoCrMo/PS Ti, cementless	[85,86]

Table 2. *Cont.*

System	Introduced	Femoral Component Material and Fixation	Acetabular Component Bearing Material/Bone-Contacting Material	References
ASR™ (DePuy Orthopaedics Inc., Warsaw, IN, USA)	2003	CoCrMo cemented	CoCrMo/CrCrMo beads + HA coating, cementless	[87,88]
Icon™ (IO International Orthopaedics Holding, Geisingen, Germany)	2004	CoCrMo cemented	CoCrMo/CrCrMo beads + HA coating	[89,90]
ReCap™ (Biomet Inc., Warsaw, IN, USA)	2004	CoCrMo cemented (cementless option PS Ti-6Al-4V)	CoCrMo/PS Ti-6Al-4V + HA coating	[91]
ACCIS™ (Van Straten Medical, The Netherlands; Implantcast, Buxtehude, Germany)	2004	TiNbN-coated CoCrMo cemented fixation, cementless from 2009	TiNbN-coated CoCrMo/ PS Ti	[92]
ADEPT® (Finsbury Orthopaedics Ltd., Leatherhead, UK)	2005	CoCrMo cemented	CoCrMo/CrCrMo beads + HA coating	[93,94]
MITCH (Stryker, Kalamazoo, MI, USA)	2006	CoCrMo cemented	CoCrMo/PS Ti + HA coating	[95]
ESKA-BIONIK® (ESKA Implants GmbH & Co., Lübeck, Germany)	2006	CoCrMo; Spongiosa Metal® (cemented option)	CoCrMo + CoCrMo insert/Spongiosa Metal®	[96–98]
ESKA-CERAM® (ESKA Implants GmbH & Co., Lübeck, Germany)	2007	CoCrMo; Spongiosa Metal® (CoCrMo with TiNb coating)	polyurethane/Al_2O_3 + polyurethane/Al_2O_3 insert/ Spongiosa Metal® (CoCrMo with TiNb coating)	[98–100]
Cormet 2000 (Corin Medical Ltd., Cirencester, UK)	2007	CoCrMo cemented	CoCrMo/PS Ti + HA coating	[84,101]
ROMAX® (Medacta, Castel San Pietro, Switzerland)	2008	CoCrMo cemented	CoCrMo/PS Ti + HA coating	[102,103]
DynaMoM (Tornier, Saint-Ismier, France)	2008	CoCrMo cemented	CoCrMo/PS Ti + HA coating	[104]
MIHR International® (Comis Orthopaedics Ltd., Birmingham, UK)	2009	CoCrMo cemented	CoCrMo/HA coating, cementless	[105]

Instead of all implants having a CoCrMo-on-CoCrMo bearing, the Advanced Ceramic Coated Implant Systems (ACCIS™) (Van Straten Medical, The Netherlands; Implantcast, Buxtehude, Germany), introduced in 2004, has titanium–niobium–nitride (TiNbN) ceramic surfaces engineered by physical vapor deposition (PVD) to minimize wear and prevent tribocorrosion and metal ion release. Total hip resurfacing represented the fastest-growing section in orthopedic surgery [106,107]. Although some early clinical follow-up studies of this system demonstrated promising results [108], there were also results reporting catastrophic failure of the prosthesis, and an unacceptably high revision rate was demonstrated due to unknown causes that led to cease implanting the ACCIS™ [109,110].

Another step towards biomimetics was the series of ESKA hip resurfacing systems developed by ESKA Implants GmbH & Co. (Lübeck, Germany) beginning in 2006. The first was a metal-on-metal ESKA-BIONIK® hip resurfacing system (also known as Biosurf® hip resurfacing), which has a unique bearing surface hydrodynamic lubrication in the bearings through the concavo–convex pattern designed to reduce abrasive wear. The pattern provided circumferentially distributed escape dimples for wear particles and improved lubrication by providing room for the lubricant [111]. The acetabular component had a spongiosa metal structured surface called Spongiosa Metal® made of a CoCrMo with a titanium niobium (TiNb) coating or an HA coating, available upon request, for cementless anchorage through osseous integration that proved to have excellent long stability in clinical trials [96,97]. The cementless anchorage via the Spongiosa Metal® was also applied in the case of the femoral component. Further, in 2007, based on the composite material ENDOCERAM®, consisting of a polyurethane matrix and a mixed-in glass ceramic powder [100,112], the ceramic-on-ceramic ESKA-CERAM® hip resurfacing system was launched. The ESKA resurfacing implants were the only designs with a cementless acetabular shell in combination with a modular ceramic insert.

Research indicates that the use of cement does not consistently ensure long-term stability of endoprostheses in bones. The most common (ca. 75% of observed) complications of currently used cement resurfacing arthroplasty include the resorption of periarticular bone tissue, loosening at the bone–cement–implant junction zone, migration of endoprosthesis components, and femoral fractures [113–116]. Additionally, stress-shielding areas near the short stem of the femoral component often lead to loosening and migration [116–121]. In cemented HRA, while cement initially anchors the femoral component, it to penetrates deeply into the femoral head's cancellous bone. Often, the area affected by cement penetration exceeds 30% of the femoral head's total volume (some studies report over 50% [122]), leading to reduced local blood flow. This, in turn, weakens the internal microstructure of the cancellous bone in the femoral head, resulting in various complications [123–129].

Currently, available HRA systems exhibit a range of survival rates over five years, from a high of 97.1% to a low of 80.9% [10]. This variability has raised safety concerns regarding certain resurfacing arthroplasty endoprostheses. As a consequence, systems such as the ASR™ by DePuy Orthopaedics Inc., based in Warsaw, IN, USA, have been withdrawn from the market due to their significant rate of postoperative complications [130].

The predominant reason for early HRA failures, accounting for approximately 35% of necessary revision surgeries, is femoral neck fracture [131,132]. Additionally, aseptic loosening of either the femoral or acetabular components is another frequent reason for HRA failures [133,134]. Aseptic bone necrosis (osteonecrosis), often linked with periprosthetic fractures or identified as a contributing factor to such fractures after HRA [116], is also seen as a consequence of using cement for endoprosthesis component fixation [135] or is due to intraoperative damage to the blood vessels supplying the femoral head [136]. The heat generated during cement polymerization can severely damage the surrounding implant tissue, leading to the collapse of the femoral head [137,138]. Moreover, as previously noted, the implantation process of the femoral HRA endoprosthesis component typically forces substantial amounts of cement into the cancellous bone of the femoral head, creating a thick cement layer [139].

Aseptic bone necrosis is often seen in the early and middle stages following hip surgery, typically linked to either reduced blood flow to the femoral head or heat damage incurred during the operation [140]. Zustin et al. [140] conducted a histological analysis of 123 bone–implant samples from various resurfacing endoprostheses systems, including ASR™ by DePuy Orthopaedics, BHR™ by Smith & Nephew, Cormet™ by Corin Group, Durom™ by Zimmer Inc., and ReCap™ by Biomet Inc. These samples were collected from patients who had diagnoses other than osteonecrosis prior to surgery. The study found that osteonecrosis occurred in 88% of the cases, often linked with periprosthetic fractures. Of the revisions examined, 85 were due to periprosthetic fractures, with 60% of these fractures exhibiting complete bone necrosis near the fracture line, which are thus classified as post-necrotic fractures. Additionally, 8% of the revisions were due to the loosening of the acetabulum, and the remaining 23% were for various other reasons, including groin pain related to the femoral component. The majority of bone–implant specimens analyzed displayed extensive aseptic necrosis histologically, identified as the reason for 46% of all failures, particularly those linked to post-necrotic periprosthetic fractures and the collapse of the femoral head [141]. Steiger et al. [142] reported that, excluding infections, the leading causes for the primary revision following HRA are hip fractures (43%), loosening/lysis (32%), metal allergic reactions (7%), and pain (6%). Therefore, in cases involving serious postoperative complications, the primary revision focused on the femoral component accounts for 62% of all such revisions in the procedures mentioned [142].

ReCap™ is the only fully porous-coated femoral component currently available. The seven-year follow-up study carried out by Gross [143] suggested that cementless femoral fixation with a short epiphyseal stem may be a viable alternative to the cement fixation of HRA and pointed out that a study on a larger number of patients should be warranted.

In Figure 1, we present, from our own experience, the roentgenogram of a patient (female, 52 y.o.) with a perceived limp requiring a cane for ambulation, showing the loosening and fracture at the femoral neck that occurred after one month.

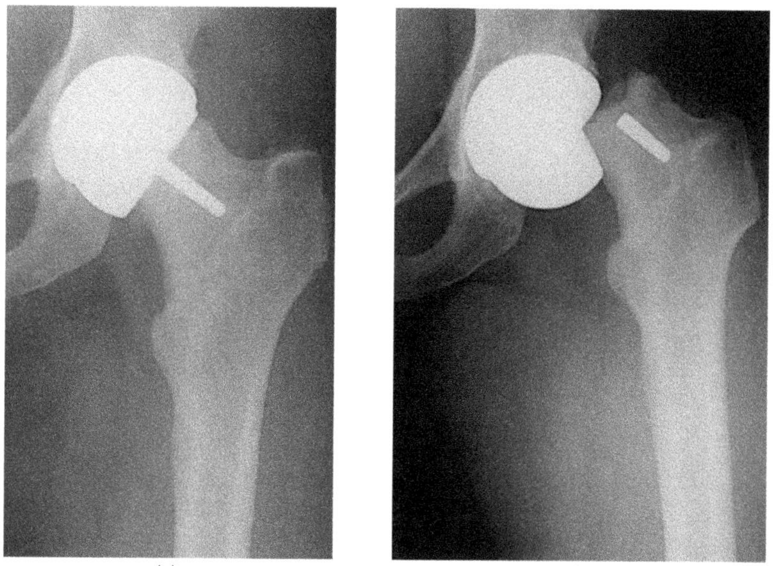

(**a**) (**b**)

Figure 1. Anteroposterior radiographs demonstrating the Birmingham Hip Resurfacing system in situ (**a**) with femoral component loosening and (**b**) femoral neck fracture after one month.

It should be noted that the cementless femoral fixation applied in the ReCap™ using a porous coating with sintered beads allows only shallow ingrowth of bone tissue into the pore space of the coating, and this requires the use of a short epiphyseal stem to ensure the proper fixation of the cap. Further studies [144,145] justify the validity of striving for more biomimetic bone ingrowth fixation methods in the case of the femoral component of hip resurfacing versus non-biomimetic approaches in terms of the cemented fixation methods.

4. First Biomimetic Resurfacing Arthroplasty Endoprosthesis with the Multi-Spiked Connecting Scaffold

A significant shift in the design philosophy for HRA endoprostheses emerged with the introduction of the original multi-spiked connecting scaffold (MSC-Scaffold) concept by Rogala [146–148]. This concept moved away from the traditional cemented hip endoprostheses with short epiphyseal stems towards a biomimetic, stemless, and entirely cementless design, which preserves periarticular bone tissue. The MSC-Scaffold employs a spike-palisade system that integrates the components of resurfacing arthroplasty HRA endoprostheses with the intertrabecular space of the periarticular cancellous bone. The prototype of the MSC-Scaffold was developed by a bioengineering clinical research team under the auspices of two research grants from the National Science Centre Poland (No.

4T07C05629 and No. N518412638, led by R. Uklejewski) and further research. This represents a pivotal innovation in the fixation of resurfacing endoprosthesis components within the periarticular trabecular bone. A comprehensive overview of the bioengineering research conducted on the biomimetic MSC-Scaffold prototype for a new generation of HRA endoprostheses can be found in a recent monograph [149].

The MSC-Scaffold's spikes are designed to imitate the natural interdigitation system of the subchondral bone, which intricately weaves into the trabeculae of the surrounding cancellous bone. This design facilitates a gradual structural and biomechanical transition between the joint's distinct morphological elements—the articular cartilage and the surrounding trabecular bone tissue of the epiphysis. Illustrated in Figure 2, the scheme shows how the subchondral bone, through its system of interdigitations, interlocks with the trabeculae of the periarticular cancellous bone. This unique interlocking mechanism forms a vital transition zone that naturally secures the articular cartilage of diarthrodial joints to the periarticular cancellous bone. Moreover, the spacing between the spikes in the MSC-Scaffold prototype is designed to allow for the growth of new bone tissue. This space acts as an osteoconductive interface, promoting the integration of periarticular trabecular bone tissue in diarthrodial human joints undergoing resurfacing arthroplasty with these endoprostheses [149].

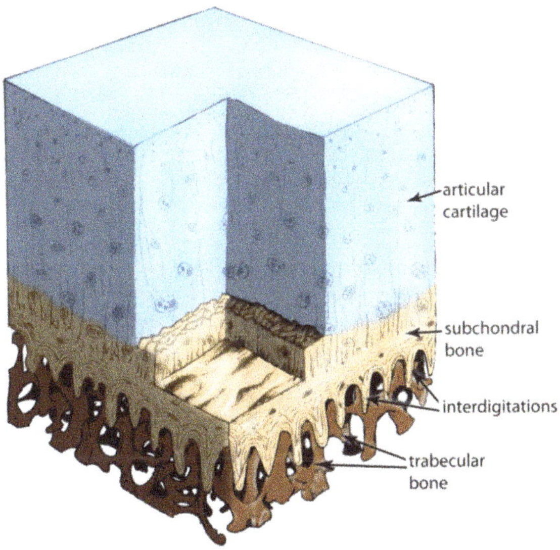

Figure 2. Histological scheme showing the hyaline cartilage and the subchondral bone with its interdigitations, anchoring among the trabeculae of the cancellous bone tissue [149].

The MSC-Scaffold facilitates the initial setting of endoprosthesis components within the bone, stimulating the bone tissue growth in its interspike spaces, leading to osseointegration. This process securely anchors the resurfacing endoprosthesis components within the periarticular bone for long-term stability. During surgery, the surgeon inserts the MSC-Scaffold spikes into the bone to a specific depth for initial mechanical stability, while the subsequent growth of new bone tissue in the spaces between the spikes provides final biological fixation during the patient's postoperative rehabilitation. The MSC-Scaffold's non-cemented approach to fixing HRA endoprostheses in a physiologically optimal manner could potentially reduce complications associated with bone cement use. When implanting the biomimetic MSC-Scaffold in the cancellous bone's intertrabecular space, the spikes in a controlled fashion disrupt the bone's trabecular microarchitecture to a beneficial osteoinductive degree. This disruption prompts adaptive remodeling and bone tissue development in the interspike areas. Furthermore, the spikes are designed to integrate with the cancellous bone's trabecular bioconstruction, effectively dampening vibrations from dynamic joint loads and enhancing the biomechanical stability of the endoprostheses in bone tissue. This integration helps prevent the spraining and loosening of the implanted components [149].

Designed according to above-described assumptions, the MSC-Scaffold prototype facilitates the in vivo infill of its interspike spaces with newly grown trabecular bone tissue, reflecting the natural microstructure, a process not achievable with cemented HRA. This biomimetic fixation device and method for the HRA endoprosthesis components within the periarticular bone suggests that the mechanical load distribution within the periarticular bone will closely resemble that in a natural hip joint. The biomimetic MSC-Scaffold prototype aims to replicate in the implant–bone interface the transfer of mechanical loads occurring in a natural joint, which leads to almost natural biodynamics and bone tissue remodeling around the implant. Consequently, effective osteoconduction is anticipated, promoting bone tissue growth into the interspike space of the MSC-Scaffold. Additionally, the macrodimensions of the femoral component's bearing part are designed to conserve the posterolateral and medial epiphyseal femoral arteries (*subcapsular arteriae retinaculares: superior and inferior*) of the femoral head. It ensures the preservation of physiological blood supply, meaning the preservation of key factors for adequate remodeling of the femoral head's trabecular bone [149].

Initial attempts at producing MSC-Scaffold prototypes through conventional subtractive manufacturing methods, such as die-sinking electrical discharge machining, were unsuccessful. This led to the conclusion that the fabrication of MSC-Scaffold for anchoring the components of joint resurfacing endoprostheses, as an integral part of these devices, necessitates the use of additive manufacturing technologies. Despite these methods being relatively obscure and not widely accessible in the mid-2000s, particularly for producing biocompatible metals and alloys [150], efforts were made to explore and assess the technological capabilities of fabricating the MSC-Scaffold using available additive technologies. Experiments with Selective Laser Sintering (SLS) and Electron Beam Melting (EBM) uncovered numerous unacceptable flaws in the early prototypes [151]. Further exploration to achieve manufacturing standards that ensured uniform density and mechanical properties akin to homogeneous materials led to the identification of Selective Laser Melting (SLM) as a suitable technology for producing prototype HRA endoprosthesis with the MSC-Scaffold [152–154]. An illustration of the HRA endoprosthesis prototype featuring the MSC-Scaffold is provided in Figure 3.

The development of the MSC-Scaffold prototype involved refining its structural and osteoconductive characteristics within the interspike space. This process led to a reevaluation of the initial design assumptions of the MSC-Scaffold, highlighting crucial insights into compensating for the revealed technological constraints of SLM technology in future design strategies. This meant focusing on enhancing pro-osteoconductive structural functionality for the effective design of subsequent MSC-Scaffold prototypes [155]. Biological verification of the functionalized MSC-Scaffold with human osteoblasts demonstrated that its interspike region offers an efficient space for the proliferation and spread of osteoblasts. These cells exhibited a propensity to form a three-dimensional intercellular network, reflecting the characteristic biostructure of lamellar bone tissue found in the trabeculae of trabecular bone [156].

Early pilot implantations of structurally functionalized MSC-Scaffold preprototypes in an animal model (swine; breed: Polish Large White) demonstrated no postoperative issues, such as implant loosening, migration, or other early complications. Histopathological analysis revealed that the majority of the interspike spaces were filled with newly formed and mineralized bone tissue. It secured the primary biological fixation of the MSC-Scaffold preprototypes in the periarticular trabecular bone [157]. These findings also highlighted the necessity for improved bone contact, suggesting the surface of the MSC-Scaffold spikes be coated with calcium phosphate (CaP) to more closely mimic the biochemical properties of native bone biomineral (HA).

Figure 3. Prototype of hip resurfacing arthroplasty endoprosthesis with the multi-spiked connecting scaffold (MSC-Scaffold) [149].

Research into electrochemical cathodic deposition of CaPs on the spike surfaces initially used constant current densities [158] and then moved to a potentiostatic electrodeposition process, followed by immersion in simulated body fluid [159]. This led to the development of efficient process parameters that ensured biomineral coverage of the bone-contacting surface, with a Ca/P molar ratio similar to that of native bone HA. Testing the CaP-coated surfaces with human osteoblasts indicated that such modifications are favorable for bone cell proliferation, alkaline phosphatase activity, and hence, the mineralization and osteoinduction/osteointegration potential of the MSC-Scaffold. The distance between the MSC-Scaffold spikes was found to be a significant factor affecting alkaline phosphatase activity. Biointegration assessments of these prototypes implanted into the knee joints of swine (breed: Polish Large White), performed eight weeks post-implantation, showed a scaffolding effect with the majority of interspike spaces filled by newly formed and remodeled bone tissue, ensuring primary biological fixation. Notably, based on calculations performed on micro-CT reconstructions of explanted knee joints, a higher percentage (about 12%) of trabeculae was observed between the spikes of the CaP-modified MSC-Scaffold [159].

Biomechanical studies on the bone–implant system relevant to the MSC-Scaffold's design were conducted, with numerical simulations identifying key geometric features that ensure physiological load transfer from the MSC-Scaffold to the surrounding bone [160]. A validated numerical model, developed and tested with micro-CT-assisted mechanical tests, simulated the MSC-Scaffold embedded in cancellous bone material [161]. Analysis of the Huber–Mises–Hencky (HMH)-reduced stress distribution from these simulations confirmed that the structural biomimicry of the MSC-Scaffold prototype enables a physiologically uniform surface transfer of the mechanical load from the spikes to the trabeculae of the periarticular trabecular bone. In conclusion, the early postoperative biomechanical load capacity (loadability) of the articular surface of the cementless and stemless HRA endoprosthesis with the MSC-Scaffold is considered to be *the crucial design criterion* for such endoprostheses. In the early postoperative rehabilitation period, the controlled loading by the patient of the joint with such endoprosthesis will enable bone tissue to grow in the MSC-Scaffold interspike spaces and will ensure the final bone–implant fixation.

Pilot studies on the MSC-Scaffold in an animal model involved the implantation of prototype partial knee resurfacing endoprostheses with the MSC-Scaffold to ten experimental animals (swine; breed: Polish Large White) [162]. Micro-CT analyses of specimens explanted from these animals eight weeks after surgery revealed (1) the interspike space occupied by newly formed and remodeled mature bone tissue, (2) initial stages of osseointegration in the form of bone tissue creeping substitution identified in deeper areas of the MSC-Scaffold and (3) the distribution of different radiological phases changing as a function of the distance from the bases of the spikes [162]. These results demonstrate that the biomimetic fixation method via the MSC-Scaffold ensures a cementless and stemless anchoring of the endoprosthesis component in the trabecular bone.

With the completion of preclinical bioengineering research on the MSC-Scaffold, the next phase of clinical surgical research in humans is set to begin, involving experimental surgical treatments of damaged knee and hip joints using prototype resurfacing endoprostheses with the biomimetic MSC-Scaffold.

5. Summary and Final Remarks

The goals of HRA are restoring joint anatomy, biomechanics, and function while prolonging the life of a patient with endoprosthesis by preserving bone stock for easy further possible revision arthroplasty in the form of traditional total hip arthroplasty. Over decades of its history, the designs of HRA endoprostheses have undergone a variety of technological innovations in terms of evolution in materials and fixation techniques, which often resulted in promising outcomes and the extension of implant survival. Although these early hip resurfacing designs showed good early functional results, they most disappointingly failed in longer follow-ups due to the inadequacy of materials, poor implant design, inadequate instrumentation, and an imprecise surgical technique. Most studies on the failures of contemporary resurfacing arthroplasties, in which practically all femoral components are fixed with cement, lead to the conclusion that femoral cement failure is the most common late cause of failure in hip resurfacing. In osteonecrotic femoral heads, there is a lack of adequate blood supply, which prevents peri-implant bone regeneration and deepens the deterioration of bone surrounding the implant leading to dysfunction in terms of load transfer in the artificial joint. This often led to femoral loosening, migration, or fractures that resulted in ceasing some of the contemporary designs of resurfacing endoprostheses. Since cement was a confirmed causative factor in femoral failures after resurfacing, the need arises for biomimetic bone–implant fixation methods that would provide a biomimetic structure and biomechanics of the artificial joint.

The newly developed fixation method for resurfacing arthroplasty endoprostheses is distinguished by the biomimetics inherent in its MSC-Scaffold, which are characterized by the following:

(1) Resemblance to the microstructure of the periarticular subchondral and cancellous bone tissue;
(2) Conservation of the femoral head's posterolateral and medial epiphyseal arteries (*subcapsular arteriae retinaculares: superior and inferior*);
(3) Facilitation of a load transfer that mimics natural bone biomechanics, reflecting the mechanical behavior observed in a natural hip joint where the load is transmitted through the trabeculae in the femoral head and neck, continuing along the femoral shaft.

Against many observed failures of the standard fixation technique of contemporary HRA endoprostheses, where the femoral component is fixed with the use of cement, our prototype of a biomimetic MSC-Scaffold can be regarded as a promising breakthrough in bone–implant advanced interfacing in joint resurfacing arthroplasty endoprostheses fixation techniques. The MSC-Scaffold prototype manufactured with modern advanced laser additive technology opens a new generation for the first biomimetic resurfacing joint endoprostheses.

This new design type of completely cementless and stemless resurfacing arthroplasty endoprostheses of the hip joint (and other joints), where endoprosthesis components are embedded in the surrounding bone via the prototype biomimetic multi-spiked connecting scaffold (MSC-Scaffold), initiates the first of all generations of biomimetic endoprostheses of diarthrodial joints.

Author Contributions: R.U.: drafted, edited, and approved the manuscript; M.W.: drafted and edited the manuscript; M.D. and P.R.: edited the manuscript. All authors have read and agreed to the published version of the manuscript.

Funding: This research received no external funding.

Institutional Review Board Statement: Not applicable.

Data Availability Statement: No new data were created or analyzed in this study. Data sharing is not applicable to this article.

Conflicts of Interest: The authors declare no conflicts of interest.

References

1. Callaghan, J.J.; Rosenberg, A.G.; Rubash, H.E.; Clohisy, J.C.; Beaulé, P.E.; DellaValle, C.J. *The Adult Hip (Two Volume Set): Hip Arthroplasty Surgery*, 3rd ed.; Lippincott Williams & Wilkins: Baltimore, MD, USA, 2015.
2. Amstutz, H.C.; Grigoris, P.; Dorey, F.J. Evolution and future of surface replacement of the hip. *J. Orthop. Sci.* **1998**, *3*, 169–186. [CrossRef]
3. Amstutz, H.C.; Sparling, E.A.; Grigoris, P.; Campbell, P.A.; Dorey, F.J. Surface Replacement: The Hip Replacement of the Future? *HIP Int.* **1998**, *8*, 187–207. [CrossRef]
4. Quesada, M.J.; Marker, D.R.; Mont, M.A. Metal-on-metal hip resurfacing: Advantages and disadvantages. *J. Arthroplast.* **2008**, *23*, 69–73. [CrossRef] [PubMed]
5. Daniel, J.; Pradhan, C.; Ziaee, H.; Pynsent, P.B.; McMinn, D.J. Results of Birmingham hip resurfacing at 12 to 15 years: A single-surgeon series. *Bone Jt. J.* **2014**, *96-B*, 1298–1306. [CrossRef] [PubMed]
6. Ford, M.C.; Hellman, M.D.; Kazarian, G.S.; Clohisy, J.C.; Nunley, R.M.; Barrack, R.L. Five to Ten-Year Results of the Birmingham Hip Resurfacing Implant in the US: A Single Institution's Experience. *J. Bone Jt. Surg.* **2018**, *100*, 1879–1887. [CrossRef]
7. Hellman, M.D.; Ford, M.C.; Barrack, R.L. Is there evidence to support an indication for surface replacement arthroplasty?: A systematic review. *Bone Jt. J.* **2019**, *101-B* (Suppl. A), 32–40. [CrossRef]
8. Amstutz, H.C.; Ball, S.T.; Le Duff, M.J.; Dorey, F.J. Resurfacing THA for patients younger than 50 year: Results of 2- to 9-year followup. *Clin. Orthop. Relat. Res.* **2007**, *460*, 159–164. [CrossRef] [PubMed]
9. Lawrie, C.M.; Barrack, R.L. Hip resurfacing arthroplasty—What has history taught us? *Ann. Jt.* **2020**, *5*, 20. [CrossRef]
10. Cadossi, M.; Tedesco, G.; Sambri, A.; Mazzotti, A.; Giannini, S. Hip Resurfacing Implants. *Orthopedics* **2015**, *38*, 504–509. [CrossRef]
11. Amstutz, H.C.; Le Duff, M.J. Hip resurfacing: History, current status, and future. *HIP Int.* **2015**, *25*, 330–338. [CrossRef]
12. Clough, E.J.; Clough, T.M. Metal on metal hip resurfacing arthroplasty: Where are we now? *J. Orthop.* **2020**, *23*, 123–127. [CrossRef]
13. Al-Jabri, T.; Ridha, M.; McCulloch, R.A.; Kayani, B.; Arif, A.; Habad, M.; Kosuge, D.; Jayadev, C.; Donaldson, J.; Skinner, J.A. Hip Resurfacing Arthroplasty: Past, Present and Future. *Orthop. Rev.* **2023**, *15*, 77745. [CrossRef]
14. Smith-Petersen, M.N. Evolution of mould arthroplasty of the hip joint. *J. Bone Jt. Surg. Br.* **1948**, *30B*, 59–75. [CrossRef]
15. Das, S.S.; Chakraborti, P. Development of Biomaterial for Total Hip Joint Replacement. *IOP Conf. Ser. Mater. Sci. Eng.* **2018**, *377*, 012177. [CrossRef]
16. Groves, E.W.H. Some contributions to the reconstructive surgery of the hip. *Br. J. Surg.* **1927**, *14*, 486–517. [CrossRef]
17. Szostakowski, B.; Jagiello, J.; Skinner, J.A. ArtiFacts: Ivory Hemiarthroplasty: The forgotten concept lives on. *Clin. Orthop. Relat. Res.* **2017**, *475*, 2850–2854. [CrossRef] [PubMed]
18. Gore, D.; Frazer, R.Q.; Kovarik, R.E.; Yepes, J.E. Vitallium. *J. Long-Term Eff. Med. Implants* **2005**, *15*, 673–686. [CrossRef] [PubMed]
19. Saadat, E.; Tiberi, J.V.; Burke, D.W.; Kwon, Y.-M. Smith-Petersen Vitallium Mold Arthroplasty: Case Report with a Fifty-One-Year Follow-Up and Histopathologic Analysis. *JBJS Case Connect.* **2013**, *3*, e112. [CrossRef] [PubMed]
20. Leng, J.; Peng, H.; Feng, B.; Weng, X. The Longest Known Follow-Up of Vitallium Mold Arthroplasty in China: A Case Report and Literature Review. *Orthop. Surg.* **2023**, *15*, 1915–1919. [CrossRef] [PubMed]
21. Law, W.A. Post-Operative Study of Vitallium Mould Arthroplasty of the Hip Joint. *J. Bone Jt. Surg.* **1948**, *30-B*, 76–83. [CrossRef]
22. Gibson, A. Vitallium-cup arthroplasty of the hip joint; review of approximately 100 cases. *J. Bone Jt. Surg.* **1949**, *31*, 861–872. [CrossRef]
23. Aufranc, O.E. Constructive Hip Surgery With the Vitallium Mold: A Report on 1000 Cases of Arthroplasty of the Hip over a Fifteen-Year Period. *J. Bone Jt. Surg.* **1957**, *39*, 237–316. [CrossRef]
24. Pramanik, S.; Agarwal, A.K.; Rai, K.N. Chronology of Total Hip Joint Replacement and Materials Development. *Trends Biomater. Artif. Organs* **2005**, *19*, 15–26.
25. Judet, R.; Judet, J. Technique and results with the acrylic femoral head prosthesis. *J. Bone Jt. Surg.* **1952**, *34-B*, 173–180.
26. Charnley, J. Surgery of the hip-joint: Present and future developments. *Br. Med. J.* **1960**, *1*, 821–826. [CrossRef] [PubMed]
27. Charnley, J. Arthroplasty of the hip. A new operation. *Lancet* **1961**, *1*, 1129–1132. [CrossRef] [PubMed]
28. Townley, C.O. Hemi and total articular replacement arthroplasty of the hip with the fixed femoral cup. *Orthop. Clin. N. Am.* **1982**, *13*, 869–894.
29. Wiles, P. The surgery of the osteoarthritic hip. *Br. J. Surg.* **1958**, *45*, 488–497. [CrossRef]

30. Haboush, E.J. A new operation for arthroplasty of the hip based on biomechanics, photoelasticity, fast-setting dental acrylic, and other considerations. *Bull. Hosp. Jt. Dis.* **1953**, *14*, 242–277.
31. Müller, M.E.; Boltzy, X. Artificial hip joints made from Protasul. *Bull. Assoc. Study Probl. Intern. Fixat.* **1968**, 1–5.
32. Müller, M.E. Total hip prostheses. *Clin. Orthop. Relat. Res.* **1970**, *72*, 46–68.
33. Müller, M.E. The benefits of metal-on-metal total hip replacements. *Clin. Orthop. Relat. Res.* **1995**, *311*, 54–59.
34. Gérard, Y.; Ségal, P.; Bedoucha, J.S. Arthroplasty of the hip with coupled cups. *Rev. Chir. Orthop. Reparatrice Appar. Mot.* **1974**, *60* (Suppl. 2), 281–289.
35. Gérard, Y. Hip arthroplasty by matching cups. *Clin. Orthop. Relat. Res.* **1978**, *134*, 25–35. [CrossRef]
36. Urist, M.R. The principles of hip-socket arthroplasty. *J. Bone Jt. Surg.* **1957**, *39-A*, 786–810. [CrossRef]
37. Nishio, A.; Eguchi, M.; Kaibara, N. Socket and cup surface replacement of the hip. *Clin. Orthop. Relat. Res.* **1978**, *134*, 53–58. [CrossRef]
38. Nishio, A.; Eguchi, M.; Ogata, K. Socket and cup surface replacement. *Orthop. Clin. N. Am.* **1982**, *13*, 843–856. [CrossRef]
39. Salzer, M.; Knahr, K.; Locke, H.; Stark, N. Cement-free bioceramic double-cup endoprosthesis of the hip joint. *Clin. Orthop. Relat. Res.* **1978**, *134*, 80–86. [CrossRef]
40. Howie, D.W.; Cornish, B.L.; Vernon-Roberts, B. Resurfacing hip arthroplasty. Classification of loosening and the role of prosthesis wear particles. *Clin. Orthop. Relat. Res.* **1990**, *255*, 144–159. [CrossRef]
41. Amstutz, H.C.; Campbell, P.; Kossovsky, N.; Clarke, I.C. Mechanism and clinical significance of wear debris-induced osteolysis. *Clin. Orthop. Relat. Res.* **1992**, *276*, 7–18. [CrossRef]
42. Grigoris, P.; Roberts, P.; Panousis, K.; Jin, Z. Hip resurfacing arthroplasty: The evolution of contemporary designs. *Proc. Inst. Mech. Eng. Part H* **2006**, *220*, 95–105. [CrossRef]
43. Trentani, C.; Vaccarino, F. The Paltrinieri-Trentani hip joint resurface arthroplasty. *Clin. Orthop. Relat. Res.* **1978**, *134*, 36–40. [CrossRef]
44. Furuya, K.; Tsuchiya, M.; Kawachi, S. Socket-cup arthroplasty. *Clin. Orthop. Relat. Res.* **1978**, *134*, 41–44. [CrossRef]
45. Freeman, M.A. Total surface replacement hip arthroplasty. *Clin. Orthop. Relat. Res.* **1978**, *134*, 2–4. [CrossRef]
46. Freeman, M.A.R. Conservative total replacement of the hip. *J. Bone Jt. Surg. Br.* **1975**, *57*, 114.
47. Capello, W.N.; Ireland, P.H.; Tramell, T.R.; Eicher, P. Conservative total hip arthroplasty: A procedure to conserve bone stock. Part I: Analysis of sixty-six patients. Part II: Analysis of failures. *Clin. Orthop. Relat. Res.* **1978**, *134*, 59–74. [CrossRef]
48. Wagner, H. Surface replacement arthroplasty of the hip. *Clin. Orthop. Relat. Res.* **1978**, *134*, 102–130. [CrossRef]
49. Amstutz, H.C.; Clarke, I.C.; Christie, J.; Graff-Radford, A. Total hip articular replacement by internal eccentric shells: The "tharies" approach to total surface replacement arthroplasty. *Clin. Orthop. Relat. Res.* **1977**, *128*, 261–284. [CrossRef]
50. Amstutz, H.C.; Graff-Radford, A.; Gruen, T.A.; Clarke, I.C. THARIES surface replacements: A review of the first 100 cases. *Clin. Orthop. Relat. Res.* **1978**, *134*, 87–101. [CrossRef]
51. Tanaka, S. Surface replacement of the hip joint. *Clin. Orthop. Relat. Res.* **1978**, *134*, 75–79. [CrossRef]
52. Howie, D.W.; Campbell, D.; McGee, M.; Cornish, B.L. Wagner resurfacing hip arthroplasty. The results of one hundred consecutive arthroplasties after eight to ten years. *J. Bone Jt. Surg.* **1990**, *72*, 708–714. [CrossRef]
53. Costi, K.; Howie, D.W.; Campbell, D.G.; McGee, M.A.; Cornish, B.L. Long-term survival and reason for revision of Wagner resurfacing hip arthroplasty. *J. Arthroplast.* **2010**, *25*, 522–528. [CrossRef]
54. Amstutz, H.C.; Le Duff, M.J. Hip resurfacing: A 40-year perspective. *HSS J.* **2012**, *8*, 275–282. [CrossRef]
55. Amstutz, H.C.; Kabo, M.; Hermens, K.; O'Carroll, P.F.; Dorey, F.; Kilgus, D. Porous surface replacement of the hip with chamfer cylinder design. *Clin. Orthop. Relat. Res.* **1987**, *222*, 140–160. [CrossRef]
56. Amstutz, H.C.; Kabo, M.; Dorey, F.J. Surface replacement arthroplasty: Evolution of today's ingrowth-fixed design. In *Osteoarthritis in the Young Adult Hip*; Reynolds, D., Freeman, M., Eds.; Churchill Livingstone: Edinburgh, UK, 1989; pp. 251–275.
57. Schmalzried, T.P.; Guttmann, D.; Grecula, M.; Amstutz, H.C. The relationship between the design, position, and articular wear of acetabular components inserted without cement and the development of pelvic osteolysis. *J. Bone Jt. Surg.* **1994**, *76*, 677–688. [CrossRef]
58. Pappas, M.J.; Makris, G.; Buechel, F.F. Titanium nitride ceramic film against polyethylene. A 48 million cycle wear test. *Clin. Orthop. Relat. Res.* **1995**, *317*, 64–70.
59. Malviya, A.; Lobaz, S.; Holland, J. Mechanism of failure eleven years following a Buechel Pappas hip resurfacing. *Acta Orthop. Belg.* **2007**, *73*, 791–794.
60. Kurtz, S.M.; Gawel, H.A.; Patel, J.D. History and systematic review of wear and osteolysis outcomes for first-generation highly crosslinked polyethylene. *Clin. Orthop. Relat. Res.* **2011**, *469*, 2262–2277. [CrossRef]
61. Amstutz, H.C. (Ed.) Surface replacement arthroplasty. In *Hip Arthroplasty*; Churchill Livingstone: New York, NY, USA, 1991; pp. 295–333.
62. Pritchett, J.W. Polyethylene for hip resurfacing—Worth a second look. *Ann. Jt.* **2020**, *5*, 10–20. [CrossRef]
63. Grigoris, P.; Roberts, P.; Panousis, K.; Bosch, H. The evolution of hip resurfacing arthroplasty. *Orthop. Clin. N. Am.* **2005**, *36*, 125–134. [CrossRef]
64. Howie, D.W.; Cornish, B.L.; Vernon-Roberts, B. The viability of the femoral head after resurfacing hip arthroplasty in humans. *Clin. Orthop.* **1993**, *291*, 171–184. [CrossRef]

65. Campbell, P.; Mirra, J.; Amstutz, H.C. Viability of femoral heads treated with resurfacing arthroplasty. *J. Arthroplast.* **2000**, *15*, 120–122. [CrossRef] [PubMed]
66. Wagner, M.; Wagner, H. Preliminary results of uncemented metal on metal stemmed and resurfacing hip replacement arthroplasty. *Clin. Orthop. Relat. Res.* **1996**, *329*, S78–S88. [CrossRef] [PubMed]
67. McMinn, D.J.W.; Treacy, R.; Lin, K.; Pynsent, P. Metal on metal surface replacement of the hip. Experience of the McMinn prosthesis. *Clin. Orthop. Relat. Res.* **1996**, *329*, S89–S98. [CrossRef] [PubMed]
68. Beaulé, P.; Le Duff, M.; Campbell, P.; Dorey, F.; Park, S.; Amstutz, H.C. Metal-on-metal surface arthroplasty with a cemented femoral component: A 7–10 year follow-up study. *J. Arthroplast.* **2004**, *19* (Suppl. 3), 17–22. [CrossRef]
69. Southgate, R.D.; Vail, T.P. Resurfacing Hip Arthroplasty: Evolution, Design, Indications, and Results. In *Surgery of the Hip*, 2nd ed.; Berry, D.J., Lieberman, J.R., Eds.; Elsevier: Amsterdam, The Netherlands, 2020; pp. 779–788.
70. Bohm, R.; Schraml, A.; Schuh, A. Long-term results with the Wagner metal-on-metal hip resurfacing prosthesis. *HIP Int.* **2006**, *16* (Suppl. 4), 58–64. [CrossRef] [PubMed]
71. McMinn, D.J.W. *Modern Hip Resurfacing*; Springer: London, UK, 2009. [CrossRef]
72. Amstutz, H.C.; Le Duff, M.J. Background of metal-on-metal resurfacing. *Proc. Inst. Mech. Eng. Part H* **2006**, *220*, 85–94. [CrossRef] [PubMed]
73. Treacy, R.B.C.; McBryde, C.W.; Shears, E.; Pynsent, P.B. Birmingham hip resurfacing: A minimum follow-up of ten years. *J. Bone Jt. Surg.* **2011**, *93-B*, 27–33. [CrossRef] [PubMed]
74. Coulter, G.; Young, D.A.; Dalziel, R.E.; Shimmin, A.J. Birmingham hip resurfacing at a mean of ten years: Results from an independent centre. *J. Bone Jt. Surg.* **2012**, *94-B*, 315–321. [CrossRef]
75. Murray, D.W.; Grammatopoulos, G.; Pandit, H.; Gundle, R.; Gill, H.S.; McLardy-Smith, P. The ten-year survival of the Birmingham hip resurfacing: An independent series. *J. Bone Jt. Surg.* **2012**, *94-B*, 1180–1186. [CrossRef]
76. Amstutz, H.C.; Le Duff, M.J.; Campbell, P.A.; Gruen, T.A.; Wisk, L.E. Clinical and radiographic results of metal-on-metal hip resurfacing with a minimum ten-year follow-up. *J. Bone Jt. Surg.* **2010**, *92*, 2663–2671. [CrossRef] [PubMed]
77. De Smet, K.; Campbell, P.; Van Der Straeten, C. (Eds.) *The Hip Resurfacing Handbook. A Practical Guide to the Use and Management of Modern Hip Resurfacings*; Woodhead Publishing Ltd.: Sawston, UK, 2013.
78. Amstutz, H.C.; Le Duff, M.J. Eleven years of experience with metal-on-metal hybrid hip resurfacing: A review of 1000 conserve plus. *J. Arthroplasty* **2008**, *23* (Suppl. 1), 36–43. [CrossRef] [PubMed]
79. McMinn, D.J. Development of metal/metal hip resurfacing. *HIP Int.* **2003**, *13*, 41–53. [CrossRef] [PubMed]
80. Amstutz, H.C. The Conserve® Plus hip resurfacing prosthesis. In *The Hip Resurfacing Handbook. A Practical Guide to the Use and Management of Modern Hip Resurfacings*; De Smet, K., Campbell, P., Van Der Straeten, C., Eds.; Woodhead Publishing Ltd.: Sawston, UK, 2013; pp. 31–38. [CrossRef]
81. Campbell, P.; De Smet, K. The Birmingham Hip Resurfacing (BHR) prosthesis. In *The Hip Resurfacing Handbook. A Practical Guide to the Use and Management of Modern Hip Resurfacings*; De Smet, K., Campbell, P., Van Der Straeten, C., Eds.; Woodhead Publishing Ltd.: Sawston, UK, 2013; pp. 25–30. [CrossRef]
82. Dhawan, R.; Young, D.A.; Van Eemeren, A.; Shimmin, A. Birmingham Hip Resurfacing at 20 years. *Bone Jt. J.* **2023**, *105-B*, 946–952. [CrossRef] [PubMed]
83. Costa, C.R.; Johnson, A.J.; Naziri, Q.; Mont, M.A. The outcomes of Cormet hip resurfacing compared to standard primary total hip arthroplasty. *Bull. NYU Hosp. Jt. Dis.* **2011**, *69* (Suppl. 1), S12–S15. [PubMed]
84. Norton, M. The Cormet™ hip resurfacing prosthesis. In *The Hip Resurfacing Handbook. A Practical Guide to the Use and Management of Modern Hip Resurfacings*; De Smet, K., Campbell, P., Van Der Straeten, C., Eds.; Woodhead Publishing Ltd.: Sawston, UK, 2013; pp. 39–43. [CrossRef]
85. Gravius, S.; Wirtz, D.; Maus, U.; Andereya, S.; Müller-Rath, R.; Mumme, T. Durom-Hip-Oberflächenersatz am Hüftgelenk: Erste klinische Ergebnisse mit dem lateralen Zugang [Durom hip resurfacing arthroplasty: First clinical experiences with a lateral approach]. *Z. Orthop. Unfall.* **2007**, *145*, 461–467. [CrossRef] [PubMed]
86. Campbell, P.; Girard, J. The Durom hip resurfacing prosthesis. In *The Hip Resurfacing Handbook. A Practical Guide to the Use and Management of Modern Hip Resurfacings*; De Smet, K., Campbell, P., Van Der Straeten, C., Eds.; Woodhead Publishing Ltd.: Sawston, UK, 2013; pp. 44–47. [CrossRef]
87. Siebel, T.; Maubach, S.; Morlock, M.M. Lessons learned from early clinical experience and results of 300 ASR hip resurfacing implantations. *Proc. Inst. Mech. Eng. Part H* **2006**, *220*, 345–353. [CrossRef] [PubMed]
88. De Smet, K. The DePuy Articular Surface Replacement (ASR™) hip resurfacing prosthesis. In *The Hip Resurfacing Handbook. A Practical Guide to the Use and Management of Modern Hip Resurfacings*; De Smet, K., Campbell, P., Van Der Straeten, C., Eds.; Woodhead Publishing Ltd.: Sawston, UK, 2013; pp. 20–24. [CrossRef]
89. Trč, T.; Šťastný, E.; Kopečný, Z.; Kos, P.; Přidal, J.; Havlas, V. Naše zkušenosti s povrchovou náhradou kyčelního kloubu ICON [Our Experience with ICON Hip Resurfacing System]. *Acta Chir. Orthop. Traumatol. Cechoslov.* **2022**, *89*, 323–331. [CrossRef]
90. Van Der Straeten, C. The ICON hip resurfacing prosthesis. In *The Hip Resurfacing Handbook. A Practical Guide to the Use and Management of Modern Hip Resurfacings*; De Smet, K., Campbell, P., Van Der Straeten, C., Eds.; Woodhead Publishing Ltd.: Sawston, UK, 2013; pp. 53–55. [CrossRef]

91. Delport, H.P. The BIOMET ReCap hip resurfacing prosthesis. In *The Hip Resurfacing Handbook. A Practical Guide to the Use and Management of Modern Hip Resurfacings*; De Smet, K., Campbell, P., Van Der Straeten, C., Eds.; Woodhead Publishing Ltd.: Sawston, UK, 2013; pp. 69–77. [CrossRef]
92. Hamelynck, K.J.; Woering, R. The advanced ceramic coated implant systems (ACCIS) hip resurfacing prosthesis. In *The Hip Resurfacing Handbook. A Practical Guide to the Use and Management of Modern Hip Resurfacings*; De Smet, K., Campbell, P., Van Der Straeten, C., Eds.; Woodhead Publishing Ltd.: Sawston, UK, 2013; pp. 3–10. [CrossRef]
93. Mancino, F.; Finsterwald, M.A.; Jones, C.W.; Prosser, G.H.; Yates, P.J. Metal-on-Metal Hips: Ten-Year Clinical and Radiographic Outcomes of the ADEPT Metal-on-Metal Hip Resurfacing and Modular Total Hip Arthroplasty. *J. Clin. Med.* **2023**, *12*, 889. [CrossRef]
94. Tuke, M. The ADEPT® hip resurfacing prosthesis. In *The Hip Resurfacing Handbook. A Practical Guide to the Use and Management of Modern Hip Resurfacings*; De Smet, K., Campbell, P., Van Der Straeten, C., Eds.; Woodhead Publishing Ltd.: Sawston, UK, 2013; pp. 11–19. [CrossRef]
95. Young, D.; Burke, J. The MITCH hip resurfacing prosthesis. In *The Hip Resurfacing Handbook. A Practical Guide to the Use and Management of Modern Hip Resurfacings*; De Smet, K., Campbell, P., Van Der Straeten, C., Eds.; Woodhead Publishing Ltd.: Sawston, UK, 2013; pp. 65–68. [CrossRef]
96. Gollwitzer, H.; Gerdesmeyer, L.; Horn, C.; Diehl, P.; Töpfer, A.; Gradinger, R. 8-year follow-up after cementless hip arthroplasty with a second generation spongy metal total hip replacement. *HIP Int.* **2009**, *19*, 359–366. [CrossRef]
97. Götze, C.; Tschugunow, A.; Wiegelmann, F.; Osada, N.; Götze, H.G.; Böttner, F. Langfristiger Einfluss der anatomisch angepassten spongiösen Endoprothese auf den periprothetischen Knochen [Long-term influence of the spongiosa metal surface prosthesis on the periprosthetic bone. A radiological and osteodensitometric analysis of implantation of the S & G (ESKA) hip prosthesis]. *Z. Orthop. Ihre. Grenzgeb.* **2006**, *144*, 192–198. (In German) [CrossRef]
98. Van Der Straeten, C. The ESKA hip resurfacing prosthesis. In *The Hip Resurfacing Handbook. A Practical Guide to the Use and Management of Modern Hip Resurfacings*; De Smet, K., Campbell, P., Van Der Straeten, C., Eds.; Woodhead Publishing Ltd.: Sawston, UK, 2013; pp. 48–52. [CrossRef]
99. Quack, G.; Willmann, G.; Krahl, H.; Grundei, H. Konzeptionelle Uberlegungen zur Verbesserung der Pfanne der ESKA-Hüftendoprothese durch die Gleitpaarung Keramik/Keramik [Design considerations for improving the acetabular cup of the ESKA hip endoprosthesis by using the ceramic/ceramic wear couple]. *Biomed. Technol.* **1996**, *41*, 253–259. [CrossRef]
100. Scholz, J.; Grundei, H.; Klingbeil, K. ESKA-Ceram®—Ein neuer Werkstoff in der Endoprothetik des Hüftgelenkes [ESKA-Ceram®—A New Material for Use in Hip Endoprostheses]. *Biomed. Technol.* **2000**, *45*, 377–379. [CrossRef]
101. Gross, T.P.; Liu, F.; Webb, L.A. Clinical outcome of the metal-on-metal hybrid Corin Cormet 2000 hip resurfacing system: An up to 11-year follow-up study. *J. Arthroplast.* **2012**, *27*, 533–538.e1. [CrossRef]
102. Giannini, S.; Laude, F.; Menge, M.; Sadri, H.; Siccardi, F. The ROMAX® hip resurfacing prosthesis. In *The Hip Resurfacing Handbook. A Practical Guide to the Use and Management of Modern Hip Resurfacings*; De Smet, K., Campbell, P., Van Der Straeten, C., Eds.; Woodhead Publishing Ltd.: Sawston, UK, 2013; pp. 78–84. [CrossRef]
103. Cadossi, M.; Tedesco, G.; Savarino, L.; Baldini, N.; Mazzotti, A.; Greco, M.; Giannini, S. Effect of acetabular cup design on metal ion release in two designs of metal-on-metal hip resurfacing. *J. Biomed. Mater. Res. Part B Appl. Biomater.* **2014**, *102*, 1595–1601. [CrossRef]
104. Girard, J. The Tornier DynaMoM hip resurfacing prosthesis. In *The Hip Resurfacing Handbook. A Practical Guide to the Use and Management of Modern Hip Resurfacings*; De Smet, K., Campbell, P., Van Der Straeten, C., Eds.; Woodhead Publishing Ltd.: Sawston, UK, 2013; pp. 85–87. [CrossRef]
105. Singh Chana, G.; Palmer, S. The MIHR International® hip resurfacing prosthesis. In *The Hip Resurfacing Handbook. A Practical Guide to the Use and Management of Modern Hip Resurfacings*; De Smet, K., Campbell, P., Van Der Straeten, C., Eds.; Woodhead Publishing Ltd.: Sawston, UK, 2013; pp. 59–64. [CrossRef]
106. Heisel, C.; Kleinhans, J.A.; Menge, M.; Kretzer, J.P. Ten different hip resurfacing systems: Biomechanical analysis of design and material properties. *Int. Orthop.* **2009**, *33*, 939–943. [CrossRef] [PubMed]
107. Australian Orthopaedic Association National Joint Replacement Registry. *Annu. Rep.* **2006**, 57–87.
108. Hamelynck, K.J.; Woering, R.G. Ceramic Surface Engineered Metal-on-Metal Hips system for Total Hip Arthroplasty and Resurfacing Hip Arthroplasty. The design rationale, pre-clinical testing and interim report on 2–7 years of clinical results. In *White Paper*; Implantcast: Buxtehude, Germany, 2009; pp. 1–8. Available online: https://silo.tips/download/ceramic-surface-engineered-metal-on-metal-hips-system-for-total-hip-arthroplasty (accessed on 19 December 2023).
109. Jemmett, P.; Parfitt, D.; Rice, R. Early clinical failure of the ACCIS® metal on metal hip arthroplasty system—A metal on metal hip with a difference. *Acta Orthop. Belg.* **2016**, *82*, 491–496. [PubMed]
110. Kim, W.Y.; Ko, M.S.; Lee, S.W.; Kim, K.S. Short-term Outcomes of Ceramic Coated Metal-on-Metal Large Head in Total Hip Replacement Arthroplasty. *HIP Pelvis* **2018**, *30*, 12–17. [CrossRef] [PubMed]
111. Gerdesmeyer, L.; Gollwitzer, H.; Al Muderis, M.; Fletcher, S.; Böhling, U. Alternative Bearing Designs for Hip Resurfacing Arthroplasty. *Tech. Orthop.* **2010**, *25*, 67–72. [CrossRef]
112. Bader, R.; Klüss, D.; Gerdesmeyer, L.; Steinhauser, E. Biomechanische Aspekte zur Implantatverankerung und Kinematik von Oberflächenersatzhüftendoprothesen [Biomechanical aspects of the implant fixation and kinematics of hip resurfacing systems]. *Orthopade* **2008**, *37*, 634–643. [CrossRef]

113. Shimmin, A.J.; Back, D. Femoral neck fractures following Birmingham hip resurfacing: A national review of 50 cases. *J. Bone Jt. Surg.* **2005**, *87*, 463–464. [CrossRef]
114. Gupta, S.; New, A.M.; Taylor, M. Bone remodelling inside a cemented resurfaced femoral head. *Clin. Biomech.* **2006**, *21*, 594–602. [CrossRef]
115. Long, J.P.; Santner, T.J.; Bartel, D.L. Hip resurfacing increases bone strains associated with short-term femoral neck fracture. *J. Orthop. Res.* **2009**, *27*, 1319–1325. [CrossRef]
116. Steffen, R.T.; Foguet, P.R.; Krikler, S.J.; Gundle, R.; Beard, D.J.; Murray, D.W. Femoral neck fractures after hip resurfacing. *J. Arthroplast.* **2009**, *24*, 614–619. [CrossRef] [PubMed]
117. Fowble, V.A.; Schuh, A.; Hoke, R.; Bitsch, R.G.; Beaulé, P.E. Clinical correlation of femoral component migration in hip resurfacing arthroplasty analyzed by Einzel-Bild-Röntgen-analyze-femoral component analysis. *Orthop. Clin. N. Am.* **2005**, *36*, 243–250. [CrossRef]
118. Amstutz, H.C.; Le Duff, M.J.; Campbell, P.A.; Dorey, F.J. The effects of technique changes on aseptic loosening of the femoral component in hip resurfacing. Results of 600 Conserve Plus with a 3 to 9 year follow-up. *J. Arthroplast.* **2007**, *22*, 481–489. [CrossRef]
119. Falez, F.; Favetti, F.; Casella, F.; Panegrossi, G. Hip resurfacing: Why does it fail? Early results and critical analysis of our first 60 cases. *Int. Orthop.* **2008**, *32*, 209–216. [CrossRef] [PubMed]
120. Zustin, J.; Hahn, M.; Morlock, M.M.; Rüther, W.; Amling, M.; Sauter, G. Femoral component loosening after hip resurfacing arthroplasty. *Skelet. Radiol.* **2010**, *39*, 747–756. [CrossRef] [PubMed]
121. Baad-Hansen, T.; Storgaard Jakobsen, S.; Soballe, K. Two-year migration results of the ReCap hip resurfacing system-a radiostereometric follow-up study of 23 hips. *Int. Orthop.* **2011**, *35*, 497–502. [CrossRef]
122. Beaulé, P.; White, C.; Lopez-Castellaro, J.; Kim, P. Acetabular component migration analysis of hip resurfacing using ein bild roentegen analyse (EBRA). *Orthop. Procs.* **2012**, *94-B*, 214. [CrossRef]
123. de Haan, R.; Buls, N.; Scheerlinck, T. Impact of implant size on cement filling in hip resurfacing arthroplasty. *Proc. Inst. Mech. Eng. Part H* **2014**, *228*, 3–10. [CrossRef]
124. Howald, R.; Kesteris, U.; Klabunde, R.; Krevolin, J. Factors affecting the cement penetration of a hip resurfacing implant: An in vitro study. *HIP Int.* **2006**, *16* (Suppl. 4), 82–89. [CrossRef] [PubMed]
125. Beaulé, P.E.; Lu, Z.; Luck, J.V.; Campbell, P. Bone thermal necrosis and cement penetration in femoral head resurfacing. *J. Bone Jt. Surg.* **2005**, *90-B*, 97. [CrossRef]
126. Beaulé, P.E.; Campbell, P.; Lu, Z.; Leunig-Ganz, K.; Beck, M.; Leunig, M.; Ganz, R. Vascularity of the arthritic femoral head and hip resurfacing. *J. Bone Jt. Surg.* **2006**, *88* (Suppl. 4), 85–96. [CrossRef]
127. Beaulé, P.E.; Campbell, P.; Shim, P. Femoral head blood flow during hip resurfacing. *Clin. Orthop. Relat. Res.* **2007**, *456*, 148–152. [CrossRef] [PubMed]
128. Beaulé, P.E.; Matar, W.Y.; Poitras, P.; Smit, K.; May, O. 2008 Otto Aufranc Award: Component design and technique affect cement penetration in hip resurfacing. *Clin. Orthop. Relat. Res.* **2009**, *467*, 84–93. [CrossRef]
129. Krause, M.; Breer, S.; Hahn, M.; Rüther, W.; Morlock, M.M.; Amling, M.; Zustin, J. Cementation and interface analysis of early failure cases after hip-resurfacing arthroplasty. *Int. Orthop.* **2012**, *36*, 1333–1340. [CrossRef]
130. Hug, K.T.; Watters, T.S.; Vail, T.P.; Bolognesi, M.P. The withdrawn ASR™ THA and hip resurfacing systems: How have our patients fared over 1 to 6 years? *Clin. Orthop. Relat. Res.* **2013**, *471*, 430–438. [CrossRef]
131. Davis, E.T.; Olsen, M.; Zdero, R.; Smith, G.M.; Waddell, J.P.; Schemitsch, E.H. Predictors of femoral neck fracture following hip resurfacing: A cadaveric study. *J. Arthroplast.* **2013**, *28*, 110–116. [CrossRef]
132. Matharu, G.S.; McBryde, C.W.; Revell, M.P.; Pynsent, P.B. Femoral neck fracture after Birmingham Hip Resurfacing Arthroplasty: Prevalence, time to fracture, and outcome after revision. *J. Arthroplast.* **2013**, *28*, 147–153. [CrossRef]
133. Haughom, B.D.; Erickson, B.J.; Hellman, M.D.; Jacobs, J.J. Do complication rates differ by gender after metal-on-metal hip resurfacing arthroplasty? A systematic review. *Clin. Orthop. Relat. Res.* **2015**, *473*, 2521–2529. [CrossRef] [PubMed]
134. Amstutz, H.C.; Le Duff, M.J. Aseptic loosening of cobalt chromium monoblock sockets after hip resurfacing. *HIP Int.* **2015**, *25*, 466–470. [CrossRef]
135. Gill, H.S.; Campbell, P.A.; Murray, D.W.; De Smet, K.A. Reduction of the potential for thermal damage during hip resurfacing. *J. Bone Jt. Surg.* **2007**, *89*, 16–20. [CrossRef] [PubMed]
136. Steffen, R.T.; Fern, D.; Norton, M.; Murray, D.W.; Gill, H.S. Femoral oxygenation during hip resurfacing through the trochanteric flip approach. *Clin. Orthop. Relat. Res.* **2009**, *467*, 934–939. [CrossRef] [PubMed]
137. Scheerlinck, T.; Delport, H.; Kiewitt, T. Influence of the cementing technique on the cement mantle in hip resurfacing: An in vitro computed tomography scan-based analysis. *J. Bone Jt. Surg.* **2010**, *92*, 375–387. [CrossRef]
138. Girard, J. Is it time for cementless hip resurfacing? *HSS J.* **2012**, *8*, 245–250. [CrossRef]
139. Baker, R.; Whitehouse, M.; Kilshaw, M.; Pabbruwe, M.; Spencer, R.; Blom, A.; Bannister, G. Maximum temperatures of 89 °C recorded during the mechanical preparation of 35 femoral heads for resurfacing. *Acta Orthop.* **2011**, *82*, 669–673. [CrossRef]
140. Zustin, J.; Sauter, G.; Morlock, M.M.; Rüther, W.; Amling, M. Association of osteonecrosis and failure of hip resurfacing arthroplasty. *Clin. Orthop. Relat. Res.* **2010**, *468*, 756–761. [CrossRef]
141. Ullmark, G.; Sundgren, K.; Milbrink, J.; Nilsson, O.; Sörensen, J. Osteonecrosis following resurfacing arthroplasty. *Acta Orthop.* **2009**, *80*, 670–674. [CrossRef]

142. de Steiger, R.N.; Miller, L.N.; Prosser, G.H.; Graves, S.E.; Davidson, D.C.; Stanford, T.E. Poor outcome of revised resurfacing hip arthroplasty. *Acta Orthop.* **2010**, *81*, 72–76. [CrossRef]
143. Gross, T.P.; Liu, F. Metal-on-metal hip resurfacing with an uncemented femoral component. A seven-year follow-up study. *J. Bone Jt. Surg.* **2008**, *90* (Suppl. 3), 32–37. [CrossRef]
144. O'Leary, R.J.; Gaillard, M.D.; Gross, T.P. Comparison of Cemented and Bone Ingrowth Fixation Methods in Hip Resurfacing for Osteonecrosis. *J. Arthroplast.* **2017**, *32*, 437–446. [CrossRef]
145. Gaillard-Campbell, D.M.; Gross, T.P. Femoral Fixation Methods in Hip Resurfacing Arthroplasty: An 11-Year Retrospective Comparison of 4013 Cases. *J. Arthroplast.* **2019**, *34*, 2398–2405. [CrossRef]
146. Rogala, P. Endoprosthesis. EU Patent EP072418 B1, 22 December 1999.
147. Rogala, P. Acetabulum Endoprosthesis and Head. U.S. Patent US5,911,759 A, 15 June 1999.
148. Rogala, P. Method and Endoprosthesis to Apply This Implantation. Canadian Patent 2,200,064, 1 April 2002.
149. Uklejewski, R.; Rogala, P.; Winiecki, M. *Prototype of a Biomimetic Multi-Spiked Connecting Scaffold for a New Generation of Resurfacing Endoprostheses*, 1st ed.; CRC Press/Taylor & Francis Group: Boca Raton, FL, USA, 2024. [CrossRef]
150. Kruth, J.-P.; Mercelis, P.; Van Vaerenbergh, J.; Froyen, L.; Rombouts, M. Binding mechanisms in selective laser sintering and selective laser melting. *Rapid Prototyp. J.* **2005**, *11*, 26–36. [CrossRef]
151. Uklejewski, R.; Winiecki, M.; Rogala, P. Technological issues of additive manufacturing of preprototypes of the multi-spiked connecting scaffold for non-cemented resurfacing arthroplasty endoprostheses. *Eng. Biomater.* **2014**, *17*, 81–82.
152. Uklejewski, R.; Rogala, P.; Winiecki, M.; Mielniczuk, J. Prototype of minimally invasive hip resurfacing endoprosthesis—Bioengineering design and manufacturing. *Acta Bioeng. Biomech.* **2009**, *11*, 65–70.
153. Uklejewski, R.; Rogala, P.; Winiecki, M.; Mielniczuk, J. Prototype of innovating bone tissue preserving THRA endoprosthesis with multi-spiked connecting scaffold manufactured in selective laser melting technology. *Eng. Biomater.* **2009**, *12*, 2–6.
154. Uklejewski, R.; Winiecki, M.; Rogala, P.; Mielniczuk, J. Selective laser melted prototype of original minimally invasive hip endoprosthesis. *Rapid Prototyp. J.* **2011**, *17*, 76–85. [CrossRef]
155. Uklejewski, R.; Winiecki, M.; Rogala, P.; Patalas, A. Structural-Geometric Functionalization of the Additively Manufactured Prototype of Biomimetic Multispiked Connecting Ti-Alloy Scaffold for Entirely Noncemented Resurfacing Arthroplasty Endoprostheses. *Appl. Bionics Biomech.* **2017**, *2017*, 5638680. [CrossRef]
156. Uklejewski, R.; Rogala, P.; Winiecki, M. Structural and Hydroxyapatite-like Functionalization of Bone Contacting Surfaces of Additively Manufactured Multispiked Connecting Scaffold Prototype for Entirely Cementless Resurfacing Arthroplasty Endoprostheses to Enhance its Osteoconduction and Osteointegration. In *Advanced Surface Engineering Materials*; Tiwari, A., Wang, R., Wei, B., Eds.; WILEY-Scrivener: Beverly, MA, USA, 2016; pp. 175–240. [CrossRef]
157. Uklejewski, R.; Rogala, P.; Winiecki, M.; Kędzia, A.; Ruszkowski, P. Preliminary Results of Implantation in Animal Model and Osteoblast Culture Evaluation of Prototypes of Biomimetic Multispiked Connecting Scaffold for Noncemented Stemless Resurfacing Hip Arthroplasty Endoprostheses. *BioMed Res. Int.* **2013**, *2013*, 689089. [CrossRef]
158. Uklejewski, R.; Rogala, P.; Winiecki, M.; Tokłowicz, R.; Ruszkowski, P.; Wołuń-Cholewa, M. Biomimetic Multi-spiked Connecting Ti-alloy Scaffold Prototype for Entirely Cementless Resurfacing Arthroplasty Endoprostheses—Exemplary Results of Implantation of the Ca-P Surface Modified Scaffold Prototypes in Animal Model and Osteoblast Culture Evaluation. *Materials* **2016**, *9*, 532. [CrossRef]
159. Uklejewski, R.; Winiecki, M.; Krawczyk, P.; Tokłowicz, R. Native Osseous CaP Biomineral Coating on a Biomimetic Multi-Spiked Connecting Scaffold Prototype for Cementless Resurfacing Arthroplasty Achieved by Combined Electrochemical Deposition. *Materials* **2019**, *12*, 3994. [CrossRef]
160. Uklejewski, R.; Winiecki, M.; Patalas, A.; Rogala, P. Numerical studies of the influence of various geometrical features of a multi-spiked connecting scaffold prototype on mechanical stresses in peri-implant bone. *Comput. Methods Biomech. Biomed. Eng.* **2018**, *21*, 541–547. [CrossRef] [PubMed]
161. Uklejewski, R.; Winiecki, M.; Patalas, A.; Rogala, P. Bone Density Micro-CT Assessment during Embedding of the Innovative Multi-Spiked Connecting Scaffold in Periarticular Bone to Elaborate a Validated Numerical Model for Designing Biomimetic Fixation of Resurfacing Endoprostheses. *Materials* **2021**, *14*, 1384. [CrossRef]
162. Rogala, P.; Uklejewski, R.; Winiecki, M.; Dąbrowski, M.; Gołańczyk, J.; Patalas, A. First Biomimetic Fixation for Resurfacing Arthroplasty: Investigation in Swine of a Prototype Partial Knee Endoprosthesis. *BioMed Res. Int.* **2019**, *2019*, 6952649. [CrossRef]

biomimetics

MDPI

Article

Long Bone Defect Filling with Bioactive Degradable 3D-Implant: Experimental Study

Arnold Popkov [1], Natalia Kononovich [1], Gleb Dubinenko [2], Elena Gorbach [1], Alexander Shastov [1], Sergei Tverdokhlebov [2] and Dmitry Popkov [1,*]

1 Ilizarov National Medical Research Center for Traumatology and Orthopedics, Kurgan 640014, Russia; popkov72@gmail.com (A.P.); n.a.kononovich@mail.ru (N.K.); gorbach.e@mail.ru (E.G.); alshastov@yandex.ru (A.S.)

2 School of Nuclear Science & Engineering, National Research Tomsk Polytechnic University, Tomsk 634050, Russia; dubinenko.gleb@gmail.com (G.D.); tverd@tpu.ru (S.T.)

* Correspondence: dpopkov@mail.ru

Abstract: Previously, 3D-printed bone grafts made of titanium alloy with bioactive coating has shown great potential for the restoration of bone defects. Implanted into a medullary canal titanium graft with cellular structure demonstrated stimulation of the reparative osteogenesis and successful osseointegration of the graft into a single bone-implant block. The purpose of this study was to investigate osseointegration of a 3D-printed degradable polymeric implant with cellular structure as preclinical testing of a new technique for bone defect restoration. During an experimental study in sheep, a 20 mm-long segmental tibial defect was filled with an original cylindrical implant with cellular structure made of polycaprolactone coated with hydroxyapatite. X-ray radiographs demonstrated reparative bone regeneration from the periosteum lying on the periphery of cylindrical implant to its center in a week after the surgery. Cellular structure of the implant was fully filled with newly-formed bone tissue on the 4th week after the surgery. The bone tissue regeneration from the proximal and distal bone fragments was evident on 3rd week. This provides insight into the use of bioactive degradable implants for the restoration of segmental bone defects. Degradable implant with bioactive coating implanted into a long bone segmental defect provides stimulation of reparative osteogenesis and osseointegration into the single implant-bone block.

Keywords: bone defect; degradable implant; polycaprolactone; hydroxyapatite; osseointegration

Citation: Popkov, A.; Kononovich, N.; Dubinenko, G.; Gorbach, E.; Shastov, A.; Tverdokhlebov, S.; Popkov, D. Long Bone Defect Filling with Bioactive Degradable 3D-Implant: Experimental Study. *Biomimetics* **2023**, *8*, 138. https://doi.org/10.3390/biomimetics8020138

Academic Editors: Ryszard Uklejewski and Mariusz Winiecki

Received: 27 February 2023
Revised: 26 March 2023
Accepted: 26 March 2023
Published: 28 March 2023

1. Introduction

Unsatisfactory outcomes of orthopedic surgery for non-unions and bone defects are not rare [1–3]. The methods of transosseous osteosynthesis developed by G. Ilizarov based on low-invasive surgery, fixation stability, preservation of blood supply, and functional loading demonstrated its efficiency in the treatment of pseudarthrosis and bone defects [4–6]. On the other hand, inconveniencies and disadvantages of external fixation are numerous: long-lasting external frame wearing, staged in-patient stay, delayed functional recovery, high rate of pin-site infections, decreased life quality of patients with external frame, and external fixation index in single-level lengthening for bone transport on tibia being over 40 days/cm in adult patients [7–9]. Recently, an idea appeared to fill bone defects with a personalized titanium implants, the design of which (inner cells and pores) is similar to the structure of spongy bone [10–12]. Experimental studies demonstrated that osseointegration of such an implant provides fast and durable restoration of a bone defect, but the titanium does not allow monitoring the process of reparative bone regeneration and its functional restructuring in clinical conditions by the methods of X-rays [13–15]. Furthermore, sometimes a metal implant remained into the bone arises concerns of a patient.

Over the past few years, there has been significant interest in the use of artificial bioresorbable scaffolds made by 3D printing to treat large bone defects [16–18]. One of the main advantages of 3D printing is the ability to personalize the scaffold, which can be designed to fit the size and shape of the bone defect with a similar structure and morphology to the tissue it will replace. Fused Deposition Modeling (FDM) is a simple and cost-effective 3D printing technique that allows for printing with thermoplastic polymers and polymer composites with a print resolution of up to 50 μm. Currently, new linear polyester-based bioresorbable materials for FDM 3D printing of personalized scaffolds for bone tissue engineering are actively studied. Polylactic acid (PLA) and PLA-based composites have been the most extensively studied, as they show good printability, a high level of biocompatibility, support cell adhesion and proliferation, and can accelerate the formation of bone tissue and osseointegration at the defect site in the combination with calcium phosphates [19–22]. An alternative to PLA is another linear polyester polycaprolactone (PCL). Compared to PLA, PCL has a much lower glass transition temperature, unique elastic properties at human body temperature, and is known for its good suturability [23]. The low melting temperature of PCL allows for adapting the strategy of FDM printing, and its high elasticity makes it easy to implant and withstand cyclic loads [24,25]. Previously, it was shown that 3D-printed PCL composite scaffolds filled with hydroxyapatite (HA) demonstrated a good level of biocompatibility, support for cell adhesion, growth, and proliferation [26,27].

Recent studies have shown that the printed infill pattern is also important for the biocompatibility of biomimetic scaffolds [28–30]. Triply periodic minimal surfaces (TPMS) have gained attention due to their high surface-to-volume ratio and interconnected porosity, which closely mimic the structure of natural materials [31]. Numerous of studies have employed gyroid-like TPMS infill patterns for 3D printing of PCL-based scaffolds and showed improved cytocompatibility, mechanical strength, and osteoblast proliferation compared to conventional cubic and rectangular infill patterns [32–38]. Studies demonstrate the potential of using TPMS structures in the design and fabrication of personalized biomimetic scaffolds for bone tissue engineering.

The purpose of this experimental study was the assessment of the reparative osteogenesis and osseointegration of a biomimetic TPMS-structured degradable cylindrical implant with bioactive coating for long bone defect filling.

2. Materials and Methods

This non-randomized study was performed on five mixed-breed sheep of feminine sex aged of 1 year in 2022. All the cheeps received the same postoperative care and same complete feeding, they were examined by veterinarian doctor 5 days per week. They were allowed to walk with no limitations in adapted vivarium.

Applied implants for bone defect filling had dimensions of 40 mm height and 18 mm diameter (Figure 1a,b). The implants were 3D printed of ε-polycaprolactone (PCL; Sigma-Aldrich, St. Louis, MO, USA; Mn 80,000 g/mol) and coated with hydroxyapatite (HA; Fluidinova, Portugal; particle size 10 ± 5 μm). The implants were fabricated via Fused Deposition Modeling (FDM) additive technology at Tomsk Polytechnic University (Tomsk, Russia). PCL pellets were melted and extruded with the use of Filabot EX2 (Filabot HQ, Barre, VT, USA) commercial single screw extruder to fabricate filament of 2.8 ± 0.15 mm diameter. The temperature of extrusion was 80 ± 2 °C and the rate of extrusion was 2 m/min. Extruded filament was used for the 3D printing of scaffolds with the use of Ultimaker S5 (Ultimaker B.V., Utrecht, The Netherlands) commercial FDM 3D printer. The temperature of the glass substrate was 35 °C, and the temperature of the printing nozzle was 200 °C. The printing was performed at printing rate of 6 mm/s. Dimensions and shape of implants were designed exactly in accordance with the transverse dimensions of the experimental animals tibial shaft and presented in Figure S1. The middle part of the implants was designed in the form of 20 mm long porous cylinder with 18 mm diameter and intramedullary cannel of 5 mm diameter. The inner space of the middle part of implants was printed in the form of a gyroid divided into cells of 1 mm diameter with walls thickness of 450 μm. Intramedullary canal of 5 mm diameter was designed in the center of the porous part. The outer walls of the cylinder had pores of 1 mm diameter to facilitate ingrowth of blood vessels to the bone defect from the side of periosteal tissues and intramedullary canal of adjacent bone fragments. The end face surfaces of the middle porous part were printed in the form of hollow cylinders of 10 mm long each with small pores of 1 mm. The external and internal surfaces of the implant were coated with a bioactive layer of HA. «Solvent/non-solvent» treatment of scaffolds was performed with the use of the mixture of toluene and ethanol at 3:7 *v/v* ratio. HA was mixed with the toluene/ethanol mixture at 10% *w/w* and stirred on magnetic stirrer for 30 min to obtain suspension. Scaffolds were dipped into the suspension for 2 min at room temperature under continuous stirring. Coated scaffolds were washed with ethanol and dried for 24 h under vacuum (1 mbar) at room temperature.

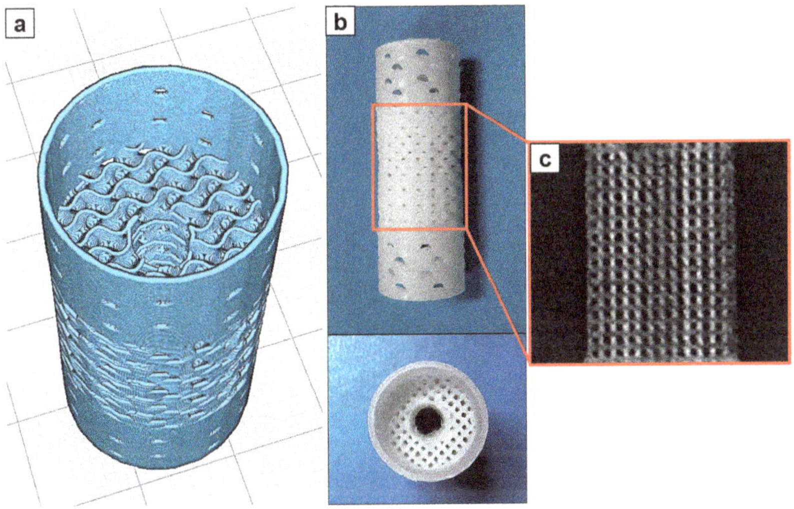

Figure 1. 3D-printed cylindrical implant: (**a**)—screenshot of 3D model of the implant from Cura 5.1.0 slicer, (**b**)—3D-printed PCL cylindrical implant; (**c**)—computer tomography of a porous part of the implant (×2.5).

Protocol. The study protocol included the following phases:

- 1–2 days before surgery: physical examination and standard AP and lateral radiographs of both tibiae including adjacent joints;
- D0: surgery;
- Postoperative period: standard AP and lateral radiographs were taken on the D1 and later once every week;
- Euthanasia at D30 (Euthanasia was performed by an intravenous injection of sodium thiopental 5%, 45 mg/kg).

Surgery. All the surgical interventions were performed under general anesthesia with a sodium thiopental 5% solution (10–15 mg/kg) with atropine as needed. The same surgeon performed all the surgeries. The first stage of the surgery consisted of the modeling of a tibial defect: a longitudinal incision was made along the anterior surface of the tibia and the shaft of the tibia was exposed. The periosteum was dissected longitudinally, and 20 mm long sector of the shaft was resected with an oscillating saw. Then, the cylindrical implant was mounted on the defect site by the integration of free ends of the bone fragments into hollow cylinders on the end face surfaces of the implant. The wound was sutured layer-by-layer, and the limb was fixed with the Ilizarov fixator. Regarding geometry of circular external fixator, the Ilizarov frame comprised two distal circular rings of 110 mm diameter made of stainless steel and two proximal rings of 110 mm diameter in three quarters connected by three 12 cm long threaded rods (Figure 2b). The gaps of proximal rings were oriented posteriorly as so allowing flexion motion in knee joint while walking. Three stain steel 1.5-mm wires were inserted in proximal tibial fragment and three Kirscher wires without olive of the same diameter were inserted in the distal tibial fragment (one of distal wires was inserted through tibia and fibula). Angulation between wires was about 90°. All wires were tensioned with 100 kg forces with graduated wire—tensioner. The external frame fixation lasted 30 days in each animal. In contrast to rigid half-pins, an external fixator with wires provides biomechanical benefit with the weight bearing on the bone regenerate while walking and does not demonstrate stiffness with negative influence on bone remodelling [39,40]. Cefazolin was administered intramuscularly 7 days after the surgery (0.5 g twice a day). In postoperative period the wounds underwent daily control, pin-site care was performed every 2 days. The experimental surgeries were performed by one surgical team.

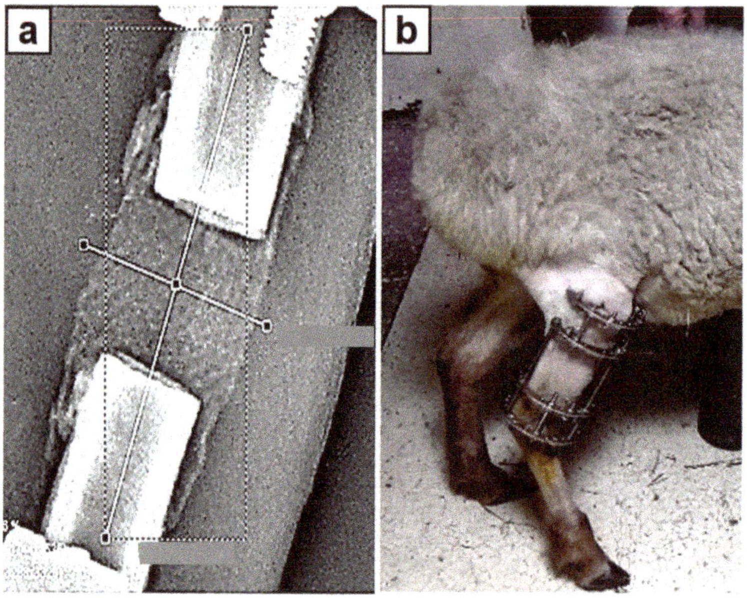

Figure 2. Filling tibia segmental defect with degradable implant: (**a**)—radiograph of a defect filled with the implant (scan lines are marked); (**b**)—photo of an experimental animal on the 4th postoperative day.

Radiographically control was performed using Premium VET X-ray system (TOSHIBA (Rotanode) Model E7239. N: 10G749, Japan) and digital radiography system with CANON CXDI-401C COMPACT flat-panel detector (Canon Inc. Medical Equipment Group, Tochigi, Japan). The optical density of the interfragmentary gap filled with implant and newly formed bone tissue was measured with original software (Certificate for the Computer Program No. 2014611777 RF) by the transverse and longitudinal scanning (Figure 2a).

Computer tomography (CT) of anatomical specimens was preformed using the PRODIS Compact table microtomography system (PRODIS. NDT Ltd., Moscow, Russia) and VG Studio 3.5 (Volume Graphics GmbH, Heidelberg, Germany) software.

For morphological study the specimens were dissected in the longitudinal direction to visually assess the state of the bone and soft tissues lying on the implant surface, and to assess the tissue substrate filling the cells of the implant and the longitudinal axial canal of the implant.

AtteStat 12.0.5 (Kurgan, Russia) software was used for the statistical analysis. The statistical values described the mean and standard deviations.

Ethics. Approval was obtained from the institutional ethics committee before the experiment (Ethics Committee of the Ilizarov National Medical Research Center for Traumatology and Orthopedics, Kurgan, Russia (protocol code 1(71), date of approval 28 April 2022). Interventions, animal care, and euthanasia conformed to the requirements of the European Convention for the Protection of Vertebrate Animals used for Experimental and other Scientific Purposes (Strasbourg, 18 March 1986), principles of laboratory animal care (NIH publication number 85-23, revised 1985), and the national laws.

3. Results

Three-dimensional-printed PCL implant and CT of porous part of the implant are presented in Figure 1b,c, respectively. Implants have smooth surface with open pores in the middle part (Figure 1b). CT images show regular porous structure of the middle part of implant with visible inner intramedullary canal (Figure 1c).

A promising approach for modifying the surface of biodegradable polyesters involves treatment of the polymer with a mixture of organic solvents to partially swell its surface [41–43]. It was shown that a mixture of toluene and ethanol has no destructive effect on the PCL scaffold structure and allows non-covalent immobilization of gelatin and iodine. The swelled surface layer of the polymer is able to adsorb molecules and particles from the contacting medium. Such mechanism of adsorption is highly suitable for the immobilization of inorganic particles. HA distributed evenly over the surface of the implant in the form of dispersed segregated particles (white particles in the SEM micrograph of implant surface) (Figure 3). EDX mapping of the HA coated surface and chemical composition of implants examined by FTIR are shown in the supporting information manuscript (Figures S2 and S3).

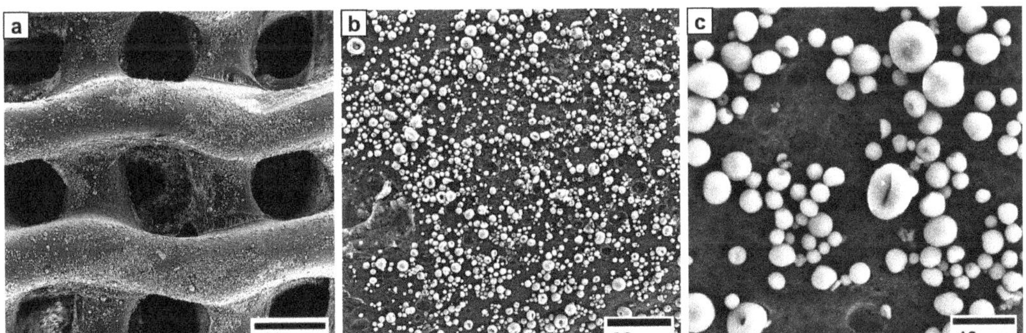

Figure 3. Microscopic appearance of the HA coated gyroid PCL implant: (**a**)—SEM micrograph of PCL implant with HA coating, (**b**)—HA coated surface at 250× magnification, (**c**)—HA coated surface at 1000× magnification.

The body weight of animals at the beginning of the study was 37.2 ± 3.2 kg (ranged from 34 to 38.5 kg). In a month after surgery the mean weight was 37.1 kg (34.5 kg to 37.9 kg). The sheep started walking with total weight-bearing on the limb that has undergone surgery since the 3th–5th day (Figure 2b). There was neither septic nor neurological complications during the study. The weight-bearing function of the experimental limb was maintained until the end of the study.

The radiography demonstrated to be enough informative for assessment of bone reparative regeneration: the PCL implant was X-ray transparent enough. X-ray transparency of the implant allowed monitoring of reparative bone regeneration and bone tissue ingrowth. On the first day of the study the cellular structure of the implant was clearly visible

(Figure 4a). The optical density inside the cells corresponded to the density of the soft tissues surrounding the bone. Less pronounced shadow was visible along the axial channel of the implant, as well as the denser shadow from the superposition of muscle tissue.

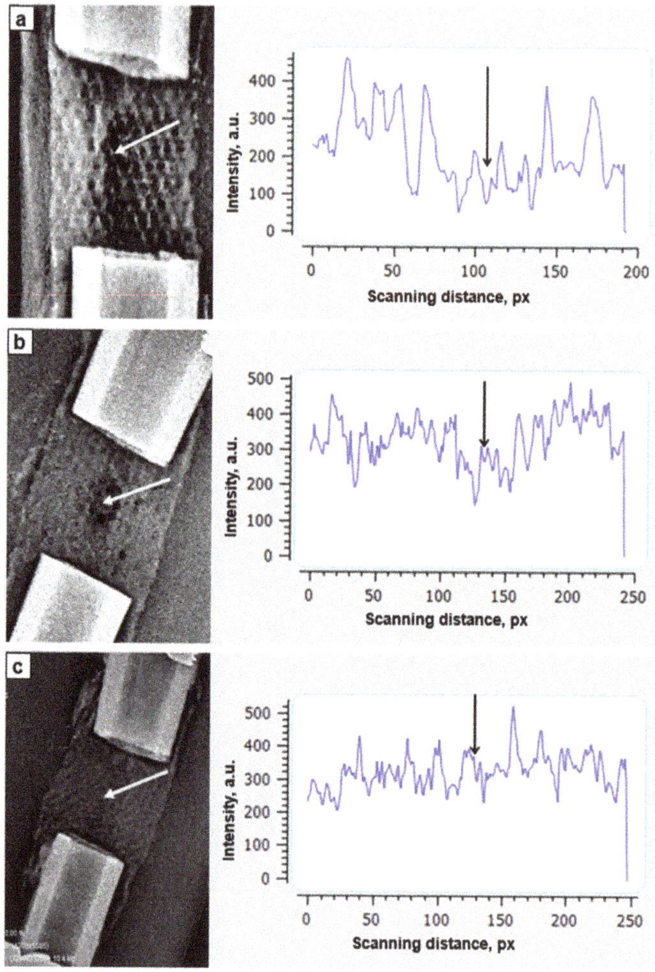

Figure 4. Optical density of the osseointegration zone of an implant (transverse scanning, white arrows indicate the zone of implant axial channel): (**a**)—day of surgery (×1.1); (**b**)—7th days of fixation (×1.0); (**c**)—21st days of fixation (zone of the cortices is expanded, zone of implant axial channel disappeared) (×0.8).

A slight periosteal reaction was noted closing to the ends of the implant at the 7th day after the surgery (Figure 4b). Densitometry revealed that all the cells of the implant were filled with the substance which optical density is higher than the density of surrounding soft tissues. An area of a low density was located in the middle of the implant along the axial line. We hypothesized that this area was not yet filled with newly forming bone tissue of low mineralization. By the 21st day of the study, the optical density in transverse scanning increased. It was no longer possible to distinguish the area of the axial canal of the implant (Figure 4c).

We compared optical density of cylindrical implant in longitudinal scanning with the optical density of the distal tibial fragment (Table 1). The optical density of the implant was increasing during the study and approaching to the values of the optical density of the distal tibial fragment.

Table 1. Dynamics of the optical density of the degradable implant on the bone defect site.

Period of Study, Days	Average Optical Density of the Bone Defect, a.u.	Optical Density of the Shaft of the Adjacent Distal Shaft End, a.u.	Coefficient of Optical Density (Relative to the Distal Bone Fragment)
D0 (surgery)	103 ± 10.7	382 ± 10.4	0.20
7	106 ± 20.4	349 ± 17.4	0.33
14	149 ± 21.2	405 ± 20.1	0.38
21	159 ± 23.1	307 ± 15.5	0.50
28	312 ± 15.9	409 ± 20.3	0.75

Longitudinal scanning of the implant zone compared to the density of the distal tibial fragment revealed increasing optical density during experimental study (Figure 5).

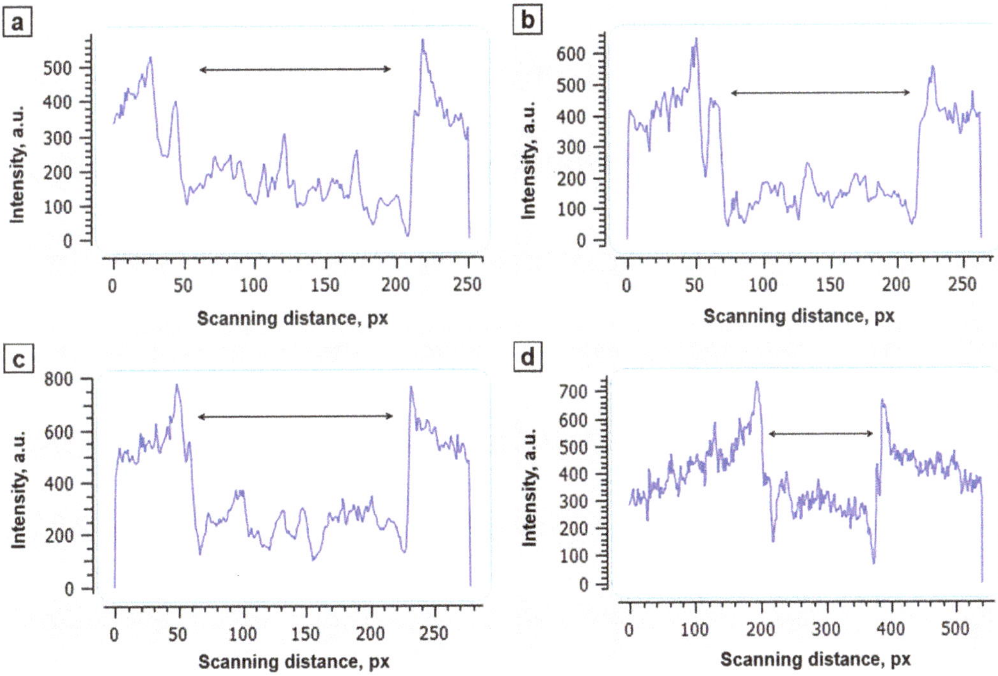

Figure 5. Dynamics of the optical density (longitudinal scanning) of bioactive implant zone (double arrow): (**a**)—the day of the surgery, (**b**)—7th day after surgery, (**c**)—14th day after surgery, (**d**)—28th day after surgery.

There was no pathological reaction from the soft tissues and signs of inflammation: subcutaneous tissue, fascia, muscles, and periosteum were of the usual structure. On the transverse and longitudinal sections of the implants after euthanasia (one month after the surgery), all cells of implants and longitudinal axial canals were filled with bone spongy substance (Figure 6a,b). Comparative analysis of the images with the use of CT reconstruction in the range of the shades of gray and color visualization revealed the presence of colored structures of high-intensity on the surface (in the outer pores of implants) and in the cells of the implant proving bone ingrowth (Figure 6c,d). In the longitudinal axial canal, images of trabecular structure of 5–6 mm long were observed (Figure 6c,e, white arrows). They were united with the proximal and distal bone fragments. The intensity of these structures corresponded to the intensity of cortical bone.

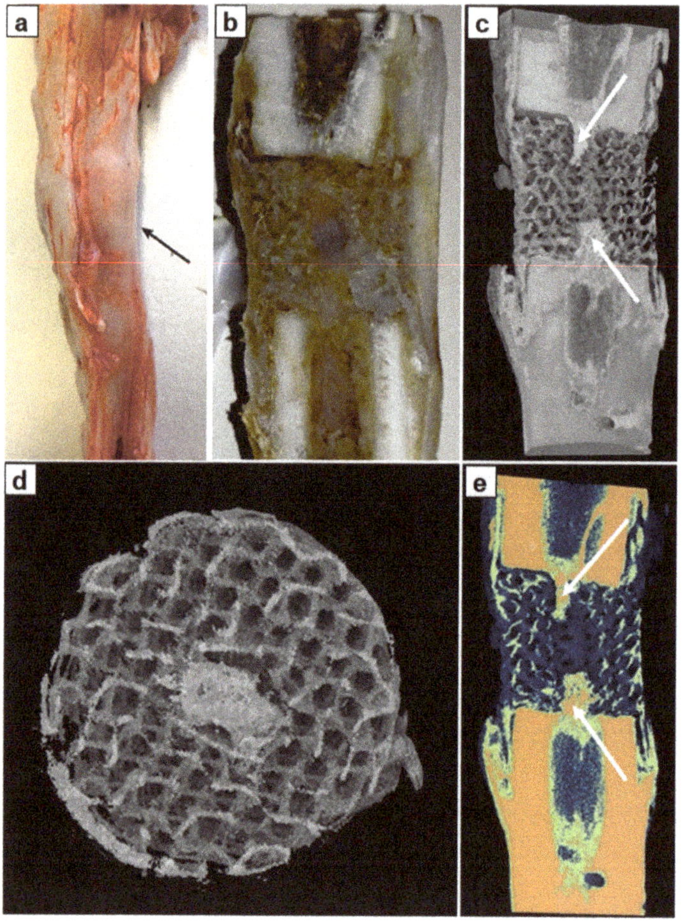

Figure 6. Bone-implant block after 30 days of the experiment: (**a**)—photo of the anatomical preparation of the tibia (the arrow indicates newly formed bone in the central canal), (**b**)—photo of the longitudinal section of the anatomical preparation, (**c**)—CT reconstruction of the anatomical preparation in the range of the shades of gray (×1.1), (**d**)—CT image of a cross-sectioned anatomical preparation in the range of the shades of gray (×2.8), (**e**)—colored visualization of a cross-sectioned anatomical preparation using the Smart colors filter (×1.1).

On the sides of the proximal and distal bone fragments, the implants were periosteally covered with cone-shaped bands of dense compact bone tissue (Figure 6e). Moreover, low mineralized bone trabeculae were found in the cells of implants (Figure 6d). There were no signs of resorption or porosity of the compact plate of a bone fragments.

Thus, the study demonstrated ingrowth of bone tissue into cylindrical PCL-HA implant both from the adjacent bone ends and periosteal structures. By the 30th postoperative day the implant and bone defect were filled in with a newly-formed bone tissue.

4. Discussion

Architecture and surface properties are among the most important characteristics for the biocompatibility of scaffolds. The ideal scaffold should provide an optimal surface for cell adhesion and interconnected porous architecture for vascularization and extracellular matrix reproduction [44,45]. Three-dimensional printing technologies implement a highly adaptable and predictable scaffold design: layer-by-layer fabrication allows the formation of various shapes and a wide range of infills from ordinary 2D grids to 3D structures with triply periodic minimal surfaces. Providing an interconnected 3D porous structure, high surface area and high-energy adsorption [46–48], TPMS appears to be biomimetic and highly desirable for the design of scaffolds. However, among various biomimetic TPMS structures, a gyroid is the most convenient to use due to the presence of the gyroid infill in advanced slicer software. The possibility to design the gyroid infill directly with the use of slicing software makes the scaffolds design highly variable and reproducible.

Ilizarov's distraction osteogenesis method demonstrates effectiveness in treatment and reconstruction of large bone defects [3–5]. However, difficulties and complications associated with the use of the Ilizarov bone transport technique have been reported: high rate of pin-site infection, prolonged time of external frame fixation, joint contractures, risk of ischemic regeneration in compromised soft tissues around a large defect, and physical and psychological suffering of patients [4,5,49,50].

Combination of Masquelet technique (induced membrane technique) followed by Ilizarov bone transport demonstrated effectiveness in the treatment of nonunions of long bones [51,52]. The bone transport through induced membrane chamber allowed improving outcomes and avoiding bone grafting. However, the healing index remained high, varying from 35 to 60 days/cm [53].

Previously, it was demonstrated that an ideal type of non-free bone grafting for bone defects filling presents a vascularized autologous osteotomized bone fragment transported gradually in the interfragmental gap within the soft tissue envelopment providing osteogenic properties [54,55]. An autologous graft allows to avoid graft harvesting but risks of delayed bone union, pin-site infection, and prolonged external frame wearing are not negligible.

Another option for treatment of extensive diaphyseal and metaphyseal bone defects, emerging for the latest 20 years, is presented by 3D printed scaffolds for hard tissue replacement [56,57]. This technique gives an alternative method to repair bone defects caused by osteomyelitis or trauma [58,59]. Materials for large bone defect reconstruction required to ensure high resistance to mechanical load. The titanium porous implants corresponded to these requirements providing high mechanical resistance to loading force [60]. On the other hand, biomaterials used in bone defect repairing should provide osteoinductivity. It is the second requirement of an ideal bone graft material for total long bone defect reconstruction. A scaffold promoting new bone tissue formation should enable conducting of bone regeneration [61]. Non-degradable materials provide only maintenance of the anatomical function rather than bone regeneration. High stiffness of metallic biomaterials results in a stress-shielding effect leading to microfractures of additional bone tissue surrounding implants [62]. Habibovic et al. [63] and Lan Levengood et al. [64] advocate necessity of an intrinsic microporosity and macropores to enhance new bone formation on the surfaces of 3D scaffold implants. Moreover, the increase in the porosity of implants decreases the stress-shielding effect. However, the strength of implants simultaneously reduces [65].

Biomaterials used in bone defect repairing should provide osteoinductivity. Three-dimensional printing of calcium-phosphates scaffolds enables biocompatibility and osteoinductive properties of implants [66]. Depending on filling the 3D-printed titanium scaffolds can provide antibacterial and osteoconductive properties [67]. However, it is still limited compared to the one of rhBMP-2 regarding osteoinduction properties. The recombinant human BMP-2 stimulates mesenchymal stem cells to differentiate into preosteoblasts and osteoblasts, and acts as a trigger for migration of osteoblasts in vivo and in vitro [68]. The use of BMP-2 for clinical purposes arises concerns: the release of rh-BMP-2, especially in surrounding soft tissues, could cause heterotopic ossifications, osteolysis and cancer [69,70]. For this reason, an osteoinductive activity of an implant has to be limited to contacting surfaces of a reconstructed bone and implant to avoid any systemic undesirable side effect.

Coating of surfaces with hydroxyapatite (HA) improves bioactivity and osteointegration of metallic implants [71]. Only implants with bioactive HA-coated surfaces ensure primary implant stability which results in an improved contact between bone and implant [72]. The osteoinductive layer of HA on the implant can be achieved by micro arc oxidation in an electrolyte with calcium and phosphate or by composite coating including fluorocarbon plastic filled with dispersed hydroxyapatite [73]. Uklejewski et al., emphasize the role of a biomineral coating with a native Ca/P ratio [74]. It can be one of crucial factors in mechanics of interacting biomaterials the innovative multi-spiked connecting scaffold [75].

On the other hand, there are certain limitations of clinical use of titanium implants in long bone defects in humans. Porous 3D scaffolds fabricated of titanium alloys induce osseointegration but material obstructs radiological control of ingrowth of newly formed bone in early postoperative period and in long-term follow up after surgery in clinics [76].

Recently, a scaffold-guided bone regeneration with use of highly porous biodegradable scaffolds providing long-term mechanical stability has been emerged [76]. Our earlier morphological studies demonstrated a positive effect of implant hydroxyapatite coating on the reparative bone formation in experimental fractures: porous surfaces contribute to a microvascular development followed by formation of dense trabecular structure on the implant surface without formation of a connective tissue envelope [77]. In cases of a significant defect in long bone shaft and the absence of bone marrow and endosteum, the periosteum ensures the main role in the reparative osteogenesis process of restoring the bone substance [13]. The formation of bone tissue is associated primarily with the intensity of angiogenesis: perivascular cells that have differentiation potential similar to mesenchymal stem/progenitor cells migrate together with the vessels into the cells of the graft differentiating into one or more of the cellular components of the vascular bed [78,79]. Possessing multipotent properties and inducibility to osteogenesis, in the presence of HA (inducer), which is present on the surface of both external and internal structures of the new implant, perivasculitis differentiate into osteoblasts [73,80].

The proposed PCL-HA cylindrical implant ensures ingrowth of bone tissue into cells of implant from the surrounding periosteum. Development of bone tissues from medullary canal of adjacent bone fragments was also achieved. The newly formed bone was observed on the implantation site in one month after implantation. Implant material (PCL) allowed to assess new bone development within the implant. In this experimental study the highly porous biodegradable construct was stabilized with Ilizarov frame.

We suppose that PCL osteoinductive biodegradable implants coated with HA allow to overcome inconveniences of stress-shielding effect and limited (even absent) radiological assessment typical for metal (titanium alloys and steels) 3D-printed implants. The use of temporary limited external fixation or degradable material providing a long-lasting mechanical resistance represents an option for mechanical resistance while bone healing.

This study did not evaluate the stability of bone fragments when the PCL-HA implant was applied for experimental defect filling. We recognize it as a weak point of the research. Thus, additional studies are required for assessment of biomechanical stability of bone fragments with such type of polymeric implants. Furthermore, morphological and biomechanical studies will clarify the remodeling of bone inside the implant in the process of polymeric matrix degradation.

5. Conclusions

In the present study, a method for the restoration of long bone segmental defect with the use of bioactive degradable 3D-printed implant was proposed. Porous biomimetic implant was 3D printed of biodegradable polymer PCL and coated with bioactive mineral HA. Experimental in vivo study with the use of laboratory sheep revealed that the porous degradable 3D implants with bioactive coating located in a long bone segmental defect provide stimulation of reparative osteogenesis and osseointegration into the single implant-bone block. Ingrowth of bone tissue into cylindrical PCL-HA implant from adjacent bone ends and periosteal structures was observed to start from day 7 of study. By the 30th postoperative day, the implant was filled in with bone tissue. Moreover, this type of implant made of degradable polymer allows radiological assessment of reparative osteogenesis in long bone defect filling.

Supplementary Materials: The following supporting information can be downloaded at: https://www.mdpi.com/article/10.3390/biomimetics8020138/s1. Figure S1. Drawing of the implant: (**a**)—front view, (**b**)—top view, (**c**)—A-A section; Figure S2. EDX elemental mapping of the HA coated implants: (**a**)—carbon (C), (**b**)—calcium (Ca), (**c**)—phosphorus (P), (**d**)—EDX spectra; Figure S3. Chemical characterization of the implants by FTIR.

Author Contributions: Conceptualization, A.P., S.T. and D.P.; methodology, N.K., D.P. and G.D.; validation, A.P. and A.S., formal analysis, G.D., E.G. and N.K.; writing—original draft preparation, A.P., G.D. and S.T.; writing—review and editing, D.P.; visualization, E.G. All authors have read and agreed to the published version of the manuscript.

Funding: This research in the part of the bioactive degradable 3D-implant manufacturing was supported by the Ministry of Science and Higher Education of the Russian Federation, project Nauka FSWW-2023-0007.

Institutional Review Board Statement: The study was conducted in accordance with the Declaration of Helsinki, and approved by the Ethics Committee of the Ilizarov National Medical Research Center for Traumatology and Orthopedics, Kurgan, Russia (protocol code 1(71), date of approval 28 April 2022).

Data Availability Statement: Data are available upon a demand to corresponding author (D.P.).

Acknowledgments: The authors would like to thank Alexey Gogolev for the kind help with CT of implants, and many thanks goes to Vadim Bocharov for his help with implants fabrication.

Conflicts of Interest: The authors declare no conflict of interest.

References

1. El-Rosasy, M.A.; Ayoub, M.A. Traumatic Composite Bone and Soft Tissue Loss of the Leg: Region-Specific Classification and Treatment Algorithm. *Injury* **2020**, *51*, 1352–1361. [CrossRef] [PubMed]
2. Lu, V.; Zhang, J.; Zhou, A.; Krkovic, M. Management of post-traumatic femoral defects with a monorail external fixator over an intramedullary nail. *Eur. J. Orthop. Surg. Traumatol.* **2022**, *32*, 1119–1126. [CrossRef] [PubMed]
3. Jia, Q.; Liu, Y.; Alimujiang, A.; Guo, J.; Chen, D.; Wang, Y.; Yusufu, A.; Ma, C. Nine-year-long complex humeral nonunion salvaged by distraction osteogenesis technique: A case report and review of the literature. *BMC Surg.* **2022**, *22*, 77. [CrossRef] [PubMed]
4. El-Alfy, B.; El-Mowafi, H.; El-Moghazy, N. Distraction osteogenesis in management of composite bone and soft tissue defects. *Int. Orthop.* **2010**, *34*, 115–118. [CrossRef]
5. Rozbruch, R.S.; Weitzman, A.M.; Watson, T.J.; Freudigman, P.; Katz, H.V.; Ilizarov, S. Simultaneous treatment of tibial bone and soft tissue defects with the Ilizarov method. *J. Orthop. Trauma* **2006**, *20*, 197–205. [CrossRef]
6. Smith, W.R.; Elbatrawy, Y.A.; Andreassen, G.S.; Philips, G.C.; Guerreschi, F.; Lovisetti, L.; Catagni, M.A. Treatment of traumatic forearm bone loss with Ilizarov ring fixation and bone transport. *Int. Orthop.* **2007**, *31*, 165–170. [CrossRef]
7. Cao, Z.; Zhang, Y.; Lipa, K.; Qing, L.; Wu, P.; Tang, J. Ilizarov Bone Transfer for Treatment of Large Tibial Bone Defects: Clinical Results and Management of Complications. *J. Pers. Med.* **2022**, *12*, 1774. [CrossRef]
8. Liodakis, E.; Kenawey, M.; Krettek, C.; Wiebking, U.; Hankemeier, S. Comparison of 39 post-traumatic tibia bone transports performed with and without the use of an intramedullary rod: The long-term outcomes. *Int. Orthop.* **2011**, *35*, 1397–1402. [CrossRef]
9. Wang, X.; Xiang, C.; Yan, C.; Chen, Q.; Chen, L.; Jiang, K.; Li, Y. Effectiveness of bone transport with a locking plate versus conventional bone transport for tibial defects. *Zhongguo Xiu Fu Chong Jian Wai Ke Za Zhi* **2022**, *36*, 983–988. (In Chinese)
10. Kumar, A.; Nune, K.C.; Misra, R.D.K. Design and biological functionality of a novel hybrid Ti-6Al-4V/hydrogel system for reconstruction of bone defects. *J. Tissue Eng. Regen. Med.* **2018**, *12*, 1133–1144. [CrossRef]
11. Wang, Z.; Han, L.; Zhou, Y.; Cai, J.; Sun, S.; Ma, J.; Wang, W.; Li, X.; Ma, L. The combination of a 3D-Printed porous Ti-6Al-4V alloy scaffold and stem cell sheet technology for the construction of biomimetic engineered bone at an ectopic site. *Mater. Today Bio.* **2022**, *16*, 100433. [CrossRef] [PubMed]

12. Ma, L.; Wang, X.; Zhou, Y.; Ji, X.; Cheng, S.; Bian, D.; Fan, L.; Zhou, L.; Ning, C.; Zhang, Y. Biomimetic Ti-6Al-4V alloy/gelatin methacrylate hybrid scaffold with enhanced osteogenic and angiogenic capabilities for large bone defect restoration. *Bioact. Mater.* **2021**, *6*, 3437–3448. [CrossRef]
13. Popkov, A.; Kononovich, N.; Gorbach, E.; Popkov, D. Osteointegration technology in long bone defect reconstruction: Experimental study. *Acta Bioeng. Biomech.* **2020**, *22*, 85–91. [CrossRef] [PubMed]
14. Hubbe, U.; Beiser, S.; Kuhn, S.; Stark, T.; Hoess, A.; Cristina-Schmitz, H.; Vasilikos, I.; Metzger, M.C.; Rothweiler, R. A fully ingrowing implant for cranial reconstruction: Results in critical size defects in sheep using 3D-printed titanium scaffold. *Biomater. Adv.* **2022**, *136*, 212754. [CrossRef]
15. Liu, B.; Hou, G.; Yang, Z.; Li, X.; Zheng, Y.; Wen, P.; Liu, Z.; Zhou, F.; Tian, Y. Repair of critical diaphyseal defects of lower limbs by 3D printed porous Ti6Al4V scaffolds without additional bone grafting: A prospective clinical study. *J. Mater. Sci. Mater. Med.* **2022**, *33*, 64. [CrossRef] [PubMed]
16. Zhang, L.; Yang, G.; Johnson, B.N.; Jia, X. Three-dimensional (3D) printed scaffold and material selection for bone repair. *Acta Biomater.* **2019**, *84*, 16–33. [CrossRef]
17. Alonzo, M.; Primo, F.A.; Kumar, S.A.; Mudloff, J.A.; Dominguez, E.; Fregoso, G.; Ortiz, N.; Weiss, W.M.; Joddar, B. Bone tissue engineering techniques, advances and scaffolds for treatment of bone defects. *Curr. Opin. Biomed Eng.* **2021**, *17*, 100248. [CrossRef]
18. Arif, Z.U.; Khalid, M.Y.; Noroozi, R.; Sadeghianmaryan, A.; Jalalvand, M.; Hossain, M. Recent advances in 3D-printed polylactide and polycaprolactone-based biomaterials for tissue engineering applications. *Int. J. Biol. Macromol.* **2022**, *218*, 930–968. [CrossRef]
19. Fu, Z.; Cui, J.; Zhao, B.; Shen, S.G.; Lin, K. An overview of polyester/hydroxyapatite composites for bone tissue repairing. *J. Orthop. Translat.* **2021**, *28*, 118–130. [CrossRef]
20. Bal, Z.; Korkusuz, F.; Ishiguro, H.; Okada, R.; Kushioka, J.; Chijimatsu, R.; Kodama, J.; Tateiwa, D.; Ukon, Y.; Nakagawa, S.; et al. A novel nano-hydroxyapatite/synthetic polymer/bone morphogenetic protein-2 composite for efficient bone regeneration. *Spine J.* **2021**, *21*, 865–873. [CrossRef]
21. Fijoł, N.; Abdelhamid, H.N.; Pillai, B.; Hall, S.A.; Thomas, N.; Mathew, A.P. 3D-printed monolithic biofilters based on a polylactic acid (PLA)—hydroxyapatite (HAp) composite for heavy metal removal from an aqueous medium. *RSC Adv.* **2021**, *11*, 32408–32418. [CrossRef]
22. Pérez, E. Mechanical Performance of in Vitro Degraded Polylactic Acid/Hydroxyapatite Composites. *J. Mater. Sci.* **2021**, *56*, 19915–19935. [CrossRef]
23. Cengiz, I.F.; Pereira, H.; Espregueira-Mendes, J.; Kwon, I.K.; Reis, R.L.; Oliveira, J.M. Suturable regenerated silk fibroin scaffold reinforced with 3D-printed polycaprolactone mesh: Biomechanical performance and subcutaneous implantation. *J. Mater. Sci. Mater. Med.* **2019**, *30*, 63. [CrossRef]
24. Evlashin, S.; Dyakonov, P.; Tarkhov, M.; Dagesyan, S.; Rodionov, S.; Shpichka, A.; Kostenko, M.; Konev, S.; Sergeichev, I.; Timashev, P.; et al. Flexible Polycaprolactone and Polycaprolactone/Graphene Scaffolds for Tissue Engineering. *Materials* **2019**, *12*, 2991. [CrossRef]
25. Chen, Y.P.; Lo, T.S.; Lin, Y.T.; Chien, Y.H.; Lu, C.J.; Liu, S.J. Fabrication of Drug-Eluting Polycaprolactone/poly(lactic-*co*-glycolic Acid) Prolapse Mats Using Solution-Extrusion 3D Printing and Coaxial Electrospinning Techniques. *Polymers* **2021**, *13*, 2295. [CrossRef] [PubMed]
26. Gerdes, S.; Mostafavi, A.; Ramesh, S.; Memic, A.; Rivero, I.V.; Rao, P.; Tamayol, A. Process-Structure-Quality Relationships of Three-Dimensional Printed Poly(Caprolactone)-Hydroxyapatite Scaffolds. *Tissue Eng. Part A* **2020**, *26*, 279–291. [CrossRef] [PubMed]
27. Porta, M.; Tonda-Turo, C.; Pierantozzi, D.; Ciardelli, G.; Mancuso, E. Towards 3D Multi-Layer Scaffolds for Periodontal Tissue Engineering Applications: Addressing Manufacturing and Architectural Challenges. *Polymers* **2020**, *12*, 2233. [CrossRef] [PubMed]
28. Olivares, A.L.; Marsal, E.; Planell, J.A.; Lacroix, D. Finite element study of scaffold architecture design and culture conditions for tissue engineering. *Biomaterials* **2009**, *30*, 6142–6149. [CrossRef]
29. Ravoor, J.; Thangavel, M.; Elsen, S.R. Comprehensive Review on Design and Manufacturing of Bio-scaffolds for Bone Reconstruction. *ACS Appl. Bio Mater.* **2021**, *4*, 8129–8158. [CrossRef]
30. Zhao, F.; Xiong, Y.; Ito, K.; van Rietbergen, B.; Hofmann, S. Porous Geometry Guided Micro-mechanical Environment Within Scaffolds for Cell Mechanobiology Study in Bone Tissue Engineering. *Front. Bioeng. Biotechnol.* **2021**, *9*, 736489. [CrossRef]
31. Siddique, S.H.; Hazell, P.J.; Wang, H.; Escobedo, J.P.; Ameri, A.A.H. Lessons from Nature: 3D Printed Bio-Inspired Porous Structures for Impact Energy Absorption—A Review. *Addit. Manuf.* **2022**, *58*, 103051. [CrossRef]
32. Hernandez, I.; Kumar, A.; Joddar, B. A Bioactive Hydrogel and 3D Printed Polycaprolactone System for Bone Tissue Engineering. *Gels* **2017**, *3*, 26. [CrossRef] [PubMed]
33. Fonseca, D.R.; Sobreiro-Almeida, R.; Sol, P.C.; Neves, N.M. Development of non-orthogonal 3D-printed scaffolds to enhance their osteogenic performance. *Biomater. Sci.* **2018**, *6*, 1569–1579. [CrossRef]
34. Hashimi, N.S.; Soman, S.S.; Govindharaj, M.; Vijayavenkataraman, S. 3D Printing of Complex Architected Metamaterial Structures by Simple Material Extrusion for Bone Tissue Engineering. *Mater. Today Commun.* **2022**, *31*, 103382. [CrossRef]
35. Wang, F.; Tankus, E.B.; Santarella, F.; Rohr, N.; Sharma, N.; Märtin, S.; Michalscheck, M.; Maintz, M.; Cao, S.; Thieringer, F.M. Fabrication and Characterization of PCL/HA Filament as a 3D Printing Material Using Thermal Extrusion Technology for Bone Tissue Engineering. *Polymers* **2022**, *14*, 669. [CrossRef] [PubMed]

36. Dong, Z.; Zhao, X. Application of TPMS structure in bone regeneration. *Eng. Regen.* **2021**, *2*, 154–162. [CrossRef]
37. Zhu, H.; Li, M.; Huang, X.; Qi, D.; Nogueira, L.P.; Yuan, X.; Liu, W.; Lei, Z.; Jiang, J.; Dai, H.; et al. 3D printed tricalcium phosphate-bioglass scaffold with gyroid structure enhance bone ingrowth in challenging bone defect treatment. *Appl. Mater. Today* **2021**, *25*, 101166. [CrossRef]
38. Van Hede, D.; Liang, B.; Anania, S.; Barzegari, M.; Verlee, B.; Nolens, G.; Pirson, J.; Geris, L.; Lambert, F. 3D-Printed Synthetic Hydroxyapatite Scaffold with In Silico Optimized Macrostructure Enhances Bone Formation In Vivo. *Adv. Funct. Mater.* **2022**, *32*, 2105002. [CrossRef]
39. Merloz, P.; Maurel, N.; Marchard, D.; Lavaste, F.; Barnole, J.; Faure, C.; Butel, J. Three-dimensional rigidity of the Ilizarov external fixator (original and modified) implanted at the femur. Experimental study and clinical deductions. *Rev. Chir. Orthop. Reparatrice L'appareil Mot.* **1991**, *77*, 65–76.
40. Glatt, V.; Samchukov, M.; Cherkashin, A.; Iobst, C. Reverse Dynamization Accelerates Bone-Healing in a Large-Animal Osteotomy Model. *J. Bone Jt. Surg. Am.* **2021**, *103*, 257–263. [CrossRef]
41. Goreninskii, S.; Stankevich, K.; Bolbasov, E.; Danilenko, N.; Filimonov, V.; Tverdokhlebov, S. Comparison of the Influence of "Solvent/Non-Solvent" Treatment for the Attachment of Signal Molecules on the Structure of Electrospun PCL and PLLA Biodegradable Scaffolds. *MATEC Web Conf.* **2016**, *79*, 01025. [CrossRef]
42. Goreninskii, S.I.; Stankevich, K.S.; Nemoykina, A.L.; Bolbasov, E.N.; Tverdokhlebov, S.I.; Filimonov, V.D. A first method for preparation of biodegradable fibrous scaffolds containing iodine on the fibre surfaces. *Bull. Mater. Sci.* **2018**, *41*, 100. [CrossRef]
43. Goreninskii, S.I.; Guliaev, R.O.; Stankevich, K.S.; Danilenko, N.V.; Bolbasov, E.N.; Golovkin, A.S.; Mishanin, A.I.; Filimonov, V.D.; Tverdokhlebov, S.I. "Solvent/non-solvent" treatment as a method for non-covalent immobilization of gelatin on the surface of poly(l-lactic acid) electrospun scaffolds. *Colloids Surf. B Biointerfaces* **2019**, *177*, 137–140. [CrossRef] [PubMed]
44. Chan, B.P.; Leong, K.W. Scaffolding in tissue engineering: General approaches and tissue-specific considerations. *Eur. Spine J.* **2008**, *17* (Suppl. S4), 467. [CrossRef] [PubMed]
45. Carotenuto, F.; Politi, S.; Ul Haq, A.; De Matteis, F.; Tamburri, E.; Terranova, M.L.; Teodori, L.; Pasquo, A.; Di Nardo, P. From Soft to Hard Biomimetic Materials: Tuning Micro/Nano-Architecture of Scaffolds for Tissue Regeneration. *Micromachines* **2022**, *13*, 780. [CrossRef] [PubMed]
46. Santos, J.; Pires, T.; Gouveia, B.P.; Castro, A.P.; Fernandes, P.R. On the permeability of TPMS scaffolds. *J. Mech. Behav. Biomed. Mater.* **2020**, *110*, 103932. [CrossRef]
47. Yang, S.D.; Lee, H.G.; Kim, J. A phase-field approach for minimizing the area of triply periodic surfaces with volume constraint. *Comput. Phys. Commun.* **2010**, *181*, 1037–1046. [CrossRef]
48. Tripathi, Y.; Shukla, M.; Bhatt, A.D. Implicit-Function-Based Design and Additive Manufacturing of Triply Periodic Minimal Surfaces Scaffolds for Bone Tissue Engineering. *J. Mater. Eng. Perform.* **2019**, *28*, 7445–7451. [CrossRef]
49. Liu, Y.; Yushan, M.; Liu, Z.; Liu, J.; Ma, C.; Yusufu, A. Complications of bone transport technique using the Ilizarov method in the lower extremity: A retrospective analysis of 282 consecutive cases over 10 years. *BMC Musculoskelet. Disord.* **2020**, *21*, 354. [CrossRef]
50. Krappinger, D.; Irenberger, A.; Zegg, M.; Huber, B. Treatment of large posttraumatic tibial bone defects using the Ilizarov method: A subjective outcome assessment. *Arch. Orthop. Trauma Surg.* **2013**, *133*, 789–795. [CrossRef]
51. Hamiti, Y.; Yushan, M.; Lu, C.; Yusufu, A. Reconstruction of massive tibial defect caused by osteomyelitis using induced membrane followed by trifocal bone transport technique: A retrospective study and our experience. *BMC Surg.* **2021**, *21*, 419. [CrossRef]
52. Borzunov, D.Y.; Kolchin, S.N.; Mokhovikov, D.S.; Malkova, T.A. Ilizarov bone transport combined with the Masquelet technique for bone defects of various etiologies (preliminary results). *World J. Orthop.* **2022**, *13*, 278–288. [CrossRef]
53. Khaled, A.; El-Gebaly, O.; El-Rosasy, M. Masquelet-Ilizarov technique for the management of bone loss post debridement of infected tibial nonunion. *Int. Orthop.* **2022**, *46*, 1937–1944. [CrossRef] [PubMed]
54. Catagni, M.A.; Azzam, W.; Guerreschi, F.; Lovisetti, L.; Poli, P.; Khan, M.S.; Di Giacomo, L.M. Trifocal versus bifocal bone transport in treatment of long segmental tibial bone defects. *Bone Jt. J.* **2019**, *101-B*, 162–169. [CrossRef] [PubMed]
55. Li, R.; Zhu, G.; Chen, C.; Chen, Y.; Ren, G. Bone Transport for Treatment of Traumatic Composite Tibial Bone and Soft Tissue Defects: Any Specific Needs besides the Ilizarov Technique? *Biomed. Res. Int.* **2020**, *2020*, 2716547. [CrossRef] [PubMed]
56. Manassero, M.; Viateau, V.; Matthys, R.; Deschepper, M.; Vallefuoco, R.; Bensidhoum, M.; Petite, H. A novel murine femoral segmental critical-sized defect model stabilized by plate osteosynthesis for bone tissue engineering purposes. *Tissue Eng. Part C Methods* **2013**, *19*, 271–280. [CrossRef]
57. Lyu, L.; Yang, S.; Jing, Y.; Zhang, C.; Wang, J. Examining trabecular morphology and chemical composition of peri-scaffold osseointegrated bone. *J. Orthop. Surg. Res.* **2020**, *15*, 406. [CrossRef]
58. Coriaty, N.; Pettibone, K.; Todd, N.; Rush, S.; Carter, R.; Zdenek, C. Titanium Scaffolding: An Innovative Modality for Salvage of Failed First Ray Procedures. *J. Foot Ankle Surg.* **2018**, *57*, 593–599. [CrossRef]
59. Gamieldien, H.; Ferreira, N.; Birkholtz, F.F.; Hilton, T.; Campbell, N.; Laubscher, M. Filling the gap: A series of 3D-printed titanium truss cages for the management of large, lower limb bone defects in a developing country setting. *Eur. J. Orthop. Surg. Traumatol.* **2022**, *33*, 497–505. [CrossRef]
60. Geetha, M.; Singh, A.K.; Asokamani, R.; Gogia, A.K. Ti-based biomaterials, the ultimate choice for orthopaedic implants. A review. *Prog. Mater. Sci.* **2009**, *54*, 397–425. [CrossRef]

61. Guarino, V.; Causa, F.; Ambrosio, L. Bioactive scaffolds for bone and ligament tissue. *Expert Rev. Med. Devices* **2007**, *4*, 405–418. [CrossRef] [PubMed]
62. Noyama, Y.; Miura, T.; Ishimoto, T.; Itaya, T.; Niinomi, M.; Nakano, T. Bone loss and reduced bone quality of the human femur after total hip arthroplasty under stress-shielding effects by titanium-based implant. *Mater. Trans.* **2012**, *53*, 565–570. [CrossRef]
63. Habibovic, P.; Yuan, H.; van der Valk, C.M.; Meijer, G.; van Blitterswijk, C.A.; de Groot, K. 3D microenvironment as essential element for osteoinduction by biomaterials. *Biomaterials* **2005**, *26*, 3565–3575. [CrossRef] [PubMed]
64. Lan Levengood, S.K.; Polak, S.J.; Poellmann, M.J.; Hoelzle, D.J.; Maki, A.J.; Clark, S.G.; Wheeler, M.B.; Wagoner Johnson, A.J. The effect of BMP-2 on micro- and macroscale osteointegration of biphasic calcium phosphate scaffolds with multiscale porosity. *Acta Biomater.* **2010**, *6*, 3283–3291. [CrossRef]
65. Chen, D.; Li, D.; Pan, K.; Gao, S.; Wang, B.; Sun, M.; Zhao, C.; Liu, X.; Li, N. Strength enhancement and modulus modulation in auxetic meta-biomaterials produced by selective laser melting. *Acta Biomater.* **2022**, *153*, 596–613. [CrossRef]
66. Le Guéhennec, L.; Van Hede, D.; Plougonven, E.; Nolens, G.; Verlée, B.; De Pauw, M.C.; Lambert, F. In vitro and in vivo biocompatibility of calcium-phosphate scaffolds three-dimensional printed by stereolithography for bone regeneration. *J. Biomed. Mater. Res. A* **2020**, *108*, 412–425. [CrossRef] [PubMed]
67. Li, S.; He, Y.; Li, J.; Sheng, J.; Long, S.; Li, Z.; Jiang, B.; Fu, H.; Weng, J.; Wu, J.; et al. Titanium scaffold loaded with strontium and copper double-doped hydroxyapatite can inhibit bacterial growth and enhance osteogenesis. *J. Biomater. Appl.* **2022**, *37*, 195–203. [CrossRef]
68. Liu, S.; Liu, Y.; Jiang, L.; Li, Z.; Lee, S.; Liu, C.; Wang, J.; Zhang, J. Recombinant human BMP-2 accelerates the migration of bone marrow mesenchymal stem cells via the CDC42/PAK1/LIMK1 pathway in vitro and in vivo. *Biomater. Sci.* **2018**, *7*, 362–372. [CrossRef]
69. Tannoury, C.A.; An, H.S. Complications with the use of bone morphogenetic protein 2 (BMP-2) in spine surgery. *Spine J.* **2014**, *14*, 552–559. [CrossRef]
70. Steib, J.-P.; Bouchaib, J.; Walter, A.; Shuller, S.; Charles, P. Could an osteoinductor result in degeneration of a neurofibroma in NF 1? *Eur. Spine J.* **2010**, *19* (Suppl. S2), S220–S225. [CrossRef]
71. Zweymuller, K.A. Bony ongrowth on the surface of HA-coated femoral implants: An x-ray analysis. *Z Orthop. Unfall.* **2012**, *150*, 27–31. [PubMed]
72. Popkov, A.V.; Gorbach, E.N.; Kononovich, N.A.; Popkov, D.A.; Tverdokhlebov, S.I.; Shesterikov, E.V. Bioactivity and osteointegration of hydroxyapatite-coated stainless steel and titanium wires used for intramedullary osteosynthesis. *Strateg. Trauma Limb Reconstr.* **2017**, *12*, 107–113. [CrossRef] [PubMed]
73. Bolbasov, E.N.; Popkov, A.V.; Popkov, D.A.; Gorbach, E.N.; Khlusov, I.A.; Golovkin, A.S.; Sinev, A.; Bouznik, V.M.; Tverdokhlebov, S.I.; Anissimov, Y.G. Osteoinductive composite coatings for flexible intramedullary nails. *Mater. Sci. Eng. C Mater. Biol. Appl.* **2017**, *75*, 207–220. [CrossRef] [PubMed]
74. Uklejewski, R.; Winiecki, M.; Krawczyk, P.; Tokłowicz, R. Native Osseous CaP Biomineral Coating on a Biomimetic Multi-Spiked Connecting Scaffold Prototype for Cementless Resurfacing Arthroplasty Achieved by Combined Electrochemical Deposition. *Materials* **2019**, *12*, 3994. [CrossRef]
75. Rogala, P.; Uklejewski, R.; Winiecki, M.; Dąbrowski, M.; Gołańczyk, J.; Patalas, A. First Biomimetic Fixation for Resurfacing Arthroplasty: Investigation in Swine of a Prototype Partial Knee Endoprosthesis. *Biomed. Res. Int.* **2019**, *2019*, 6952649. [CrossRef]
76. Laubach, M.; Kobbe, P.; Hutmacher, D.W. Biodegradable interbody cages for lumbar spine fusion: Current concepts and future directions. *Biomaterials* **2022**, *288*, 121699. [CrossRef]
77. Popkov, A.V.; Popkov, D.A.; Kononovich, N.A.; Gorbach, E.N.; Tverdokhlebov, S.I.; Bolbasov, E.N.; Darvin, E.O. Biological activity of the implant for internal fixation. *J. Tissue Eng. Regen. Med.* **2018**, *12*, 2248–2255. [CrossRef]
78. Valarmathi, M.T.; Davis, J.M.; Yost, M.J.; Goodwin, R.L.; Potts, J.D. A three-dimensional model of vasculogenesis. *Biomaterials* **2009**, *30*, 1098–1112. [CrossRef]
79. Siow, R.C.; Mallawaarachchi, C.M.; Weissberg, P.L. Migration of adventitial myofibroblasts following vascular balloon injury: Insights from in vivo gene transfer to rat carotid arteries. *Cardiovasc. Res.* **2003**, *59*, 212–221. [CrossRef]
80. Barrère, F.; van der Valk, C.M.; Dalmeijer, R.A.; Meijer, G.; van Blitterswijk, C.A.; de Groot, K.; Layrolle, P. Osteogenecity of octacalcium phosphate coatings applied on porous metal implants. *J. Biomed. Mater. Res. A* **2003**, *66*, 779–788. [CrossRef]

biomimetics

MDPI

Article

Preparation, Characterization, and Drug Delivery of Hexagonal Boron Nitride-Borate Bioactive Glass Biomimetic Scaffolds for Bone Tissue Engineering

Mertcan Ensoylu [1], Aylin M. Deliormanlı [1,*] and Harika Atmaca [2]

1 Department of Metallurgical and Materials Engineering, Manisa Celal Bayar University, 45140 Yunusemre, Manisa, Turkey
2 Department of Biology, Manisa Celal Bayar University, 45140 Yunusemre, Manisa, Turkey
* Correspondence: aylin.deliormanli@cbu.edu.tr; Tel.: +90-0236-2012405

Abstract: In this study, biomimetic borate-based bioactive glass scaffolds containing hexagonal boron nitride hBN nanoparticles (0.1, 0.2, 0.5, 1, and 2% by weight) were manufactured with the polymer foam replication technique to be used in hard tissue engineering and drug delivery applications. To create three-dimensional cylindrical-shaped scaffolds, polyurethane foams were used as templates and covered using a suspension of glass and hBN powder mixture. Then, a heat treatment was applied at 570 °C in an air atmosphere to remove the polymer foam from the structure and to sinter the glass structures. The structural, morphological, and mechanical properties of the fabricated composites were examined in detail. The in vitro bioactivity of the prepared composites was tested in simulated body fluid, and the release behavior of gentamicin sulfate and 5-fluorouracil from glass scaffolds were analyzed separately as a function of time. The cytotoxicity was investigated using osteoblastic MC3T3-E1 cells. The findings indicated that the hBN nanoparticles, up to a certain concentration in the glass matrix, improved the mechanical strength of the glass scaffolds, which mimic the cancellous bone. Additionally, the inclusion of hBN nanoparticles enhanced the in vitro hydroxyapatite-forming ability of bioactive glass composites. The presence of hBN nanoparticles accelerated the drug release rates of the system. It was concluded that bioactive glass/hBN composite scaffolds mimicking native bone tissue could be used for bone tissue repair and regeneration applications.

Keywords: bioactive glass; hBN nanoparticles; biomimetic; scaffolds; drug delivery

Citation: Ensoylu, M.; Deliormanlı, A.M.; Atmaca, H. Preparation, Characterization, and Drug Delivery of Hexagonal Boron Nitride-Borate Bioactive Glass Biomimetic Scaffolds for Bone Tissue Engineering. *Biomimetics* **2023**, *8*, 10. https://doi.org/10.3390/biomimetics8010010

Academic Editors: Ryszard Uklejewski and Mariusz Winiecki

Received: 17 November 2022
Revised: 15 December 2022
Accepted: 16 December 2022
Published: 26 December 2022

1. Introduction

Bone tissue may be deformed and unable to function because of aging, disease, trauma, or injury. Today, new methods are being developed for the treatment and regeneration of damaged bone tissue [1]. Ceramic, polymer, and metal-based biomaterials can be utilized in bone tissue engineering applications. Another type of biomedical material utilized in the same application area is bioactive glass. Bioactive glasses are materials that can react with physiological fluids and bond to the bone surface by forming hydroxyapatite on their surface [2]. They exhibit calcium phosphate nucleation and mimic bone mineral maturation, also demonstrating attractive characteristics for bone tissue engineering [2–4]. The 45S5 coded glass synthesized by Hench et al. [2] and known as Bioglass contains P_2O_5–SiO_2–CaO–Na_2O. Similarly, the 13-93 composition (wt%), another bioactive glass used in biomedical applications, has 53% SiO_2, 6 Na_2O, 12 K_2O, 5 MgO, 20 CaO, and 4 P_2O_5 contents [5]. Likewise, borate-based 13-93B3 bioactive glass has been developed by replacing SiO_2 in the 13-93 bioactive glass composition with B_2O_3, and it has higher bioactivity compared to silicate-based bioactive glasses [6,7]. However, the primary disadvantage of borate glasses is their lower mechanical strength compared to their silicate-based counterparts. To overcome this limitation, the preparation of the bioactive glass matrix composites in the presence of two dimensional nanomaterials and also polymer coatings are commonly employed. Previously, the inclusion of pristine graphene [8–11], graphene oxide [12,13], and tungsten disulfide [14,15] in bioactive glasses has been reported. In general, the use of the aforementioned two dimensional materials in the glass network improved the mechanical properties of the composites.

Hexagonal boron nitride is a two dimensional material that is similar to graphene both physically and chemically and is therefore known as white graphene. It has the same crystallographic appearance; however, in contains boron and nitrogen atoms instead of carbon [16–18]. On the other hand, unlike graphene, it is an insulator. Studies in the field of biomedical applications related to hBN have revealed that boron nitride nanotubes are biocompatible and do not have a toxic effect [19,20]. On the other hand, there are also few studies reporting the cytotoxic influence of boron nitride nanotubes on certain cell types. Their toxicity was found to be highly dependent on the cellular accumulation enhanced for straight nanotubes [21]. For this reason, many composite structures, including boron nitride and, especially, hydroxyapatite–boron nitride composites, have been prepared and demonstrated as a scaffolding material for tissue engineering applications as well as a drug delivery vehicle [22–24]. In a former study, the influence of hexagonal boron nitride nanoparticles incorporated into a PCL and PCL-PLGA matrix, which was coated on the surface of borate bioactive glass scaffolds, have been investigated [25]. Results showed that the incorporation of hBN nanoparticles inside the polymer matrix improved the compressive strength of the bioactive glass composite scaffolds as well as their in vitro bioactivity and biocompatibility [25]. The function of boron nitride nanosheets as the reinforcing material on the mechanical strength of borosilicate glass matrix was also investigated by Saggar et al. [26]. Results indicated that fracture toughness and the flexural strength of the glass composites raised as a function of boron nitride concentration in the glass. Although the use of boron nitride-based systems in the biomedical field and the preparation of biocomposites have been studied, the inclusion of hBN nanoparticles directly inside the bioactive glass matrix on the structural, mechanical, and biological performance and drug-release behavior has not yet been published. This study aimed to fabricate borate-based bioactive glass scaffolds containing hexagonal boron nitride (0.1, 0.2, 0.5, 1, and 2 weight percent) nanoparticles for bone tissue engineering applications using the polymer foam replication method. In this context, the structural and morphological characteristics of hBN-containing bioactive glass composite scaffolds, compressive strength, in vitro mineralization in simulated body fluid, cytotoxicity against pre-osteoblast MC3T3-E1 cells, and the drug delivery properties of gentamicin and fluorouracil-loaded scaffolds were investigated.

2. Experimental Studies

2.1. Materials

In the study, 13-93B3 bioactive glass powders (5.5 Na_2O, 11.1 K_2O, 4.6 MgO, 18.5 CaO, 3.7 P_2O_5, 56.6 B_2O_3 wt.%) synthesized by the melt-quenching method (d_{50}: 2.5 µm and a density of 2.5 g/cm^3) were used. The hexagonal boron nitride (hBN) nanopowders (99.85%+ purity, 65–75 nm, density: 2.3 g/cm^3) used in the preparation of the composites were obtained from Nanografi Nanotechnology, Ankara, Turkey. They contain 0.03% Fe_2O_3, 0.002% CaO, 0.04% MgO, and 0.1% B_2O_3 as an impurity. Anhydrous ethanol ($\geq$99.9% purity) and ethyl cellulose (d = 1.14 g/mL) were purchased from Sigma-Aldrich (Steinheim, Germany) to be used in scaffold manufacture using the polymer foam replication method.

2.2. Porous Biomimetic Scaffold Manufacture

The polymer foam replication method, which was developed within the scope of another study, was used to manufacture three-dimensional bioactive glass scaffolds that mimic the cancellous bone [10,25]. In the method, poly(urethane) foams with a pore density of 60 pores per inch were cut to a diameter of 12 mm and a length of 35 mm and coated by dipping method using a homogeneously prepared bioactive glass-based suspension (40 vol.% bioactive glass powder, ethanol, 4% ethyl cellulose). Likewise, polymer foams coated with the same glass suspension, but also containing 0.1, 0.2, 0.5, 1, and 2 hBN nanoparticles by weight, were prepared and left to dry at room temperature and then heat-treated at 570 °C (1 h, heating rate < 1 °C/min below 350 °C) to remove the polyurethane foam in the structure and fabricate a dense three-dimensional bioactive glass scaffold.

2.3. Instrumentation

The morphological characteristics of the prepared glass scaffolds were examined with a stereo microscope (Nikon, SMZ745T, Tokyo, Japan) to observe the distribution of the additive nanoparticles with the bioactive glass and to determine the changes in the pore structure of the prepared scaffolds. In addition, total porosity measurement was performed using Archimedes' principle to determine the porosity of the scaffolds.

FTIR spectroscopy (Thermo Scientific, Nicolet, IS20, Waltham, MA USA) was used for structural analysis, and measurements were made by using an ATR module in the wavelength range of 550–4000 cm^{-1}. In XRD analysis, the Malvern Pan-Analytical brand, Empyrean model diffractometer was used. A Cu-Kα X-ray tube was utilized in the measurement, and samples were analyzed in the range of 10°–90° at a scanning speed of 0.01°/min.

The effect of hBN nanoparticles used in the study on the mechanical properties of bioactive glass scaffolds was analyzed by compression test. Cylindrical bioactive glass scaffolds with a height of ~6–8 mm and a diameter of ~6 mm were used for the compression test. Measurements were carried out with a Shimadzu brand Autograph AG-IS model test device using a deformation rate of 0.5 mm/min. Measurements were performed for 5 different samples and results were averaged. Before mechanical testing, the contact surfaces of each sample were ground to produce parallel surfaces.

2.4. In Vitro Mineralization

The bioactivity of the samples was tested in simulated body fluid (SBF). For the preparation of the simulated body fluid, the protocol developed by Kokubo et al. [27] was followed. The chemicals NaCl, $NaHCO_3$, KCl, $K_2HPO_4.3H_2O$, $MgCl_2.6H_2O$, $CaCl_2$, and Na_2SO_4 (Sigma-Aldrich, Steinheim, Germany) were dissolved in deionized water and buffered at a pH of 7.40 with tris(hydroxymethyl)aminomethane ($(CH_2OH)_3CNH_2$) and 1 M hydrochloric acid (Fisher Scientific Inc., USA) at 37 °C. A total 500 mL of SBF per 1 g of sample was used and the scaffolds were disinfected with ethanol before immersion in SBF. Tissue scaffolds immersed in SBF were kept in an incubator at 37 °C for 7, 14, and 30 days. When the specified time expired, the scaffolds were removed from the incubator and left to dry after washing with deionized water and ethanol. The changes in the samples as a result of holding in SBF were examined using a scanning electron microscope (ZEISS, GeminiSEM 560) and FTIR spectroscopy under the conditions described previously.

2.5. Drug Delivery Studies

2.5.1. Gentamicin

Gentamicin (Genta ampoule, İbrahim Etem Ulagay İlaç, İstanbul, Turkey, containing 80 mg/2 mL gentamicin, 124.8 mg in the form of gentamicin sulfate) was used as the first drug to investigate drug release behavior from bioactive glass composite scaffolds prepared in the study. Before the drug release studies, the samples were disinfected by soaking in ethanol, and then 50 microliters of Genta solution (40 mg/mL gentamicin) were dropped onto the sample surface with the help of a micropipette. After drying for 24 h, drug-loaded samples were soaked in a 5 mL of phosphate-buffered saline (PBS) solution and the amount of drug released into the PBS was measured with a UV–Vis spectrophotometer at 256 nm (Thermo Scientific, Evolution 201, Waltham, MA USA) for up to 96 h.

2.5.2. Fluorouracil (5-FU)

For the 5-FU loading studies, bioactive glass composite scaffolds were immersed in a 5 mL of 5 mg/100 mL drug solution (5-FU, Sigma-Aldrich, Steinheim Germany), which is prepared in PBS (pH 7.4) for 48 h at 37 °C in a dark environment. In the adsorption study, at the end of 48 h, 2 mL of the drug solution mixture was taken, and the absorbance was measured at 266 nm with a UV–Visible spectrophotometer (Thermo Scientific, Evolution 201, Waltham, MA, USA). Bioactive glass scaffolds were removed from the drug solution and dried at 40 °C for 48 h before release experiments.

For drug release studies, the drug-loaded bioactive glass scaffolds were immersed in 10 mL of PBS at pH 7.4. The release experiments were made at 37 °C under static conditions. For each time interval, 2 mL of the drug solution sample was taken and replaced with the same amount of fresh PBS solution. The absorbance values of the samples were recorded at 266 nm using the spectrophotometer, and the amount released from the calibration curve was obtained. Drug loading and delivery experiments were performed in triplicate and results were averaged. Through the experiments, the direct light contact of the drug solution was prevented and drug solutions were freshly prepared.

In the study, the drug release kinetics were also investigated. For this purpose, the obtained release profiles were analyzed using zero order, first order, and Higuchi kinetic models [28].

2.6. Cytotoxicity

The in vitro cytotoxicity of the biomimetic scaffolds was examined using the osteoblastic cells (MC3T3-E1, Subclone-4, ATCC, CRL-2593, Manassas, VA, USA) using MTT (3-[4,5-dimethylthiazole-2-yl]-2,5-diphenyltetrazolium bromide) (Sigma-Aldrich, Steinheim, Germany) assay which is a colorimetric method used to understand the metabolic activity of living cells. Scaffolds were sterilized at 350 °C before cell culture experiments. Osteoblastic cells were cultured in a growth medium containing Alpha-Minimum Essential Medium with L-glutamine with 10% fetal bovine serum and 100 U/mL penicillin–100 mg/mL streptomycin. For this purpose, MC3T3-E1 cells (5×10^4) were seeded onto each scaffold in the presence of 1.9 mL of culture media and cultured for 72 h at 37 °C in a 5% CO_2 incubator. Following this, MTT solution was added and the cells were cultured for a further 4 h. The formazan crystals formed at this stage were dissolved by dimethyl sulfoxide (DMSO, Sigma-Aldrich, Steinheim, Germany). A multi-plate reader (Thermo-Scientific, Waltham, MA, USA) was used to measure the color change at a wavelength of 570 nm that is directly related to the amount of formazan. The morphology of the osteoblastic cells was observed after culturing with the glass scaffolds for 72 h using an optical microscope.

The statistical analyses for the MTT test results were carried out with Graph Pad Prism 5. Results were analyzed by using one-way ANOVA. Values with $p \leq 0.05$ (*), $p \leq 0,01$ (**) were considered statistically significant.

3. Results and Discussion

The SEM images of hBN nanoparticles used in the study and the digital images of the fabricated bioactive glass scaffolds are given in Figure 1a–c. Based on the SEM micrographs and the report of the manufacturer company, the average particle size of the hBN nanoparticles is 65–75 nm, and it can be seen that they exist in a platelet structure with round morphology and also tend to agglomerate.

The digital images of porous bioactive glass scaffolds (6 mm diameter, 3 mm height) produced using the polymer foam replication method and containing hBN nanoparticles at different volume ratios show no significant change in the morphology of the scaffolds at varying hBN concentrations. This result is also supported by the optical microscope images given in Figure 2. Optical microscope images demonstrate that the prepared scaffolds have an interconnected, open pore structure, and the addition of hBN does not change the pore structure significantly. It is understood from the optical microscope images given at high magnification that the average pore diameter is ~500 μm.

Figure 1. (**a**,**b**) SEM micrographs of the as-received hBN nanopowders; (**c**) digital image of the fabricated glass scaffolds.

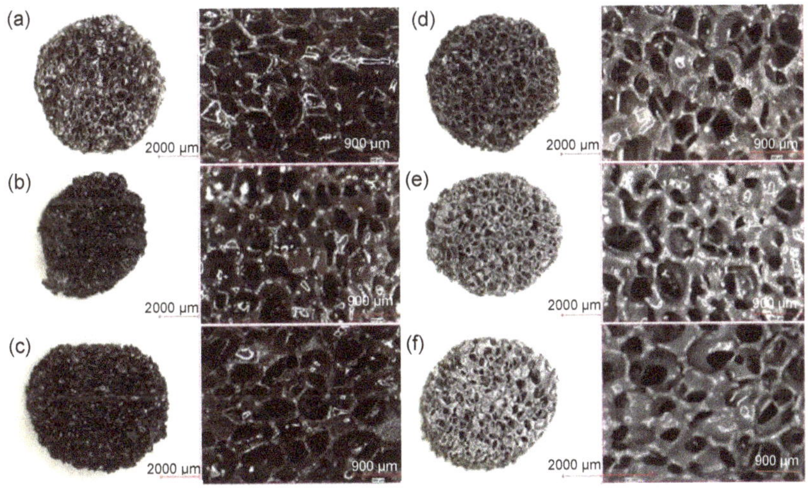

Figure 2. Optical microscope images of the (**a**) bare B3, (**b**) 0.1 hBN, (**c**) 0.2 hBN, (**d**) 0.5 hBN, (**e**) 1 hBN, and (**f**) 2 hBN-containing borate bioactive glass scaffolds. Low magnification image scale bar: 2000 μm; high magnification image scale bar: 900 μm.

The XRD patterns of the prepared scaffolds after sintering at 570 °C are shown in Figure 3a. It can be seen that the glass samples subjected to heat treatment maintain their amorphous structure up to 0.5% hBN concentration. The characteristic peak formation of hBN in the structure is observed with bioactive glass samples at higher boron nitride concentrations. The intense peak observed in Figure 3a corresponds to the hBN peak of (2θ-27°) the 002 plane in the XRD pattern (JCPDS 034-0421). This characteristic peak is similar to the 2θ-26° peak seen in the XRD pattern of graphene. In addition, although the characteristic peak observed overlaps with the hydroxylated boron nitride (BNO) peak, it is known that BNO conversion occurs at 1000 °C [29]. It is known that the oxidation character of boron nitride nanostructures is affected by the specific surface area, and boron nitride nanocrystals (for 210 nm edge length, 270 nm thickness) maintain their thermal stability in the air up to 900 °C. [30]. In addition, the low-intensity peaks observed in the pattern at ~2θ-29° belong to B_2O_3 [31].

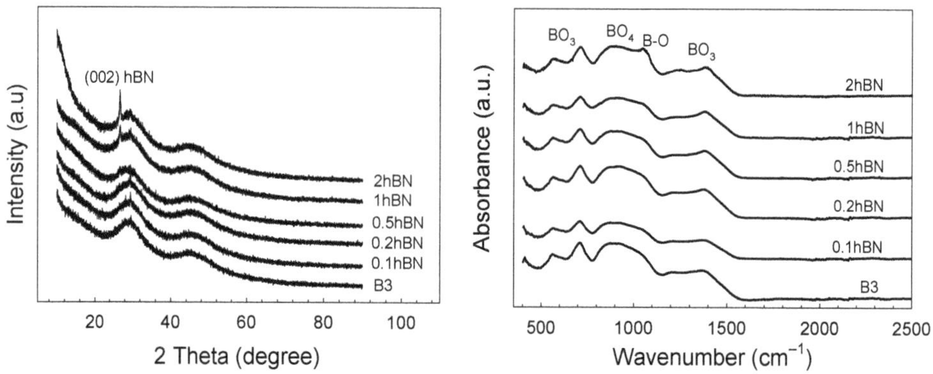

(a) (b)

Figure 3. (**a**) XRD pattern; (**b**) FTIR spectra of the prepared bioactive scaffolds containing hBN nanopowders.

According to the FTIR spectra given in Figure 3b, it is seen that the inclusion of hBN nanoparticles in the bioactive glass structure does not cause a significant change in the molecular structure of the composite scaffolds. Accordingly, peaks at 1300–1500 cm^{-1} and 720 cm^{-1} wavenumbers are due to the presence of B_2O_3 groups. The broad peak in the 1300–1500 cm^{-1} wavenumber range belongs to the bending and stretching vibrations of the B–O–B bonds in the BO_3 triple system and the low-intensity peak at ~726 cm^{-1} is due to B−O−B linkages. On the other hand, the broad band in the range of 900–1100 cm^{-1} also belongs to the stretching vibrations of $BO_4{}^-$ groups in the structure [32]. In the spectrum of the 2hBN-B3 glass, the shoulder observed at 1100 cm^{-1} corresponds to the B–O–H in-plane bending [33].

In Figure 4a, according to the graph showing the total porosity values of the scaffolds, it was determined that there was a slight decrease in the porosity values with increasing hBN concentration. While the mean porosity value of the scaffolds without hBN was 77 ± 3.6%, this value was measured as 72 ± 2.8% in the scaffold with the highest hBN concentration.

In Figure 4b, the compressive strength values of the scaffolds are given. Accordingly, while the compressive strength value of the 13-93B3 scaffold without additives was 0.79 ± 0.2 MPa, the compressive strength of the sample containing 0.2% hBN was measured to be 2.22 ± 0.3 MPa. A decrease in compressive strength values was observed at hBN concentrations higher than this value presumably due to the agglomeration of hBN nanopowders added to the structure at high concentrations.

In a previous study by Turk and Deliormanlı [10], the compressive strength of 13-93B3 bioactive glass scaffolds prepared using the polymer foam replication method containing graphene nanopowders at different concentrations (1, 3, 5, and 10 wt%) was investigated. The results showed that the highest compressive strength value was obtained in the sample containing 5% graphene as 1.86 ± 0.7 MPa. In the current study, the maximum compressive strength value was obtained for the hBN-containing 13-93B3 bioactive glass scaffolds at 0.2% hBN concentration, and the strength of these scaffolds is approximately 19% higher than that of graphene-bioactive glass scaffolds. The special layer-stacking structure of hBN, the partial ionicity of boron and nitrogen atoms, and the polarity of the orbitals in the structure may influence the observed increase in the compressive strength of the scaffolds. Due to the partial ionicity of the B and N atoms in the hBN structure, the layers are aligned by the overlapping of the positive B and negative N atoms. In this particular case, the hBN atoms have an AÁ stacking structure, while the graphite structure has an AB (Bernall stacking) stacking. In this case, only some of the carbon atoms are located directly above or below the neighbor. Unlike graphene, in hexagonal boron nitride, the layer thickness does not affect the mechanical properties much. As the layer thickness increases in the graphene structure, the deviation of the mechanical property values is related to the inhomogeneous deformation and stacking structure. These mentioned factors cause the shift of the interlayer and energy loss during the loading and unloading cycle. The main reason for this difference in graphene is the spontaneous sliding of the graphene layers on the graphene surface due to the negative increase in shear energy as a result of large in-plane stress and out-of-plane pressure applications and warped layer (AB) stacking. This is because of the overlapping 2Pz orbitals in the graphene structure. On the other hand, the more polar orbitals in hBN become localized in the same stress condition to positively increase the sliding energy barrier, thus making the hBN resistant to interlayer shifting [34–39]. In the current study, the improvement obtained in the compressive strength of bioactive glass-based scaffolds in the presence of hBN nanoparticles may be attributed to the uniform dispersion of the nanoparticles in the glass matrix and the stress transfer between the nanoparticles and matrix. A higher level of nanoparticle loading presumably reduced the load transfer between matrix and filler due to the agglomeration of nanoparticles, which in turn caused a decrease in the compressive strength.

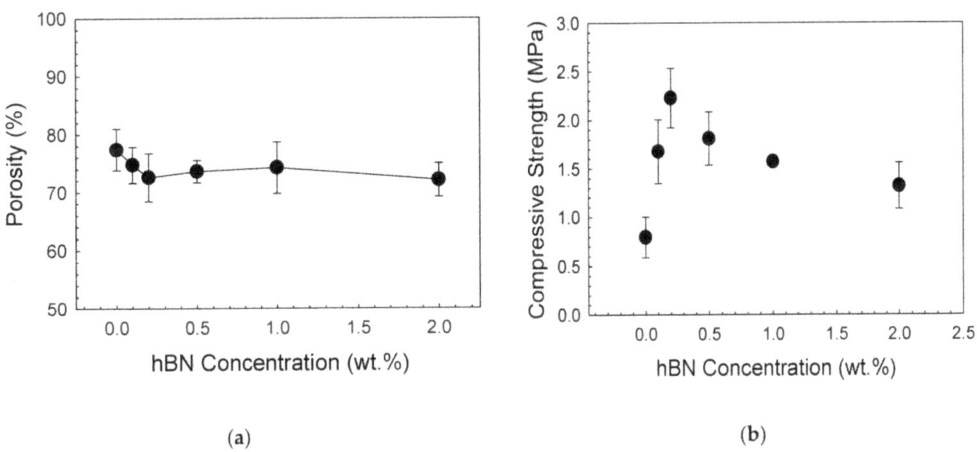

Figure 4. Graphs showing (**a**) the total porosity and (**b**) the compressive strength of the composite scaffolds.

In another study [40], it was reported that boron nitride nanolayers added to the akermanite matrix improved the mechanical strength of akermanite. In that study, the compressive strength of akermanite-boron nitride nanolayer composites increased with 0.5% and 1% boron nitride additives by weight, while the compressive strength decreased at higher concentrations.

The FTIR spectra of bioactive glass scaffolds kept in SBF for 7, 14, and 30 days are given in Figure 5a–c. When the results of the SBF-treated (7 days) scaffolds were examined, no significant difference was observed in the spectra of the scaffolds at low hBN concentrations, whereas the formation of PO_4^{3-} (at 550 cm^{-1} and 1000 cm^{-1}) and CO_3^- (at 1389 cm^{-1}) groups starting from 1% hBN were observed. This is particularly evident in glass samples containing 2% hBN. When the scaffolds kept in SBF for 14 days were examined, the peaks representing hydroxyapatite formation on the surfaces of the samples were seen on the spectrum. Accordingly, the peak at 1022 cm^{-1} was assigned to the PO_4^{3-} group v_3 vibration, and the peak at 554 cm^{-1} corresponded to the v_4 vibration of the same group. The split peak at 550 cm^{-1} and 604 cm^{-1} observed in the glass samples kept in SBF for 30 days belongs to the bending mode of orthophosphate and demonstrates that hydroxyapatite is formed in the structure. The peak observed at 964 cm^{-1} represents the PO_4^{3-} v_1 vibrations and the peaks at ~874 cm^{-1} and ~1400 cm^{-1} may be due to the presence of the CO_3^- group [41,42]. Results revealed that peak intensities representing crystalline hydroxyapatite formation increased as a function of immersion time in SBF.

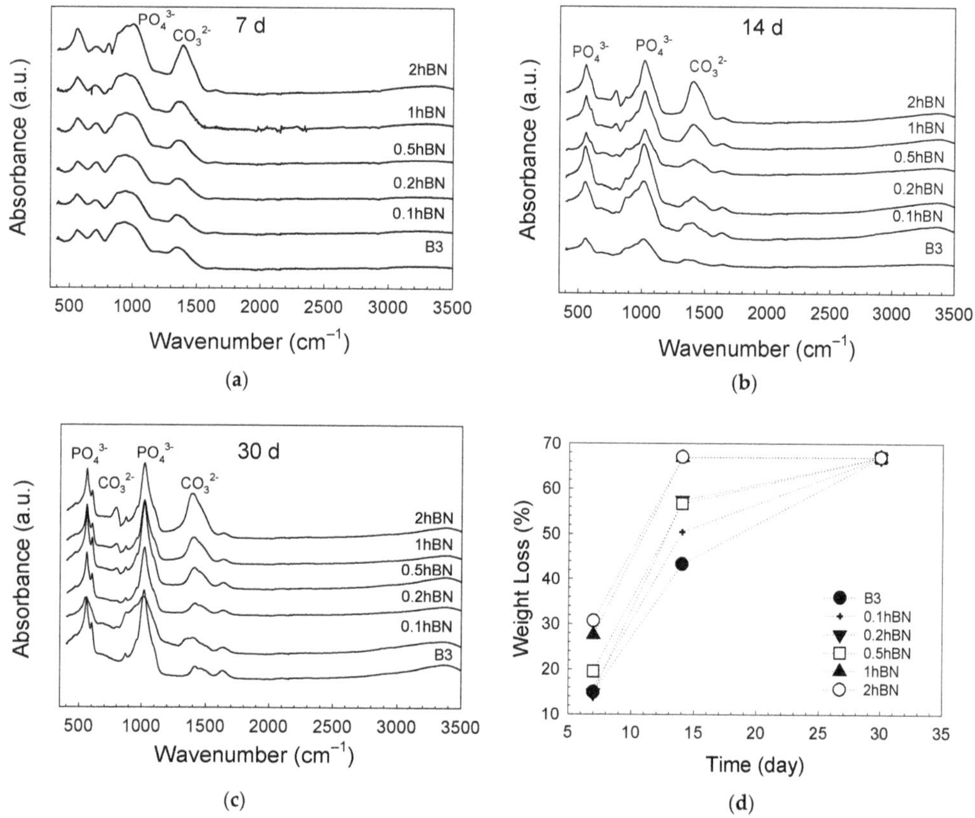

Figure 5. FTIR spectra of the SBF treated bioactive glass scaffolds for (**a**) 7d, (**b**) 14 d, and (**c**) 30 d; (**d**) graph showing the weight loss of the scaffolds as a function of SBF immersion time.

In general, the presence of hBN nanoparticles appeared to enhance the HA-forming ability of bioactive glass starting from certain concentrations, presumably due to their high surface area. Similarly, in a different study, hydroxyapatite precipitation was reported on the surface of boron nitride nanoparticles in 5 days when immersed in simulated body fluid [43]. This result supports the increase in bioactivity seen in bioactive glass scaffolds containing hBN nanoparticles.

The weight loss values of the fabricated bioactive glass scaffolds after soaking in SBF at 37 °C are given in Figure 5d. Accordingly, it was observed that the bare borate glass scaffolds kept in SBF for 7 days lost approximately 15% of their original weight, while this value was calculated as ~31% in the scaffolds containing 2% hBN. At the end of 30 days, it was observed that all of the glass scaffolds under investigation lost 67% of their weight. It was observed that the percentage of weight loss increased with increasing hBN concentration. This behavior may be correlated with the loose network of the borate glasses. Unlike silica, the coordination number of boron prevents the full formation of the 3D network structure, causing the boron-based glass to have lower chemical stability [43]. It is also known that ions such as Na^+, K^+, Mg^{2+}, and $(BO_3)^{-3}$ dissolve in solution and all

CaO in the glass reacts with phosphate ions in SBF to form HA and the theoretical weight loss of fully transformed 13-93B3 bioactive glass scaffolds is 67% [3,44].

The SEM images of glass scaffolds kept in SBF for 7 days and 30 days are given in Figures 6 and 7, respectively. Accordingly, it is understood that a new substance formation occurred on the surface of the scaffolds, which were kept in SBF for both 7 days and 30 days. At the end of 30 days, HA formation increased significantly and formed a thick layer on the surface of the scaffolds. Plate-like HA formations came together to form spherical aggregates. It was determined that the second phase material observed in the micrographs was compatible with the HA morphology [45,46].

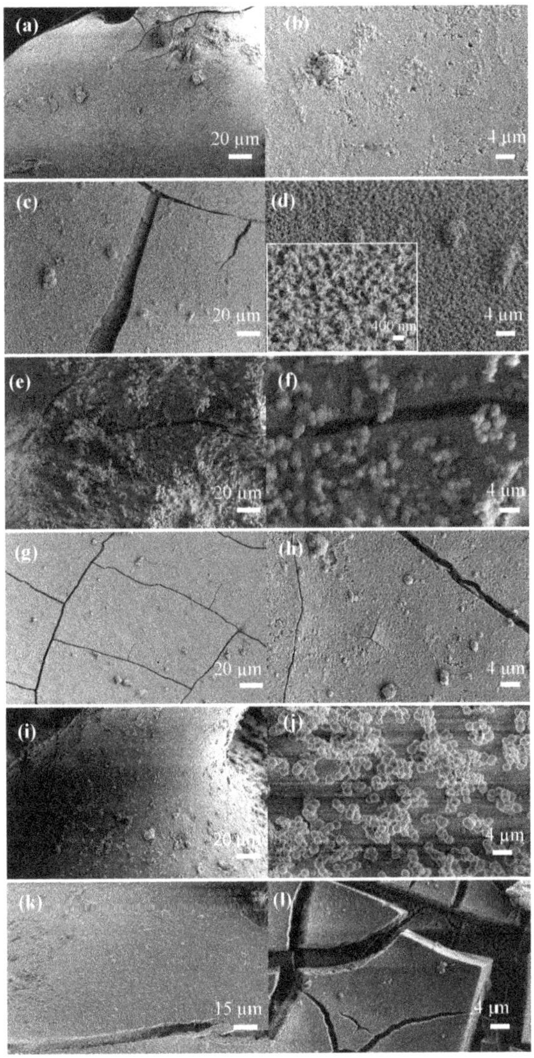

Figure 6. SEM micrographs of the SBF-treated bioactive glass scaffolds for 7 d (**a**,**b**) B3; (**c**,**d**) 0.1 hBN-B3; (**e**,**f**) 0.2 hBN-B3; (**g**,**h**) 0.5 hBN-B3; (**i**,**j**) 1 hBN-B3; (**k**,**l**) 2 hBN-B3.

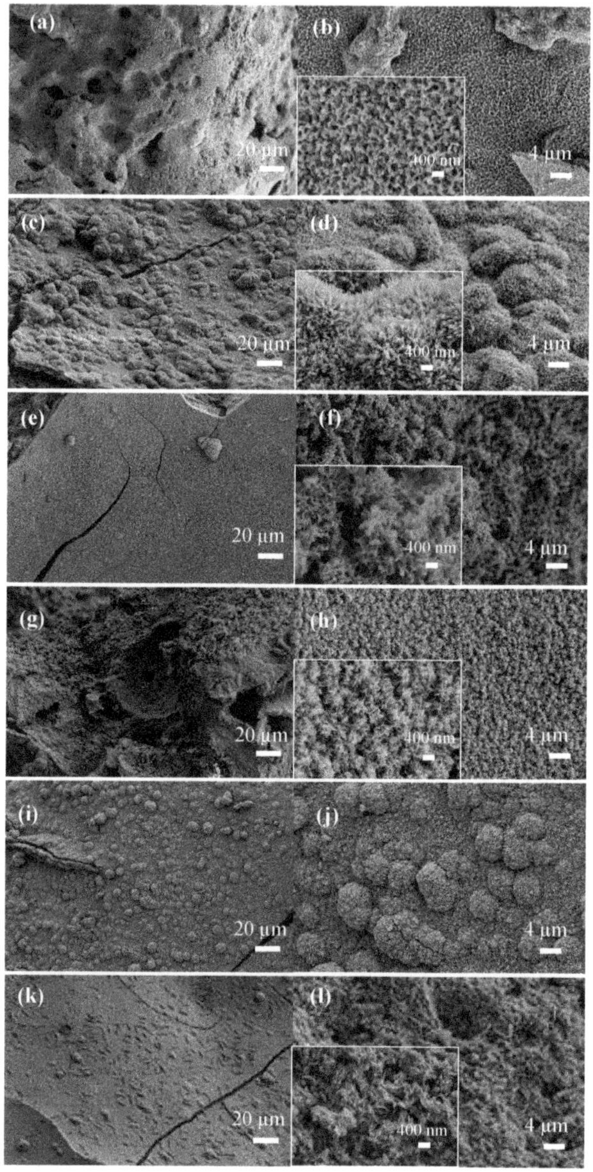

Figure 7. SEM micrographs of the SBF-treated bioactive glass scaffolds for 30 d (**a**,**b**) B3; (**c**,**d**) 0.1 hBN-B3; (**e**,**f**) 0.2 hBN-B3; (**g**,**h**) 0.5 hBN-B3; (**i**,**j**) 1 hBN-B3; (**k**,**l**) 2 hBN-B3.

The results of gentamicin release experiments from bioactive glass scaffolds containing hBN are demonstrated in Figure 8. In this study, drug delivery from the scaffolds loaded with gentamicin sulfate was monitored for up to 96 h. The results revealed that the gentamicin release from the scaffolds was very rapid and the cumulative release amount reached 100% in up to 24 h. This may be attributed to the weak physical adsorption of the gentamicin sulfate to the glass surface. Drug release kinetic studies revealed that the highest correlation coefficient (R^2) value among gentamicin release profiles was obtained in the first order model (see Table 1). The logarithm of the percentage of the drug remaining in this kinetic model [47] versus the time (hour) plot gives a straight line. It can be seen that the drug release behaviors of the bare and hBN-containing bioactive glass samples (at all concentrations) were similar.

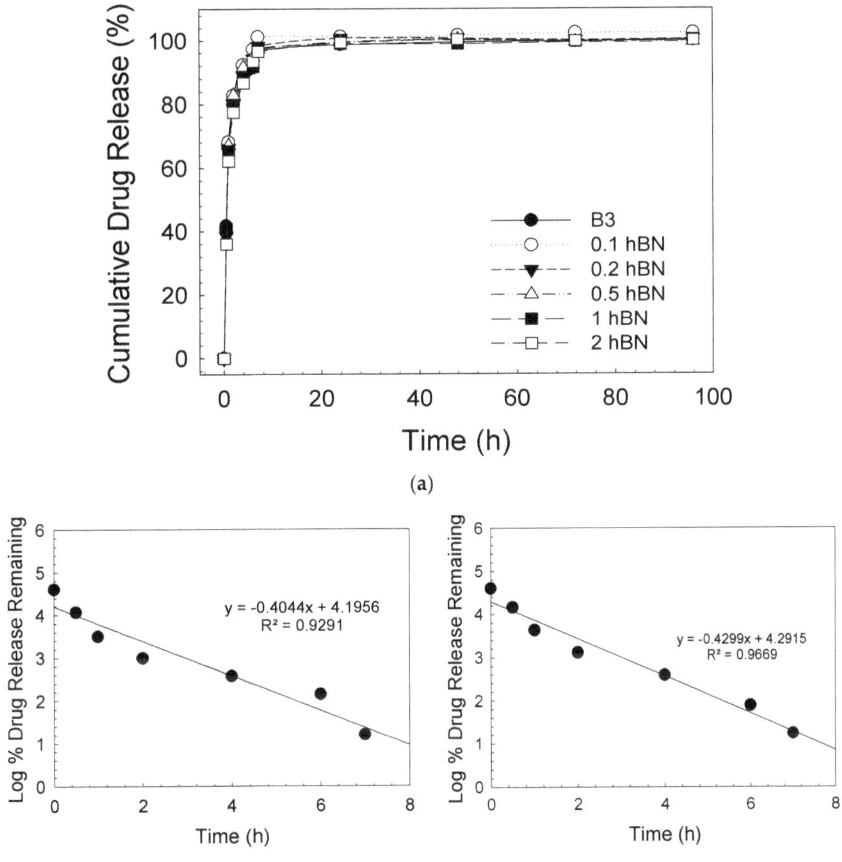

(**a**) (**b**) (**c**)

Figure 8. (**a**) Cumulative gentamicin sulfate release curve; graphs showing the first order drug release for (**b**) bare B3, (**c**) 2 hBN-B3.

Table 1. Gentamicin sulfate release kinetics model parameters. K_0, K_1, and K_H are zero order, first order, and Higuchi model rate constants. t is time, and C_0 is the initial concentration of the drug.

Sample	Zero Order $C_t = C_0 + K_0.t$		First Order $logC = logC_0 - K_1.t/2.303$		Higuchi $Q = K_H\sqrt{t}$	
	R^2	K_0	R^2	K_1	R^2	K_H
B3	0.6478	10.024	0.9291	0.4044	0.8682	33.372
0.1 hBN-B3	0.6729	10.901	0.9809	0.5964	0.8855	35.956
0.2 hBN-B3	0.6754	10.595	0.9526	0.5013	0.886	34.895
0.5 hBN-B3	0.6462	10.389	0.9501	0.4641	0.8681	34.626
1 hBN-B3	0.6671	10.229	0.9357	0.4301	0.8819	33.82
2 hBN-B3	0.7027	10.676	0.9669	0.4299	0.9041	34.824

The other drug tested in the study was fluorouracil. The antimetabolite drug fluoropyrimidine 5-fluorouracil is frequently used to treat cancer. 5-FU has anticancer properties, since it inhibits thymidylate synthase and integrates its metabolites into RNA and DNA [48–50]. Figure 9 demonstrates the results of the fluorouracil loading and the release experiments into a PBS medium. Accordingly, the drug adsorption percentage to the bioactive glass scaffolds was in the range of 25% to 30%. Drug release experiments showed that after 24 h, the cumulative drug release amount was calculated to be 11% to 13%. After 196 h immersion in PBS, 20 to 26% of the drug was released from the scaffolds. The difference obtained in the cumulative release rates between gentamicin and the 5-FU may be attributed to the chemical structure of the drugs and the loading method followed in the experiments. In a past study, Dehaghani et al. [51] investigated the encapsulation of 5-FU into carbon nanotubes and boron nitride nanotubes (BNT) theoretically. Results revealed that due to the raised van der Waals contact energy between the drug and the BNT, 5-FU was adsorbed into the cavity of the BNT more quickly than the CNT. Similarly, the electrical response of BN nanocones to 5-FU was studied by Wang et. al. [52] using density functional theory. Results indicated that the boron nitride nanocones may be suitable candidates for the detection of 5-FU and can be utilized in electronic sensors. In the current study, the drug release kinetics studies for 5-FU showed that the release of the drug followed Higuchi kinetic model (R^2 = 0.97 and 0.98 bare and 2% hBN-containing glass scaffolds, respectively), which describes the diffusion-based delivery process from a porous vehicle (Table 2). Figure 10 depicts the drug release graphs fitted by the Higuchi kinetic model. El-Kady et al. [53] also reported that 5-FU release from silicate-based sol-gel-derived bioactive glass nanoparticles was fitted using the Higuchi model (square root of time) and followed by a diffusion-controlled release mechanism.

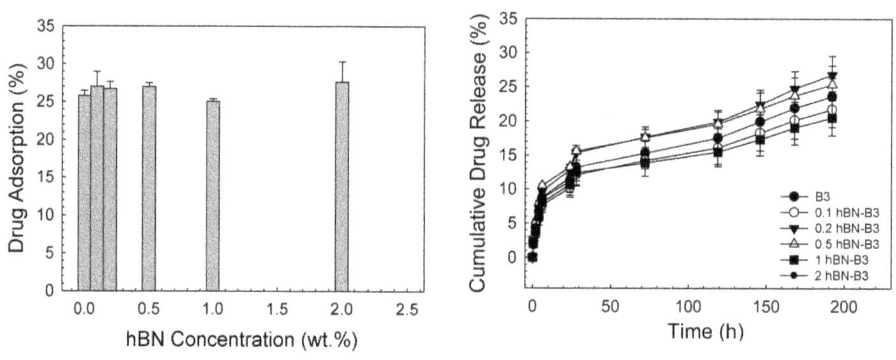

(a) (b)

Figure 9. (**a**) 5-FU adsorption curve; (**b**) cumulative 5-FU release curve for studied bioactive glass scaffolds.

Table 2. 5-FU release kinetics model parameters. K_0, K_1, and K_H are zero order, first order, and Higuchi model rate constants. T is time, and C_0 is the initial concentration of the drug.

Sample	**Zero Order** $C_t = C_0 + K_0.t$		**First Order** $logC = logC_0 - K_1.t/2.303$		**Higuchi** $Q = K_H\sqrt{t}$	
	R^2	K_0	R^2	K_1	R^2	K_H
B3	0.896	0.0923	0.9138	0.0011	0.9661	1.415
0.1 hBN-B3	0.893	0.0866	0.9097	0.001	0.9669	1.331
0.2 hBN-B3	0.887	0.1048	0.9079	0.0012	0.9632	1.613
0.5 hBN-B3	0.59	0.0951	0.8814	0.0011	0.9487	1.476
1 hBN-B3	0.859	0.782	0.8767	0.0009	0.948	1.214
2 hBN-B3	0.908	0.1063	0.9268	0.0013	0.9759	1.627

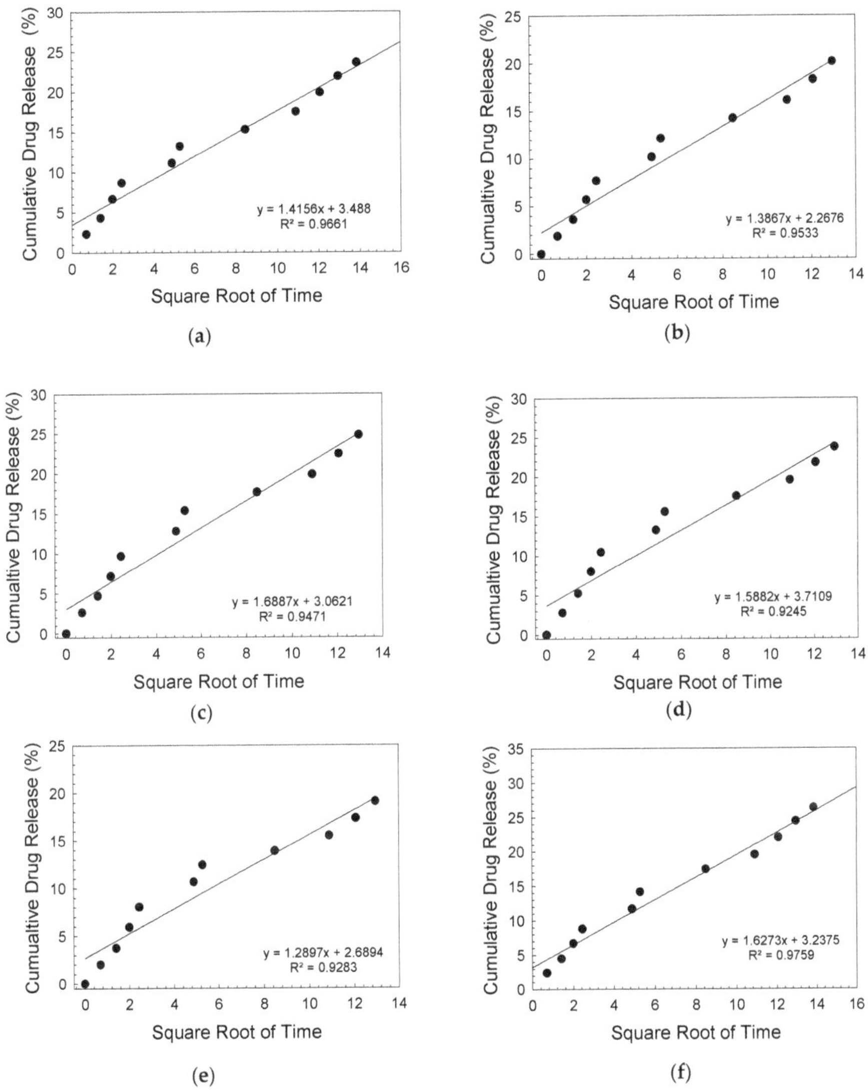

Figure 10. Graphs showing the Higuchi model 5-FU release for (**a**) bare B3, (**b**) 0.1 hBN-B3, (**c**) 0.2 hBN-B3, (**d**) 0.5 hBN-B3, (**e**) 1 hBN-B3, and (**f**) 2 hBN-B3.

The results of the MTT assay, which was made to understand the cytotoxicity of the fabricated samples, are given in Figure 11. Accordingly, starting from a 0.5% hBN content, a significant decrease (~50%) was observed in cell viability after 72 h. However, cell viability was calculated to be 86% for 0.2% hBN-containing glass scaffolds under the same conditions. This may be attributed to the increase obtained from the in vitro degradation of the glass scaffolds as the hBN concentration increased. Recently, it was seen that when they are used as a coating material inside a polymeric matrix over borate-based 13-93B3 bioactive glass scaffolds, hBN nanoparticles did not show any toxic effect for the same type of cells [25]. However, in the current study, hBN nanoparticles were embedded directly inside the glass matrix, and following the degradation of the glass in the cell culture medium, 2D nanoparticles may diffuse into the cell membrane more easily. Similarly, a recent study by Khalid et al. [54] reported the toxicological effects of boron nitride nanotubes. Results showed that the cytotoxicity of boron nitride nanotubes was higher in HeLa cancer cells (80%) compared to the HEK-293 normal cells (60%) for 48 h incubation.

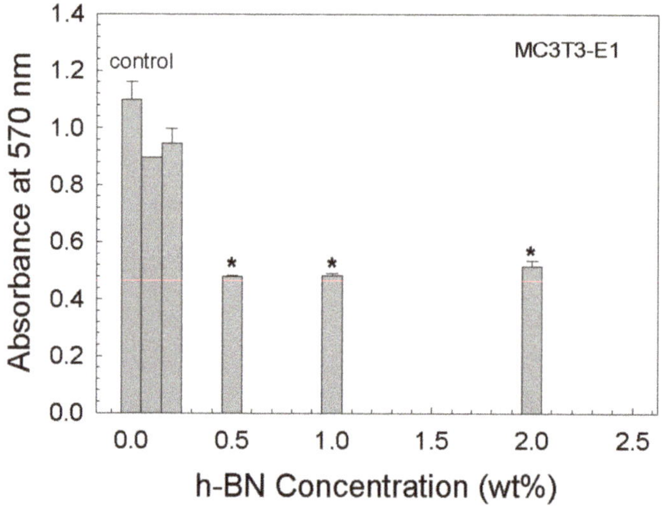

Figure 11. Graph showing the MTT assay results after incubation for 72 h. * indicates the statistical significance, $p < 0.05$.

Figure 12 shows the optical microscope images of the osteoblastic MC3T3-E1 cells after incubation for 72 h in the presence of fabricated bioactive glass composite scaffolds. Based on the optical microscope images there was no significant difference in cell morphology depending on the hBN concentration in the glass matrix.

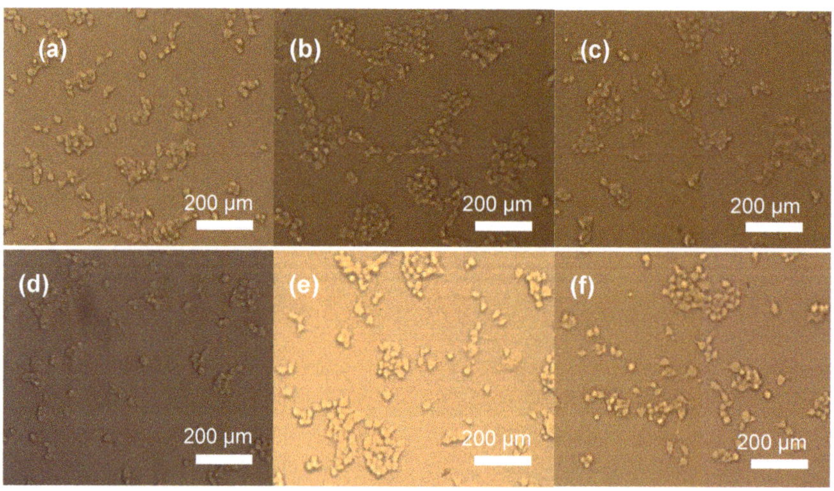

Figure 12. Optical microscope images of the MC3T3-E1 cells after incubation for 72 h in the presence of composite bioactive glass scaffolds containing (**a**) 0, (**b**) 0.1%, (**c**) 0.2%, (**d**) 0.5%, (**e**) 1%, and (**f**) 2% hBN nanopowders. Magnification at 100×.

4. Conclusions

Within the scope of this study, biomimetic borate bioactive glass scaffolds containing hBN nanopowders were produced and in vitro characterizations were carried out. According to XRD analysis results, it was seen that hBN generally remained stable in the bioactive glass matrix. The applied mechanical test results showed that the hBN nanoparticle addition (depending on the concentration) improved the compressive strength of the bioactive glass scaffolds. In vitro bioactivity studies in SBF indicated that the incorporation of hBN improved the hydroxyapatite-forming ability of bioactive glass scaffolds. According to the results of drug release experiments of antibiotic-loaded scaffolds in PBS medium, it was understood that all scaffolds showed burst drug release behavior, presumably due to the rapid degradation of 13-93B3 glass and weak interaction of the drug molecule with the glass surface. It was observed that the drug release kinetics of the scaffolds fit the first order kinetic model. On the other hand, the cumulative release of 5-FU from the scaffolds was measured to be 20% to 26% after 196 h. Fluorouracil showed sustained release behavior from the borate glass scaffolds and the release kinetics were fitted by the Higuchi model. In vitro cytotoxicity experiments revealed that the presence of hexagonal boron nitride nanopowders in a bioactive glass matrix up to 0.2% did not cause any toxicity in mouse calvarial pre-osteoblast cells after 72 h of incubation.

Author Contributions: Conceptualization, A.M.D.; methodology, A.M.D. and H.A.; validation, A.M.D. and H.A.; formal analysis, M.E. and H.A.; investigation, M.E.; data curation, M.E and H.A.; writing—original draft preparation, A.M.D. and M.E.; writing—review and editing, A.M.D. and H.A.; visualization, A.M.D. and M.E.; supervision, A.M.D.; project administration, A.M.D.; funding acquisition, A.M.D. All authors have read and agreed to the published version of the manuscript.

Funding: This research was funded by The Scientific and Technological Research Council of Turkey (TÜBİTAK), grant number 119M935.

Institutional Review Board Statement: Not applicable.

Data Availability Statement: Not applicable.

Acknowledgments: The authors thank TÜBİTAK for the support provided.

Conflicts of Interest: The authors declare that they have no conflicts of interest.

References

1. Wang, E.T.; Zhao, M. Regulation of tissue repair and regeneration by electric fields. *Chin. J. Traumatol.* **2010**, *13*, 55–61. [PubMed]
2. Hench, L.L. Bioceramics. *J. Am. Ceram. Soc.* **1998**, *81*, 1705–1728. [CrossRef]
3. Rahaman, M.N.; Day, D.E.; Bal, B.S.; Fu, Q.; Jung, S.B.; Bonewald, L.F.; Tomsia, A.P. Biactive glass in tissue engineering. *Acta Biomater.* **2011**, *7*, 2355–2373. [CrossRef]
4. Fu, Q.; Rahaman, M.N.; Fu, H.; Liu, X. Silicate, borosilicate, and borate bioactive glass scaffolds with controllable degradation rate for bone tissue engineering applications. I. Preparation and in vitro degradation. *J. Biomed. Mater. Res. Part A* **2010**, *95*, 164–171. [CrossRef] [PubMed]
5. Brink, M.; Turunen, T.; Happonen, R.P.; Yli-Urpo, A. Compositional dependence of bioactivity of glasses in the system Na2O-K2O-MgO-CaO-B2O3-P2O5-SiO2. *J. Biomed. Mater. Res.* **1997**, *37*, 114–121. [CrossRef]
6. Fu, H.; Fu, Q.; Zhou, N.; Huang, W.; Rahaman, M.N.; Wang, D.; Liu, X. In vitro evaluation of borate-based bioactive glass scaffolds prepared by a polymer foam replication method. *Mater. Sci. Eng. C* **2009**, *29*, 2275–2281. [CrossRef]
7. Bi, L.; Rahaman, M.N.; Day, D.E.; Brown, Z.; Samujh, C.; Liu, X.; Mohammadkhah, A.; Dusevich, V.; Eick, J.D.; Bonewald, L.F. Effect of bioactive borate glass microstructure on bone regeneration, angiogenesis, and hydroxyapatite conversion in a rat calvarial defect model. *Acta Biomater.* **2013**, *9*, 8015–8026. [CrossRef]
8. Porwal, H.; Grasso, S.; Cordero-Arias, L.; Li, C.; Boccaccini, A.R.; Reece, M.J. Processing and bioactivity of 45S5 Bioglass®-graphene nanoplatelets composites. *J. Mater. Sci. Mater. Med.* **2014**, *25*, 1403–1413. [CrossRef]
9. Gao, C.; Liu, T.; Shuai, C.; Peng, S. Enhancement mechanisms of graphene in nano-58S bioactive glass scaffold: Mechanical and biological performance. *Sci. Rep.* **2014**, *4*, 4712. [CrossRef]
10. Turk, M.; Deliormanlı, A.M. Electrically conductive borate-based bioactive glass scaffolds for bone tissue engineering applications. *J. Biomater. Appl.* **2017**, *32*, 28–39. [CrossRef]
11. Türk, M.; Deliormanlı, A.M. Graphene-containing PCL- coated Porous 13-93B3 Bioactive Glass Scaffolds for Bone Regeneration. *Mater. Res. Express* **2018**, *5*, 045406. [CrossRef]

12. Ilyas, K.; Zahid, S.; Batool, M.; Chaudhry, A.A.; Jamal, A.; Iqbal, F.; Nawaz, M.H.; Goerke, O.; Gurlo, A.; Shah, A.T.; et al. In-vitro investigation of graphene oxide reinforced bioactive glass ceramics composites. *J. Non-Cryst. Solids* **2019**, *505*, 122–130. [CrossRef]
13. Ahn, J.H.; Kim, I.R.; Kim, Y.; Kim, D.H.; Park, S.B.; Park, B.S.; Bae, M.K.; Kim, Y.I. The Effect of Mesoporous Bioactive Glass Nanoparticles/Graphene Oxide Composites on the Differentiation and Mineralization of Human Dental Pulp Stem Cells. *Nanomaterials* **2020**, *10*, 620. [CrossRef] [PubMed]
14. Deliormanlı, A.M.; Ensoylu, M.; Issa, S.A.; Elshami, W.; Al-Baradi, A.M.; Al-Buriahi, M.S.; Tekin, H.O. WS2/bioactive glass composites: Fabrication, structural, mechanical and radiation attenuation properties. *Ceram. Int.* **2021**, *47*, 29739–29747. [CrossRef]
15. Ensoylu, M.; Atmaca, H.; Deliormanlı, A.M. Fabrication and in vitro characterization of macroporous WS2/ bioactive glass scaffolds for biomedical applications. *J. Aust. Ceram Soc.* **2022**, *58*, 397–409. [CrossRef]
16. Wang, J.; Ma, F.; Sun, M. Graphene, hexagonal boron nitride, and their heterostructures: Properties and applications. *RSC Adv.* **2017**, *7*, 16801–16822. [CrossRef]
17. Yankowitz, M.; Ma, Q.; Jarillo-Herrero, P.; LeRoy, B.J. van der Waals heterostructures combining graphene and hexagonal boron nitride. *Nat. Rev. Phys.* **2019**, *1*, 112–125. [CrossRef]
18. Lorrette, C.; Weisbecker, P.; Jacques, S.; Pailler, R.; Goyhénèche, J.M. Deposition and characterization of hex-BN coating on carbon fibres using tris (dimethylamino) borane precursor. *J. Eur. Ceram. Soc.* **2007**, *27*, 2737–2743. [CrossRef]
19. Jedrzejczak-Silicka, M.; Trukawka, M.; Dudziak, M.; Piotrowska, K.; Mijowska, E. Hexagonal Boron Nitride Functionalized with Au Nanoparticles—Properties and Potential Biological Applications. *Nanomaterials* **2018**, *8*, 605. [CrossRef]
20. Lu, T.; Wang, L.; Jiang, Y.; Huang, C. Hexagonal boron nitride nanoplates as emerging biological nanovectors and their potential applications in biomedicine. *J. Mater. Chem. B* **2016**, *4*, 6103–6110. [CrossRef]
21. Horvath, L.; Magrez, A.; Golberg, D.; Zhi, C.; Bando, Y.; Smajda, R.; Horvath, E.; Forro, L.; Schwaller, B. In Vitro Investigation of the Cellular Toxicity of Boron Nitride Nanotubes. *ACS Nano* **2011**, *5*, 3800–3810. [CrossRef] [PubMed]
22. Kartal, İ.; Boztoprak, Y. Bor nitrür partiküllеriyle takviye edilmiş vinil ester matrisli kompozitlerin mekanik özelliklerinin incelenmesi. *El-Cezerî J. Sci. Eng.* **2019**, *6*, 43–50.
23. Lahiri, D.; Singh, V.; Benaduce, A.P.; Seal, S.; Kos, L.; Agarwal, A. Boron nitride nanotube reinforced hydroxyapatite composite: Mechanical and tribological performance and in-vitro biocompatibility to osteoblasts. *J. Mech. Behav. Biomed. Mater.* **2011**, *4*, 44–56. [CrossRef] [PubMed]
24. Lu, F.; Wang, F.; Cao, L.; Kong, C.Y.; Huang, X. Hexagonal Boron Nitride Nanomaterials: Advances towards Bioapplications. *Nanosci. Nanotechnol. Lett.* **2012**, *4*, 949–961. [CrossRef]
25. Ensoylu, M.; Deliormanlı, A.M.; Atmaca, H. Hexagonal Boron Nitride/PCL/PLG Coatings on Borate Bioactive Glass Scaffolds for Bone Regeneration. *J. Inorg. Organomet. Polym. Mater.* **2022**, *32*, 1551–1566. [CrossRef]
26. Saggar, R.; Porwal, H.; Tatarko, P.; Dlouhý, I.; Reece, M.J. Boron nitride nanosheets reinforced glass matrix composites. *Adv. Appl. Ceram.* **2015**, *114* (Suppl. 1), S26–S33. [CrossRef]
27. Kokubo, T.; Takadama, H. How Useful is SBF in Predicting in vivo Bone Bioactivity? *Biomaterials* **2006**, *27*, 2907–2915. [CrossRef]
28. Wojcik-Pastuszka, D.; Krzak, J.; Macikowski, B.; Berkowski, R.; Osiński, B.; Musiał, W. Evaluation of the Release Kinetics of a Pharmacologically Active Substance from Model Intra-Articular Implants Replacing the Cruciate Ligaments of the Knee. *Materials* **2019**, *12*, 1202. [CrossRef]
29. Songfeng, E.; Ye, X.; Wang, M.; Huang, J.; Ma, Q.; Jin, Z.; Ning, D.; Lu, Z. Enhancing the tribological properties of boron nitride by bioinspired polydopamine modification. *Appl. Surf. Sci.* **2020**, *529*, 147054.
30. Wang, L.; Hang, R.; Xu, Y.; Guo, C.; Qian, Y. From ultrathin nanosheets, triangular plates to nanocrystals with exposed (102) facets, a morphology and phase transformation of sp2 hybrid BN nanomaterials. *RSC Adv.* **2014**, *4*, 14233–14240. [CrossRef]
31. Ojha, P.K.; Maji, R.; Karmakar, S. Effect of crystallinity on droplet regression and disruptive burning characteristics of nanofuel droplets containing amorphous and crystalline boron nanoparticles. *Combust. Flame* **2018**, *188*, 412–427. [CrossRef]
32. Balachander, L.; Ramadevudu, G.; Shareefuddin; Sayanna, R.; Venudhar, Y.C. IR analysis of borate glasses containing three alkali oxides. *ScienceAsia* **2013**, *39*, 278–283. [CrossRef]
33. Sudeep, P.M.; Vinod, S.; Ozden, S.; Sruthi, R.; Kukovecz, A.; Konya, Z.; Vajtai, R.; Anantharaman, M.R.; Ajayan, P.M.; Narayanan, T.N. Functionalized boron nitride porous solids. *RSC Adv.* **2015**, *5*, 93964. [CrossRef]
34. Geick, R.; Perry, C.H.; Rupprecht, G. Normal modes in hexagonal boron nitride. *Phys. Rev.* **1966**, *146*, 543–547. [CrossRef]
35. Pierson, H.O. *Handbook of Carbon, Graphite, Diamond, and Fullerenes: Properties, Processing, and Applications*; Noyes Publications: Devon, UK, 1993.
36. Riaz, I. Graphene and Boron Nitride: Members of Two Dimensional Material Family. Doctor's Thesis, School of Physics and Astronomy, Manchester Üniversitesi, Manchester, UK, 2012, 144p.
37. Wei, X.; Meng, Z.; Ruiz, L.; Xia, W.; Lee, C.; Kysar, J.W.; Hone, J.C.; Keten, S.; Espinosa, H.D. Recoverable slippage mechanism in multilayer graphene leads to repeatable energy dissipation. *ACS Nano* **2016**, *10*, 1820–1828. [CrossRef]
38. Kim, J.H.; Jeong, J.H.; Kim, N.; Joshi, R.; Lee, G.H. Mechanical properties of two-dimensional materials and their applications. *J. Phys. D Appl. Phys.* **2018**, *52*, 083001. [CrossRef]
39. Golberg, D.; Bando, Y.; Huang, Y.; Terao, T.; Mitome, M.; Tang, C.; Zhi, C. Boron Nitride Nanotubes and Nanosheets. *ACS Nano* **2010**, *4*, 2979–2993. [CrossRef]
40. Shuai, C.; Han, Z.; Feng, P.; Gao, C.; Xiao, T.; Peng, S. Akermanite scaffolds reinforced with boron nitride nanosheets in bone tissue engineering. *J. Mater. Sci. Mater. Med.* **2015**, *26*, 188. [CrossRef] [PubMed]

41. Berzina-Cimdina, L.; Borodajenko, N. Research of calcium phosphates using Fourier transform infrared spectroscopy. *Infrared Spectrosc.-Mater. Sci. Eng. Technol.* **2012**, *12*, 251–263.
42. Lahiri, D.; Singh, V.; Keshri, A.K.; Seal, S.; Agarwal, A. Apatite formability of boron nitride nanotubes. *Nanotechnology* **2011**, *22*, 205601. [CrossRef]
43. Schuhladen, K.; Pantulap, U.; Engel, K.; Jeleń, P.; Olejniczak, Z.; Hupa, L.; Sitarz, M.; Boccaccini, A.R. Influence of the replacement of silica by boron trioxide on the properties of bioactive glass scaffolds. *Int J Appl Glass Sci.* **2021**, *12*, 293–312. [CrossRef]
44. El-Batal, F.; El-Kheshen, A.A.; El-Bassyouni, G.T.; Abd El Aty, A.A. In Vitro Bioactivity Behavior of some Borate Glasses and their Glass-Ceramic Derivatives Containing Zn2+, Ag+ or Cu2+ by Immersion in Phosphate Solution and their Anti-Microbial Activity. *Silicon* **2018**, *10*, 943–957. [CrossRef]
45. Ksouri, D.; Khireddine, H.; Aksas, A.; Valente, T.; Bir, F.; Slimani, N.; Cabal, B.; Torrecillas, R.; Santos, J.D. Synthesis of ternary bioactive glass derived aerogel and xerogel: Study of their structure and bioactivity. *Nova Biotechnol. Chim.* **2018**, *17*, 150–159. [CrossRef]
46. Nawaz, A.; Bano, S.; Yasir, M.; Wadood, A.; Rehman, M.A.U. Ag and Mn-doped mesoporous bioactive glass nanoparticles incorporated into the chitosan/gelatin coatings deposited on PEEK/bioactive glass layers for favorable osteogenic differentiation and antibacterial activity. *Mater. Adv.* **2020**, *1*, 1273–1284. [CrossRef]
47. Bruschi, M.L. Chapter 5—Mathematical models of drug release. In *Strategies to Modify the Drug Release from Pharmaceutical Systems*; Woodhead Publishing: Sawston, UK, 2015; pp. 63–86. [CrossRef]
48. Longley, D.; Harkin, D.; Johnston, P. 5-Fluorouracil: Mechanisms of action and clinical strategies. *Nat. Rev. Cancer* **2003**, *3*, 330–338. [CrossRef] [PubMed]
49. He, L.; Zhu, H.; Zhou, S.; Wu, T.; Wu, H.; Yang, H.; Mao, H.; SekharKathera, C.; Janardhan, A.; Edick, A.M.; et al. Wnt pathway is involved in 5-FU drug resistance of colorectal cancer cells. *Exp. Mol. Med.* **2018**, *50*, 1–12. [CrossRef] [PubMed]
50. Sara, J.D.; Kaur, J.; Khodadadi, R.; Rehman, M.; Lobo, R.; Chakrabarti, S.; Herrmann, J.; Lerman, A.; Grothey, A. 5-fluorouracil and cardiotoxicity: A review. *Ther. Adv. Med. Oncol.* **2018**, *10*, 1758835918780140. [CrossRef]
51. Zarghami Dehaghani, M.; Yousefi, F.; Sajadi, S.M.; Tajammal Munir, M.; Abida, O.; Habibzadeh, S.; Mashhadzadeh, A.H.; Rabiee, N.; Mostafavi, E.; Saeb, M.R. Theoretical Encapsulation of Fluorouracil (5-FU) Anti-Cancer Chemotherapy Drug into Carbon Nanotubes (CNT) and Boron Nitride Nanotubes (BNNT). *Molecules* **2021**, *26*, 4920. [CrossRef]
52. Wang, C.; Shen, L.; Wu, L. Adsorption and sensing of an anticancer drug on the boron nitride nanocones; a computational inspection. *Comput. Methods Biomech. Biomed. Eng.* **2021**, *24*, 151–160. [CrossRef]
53. El-Kady, A.; Farag, M. Bioactive Glass Nanoparticles as a New Delivery System for Sustained 5-Fluorouracil Release: Characterization and Evaluation of Drug Release Mechanism. *J. Nanomater.* **2015**, *16*, 399. [CrossRef]
54. Khalid, A.; Ahmad, P.; Khan, A.; Khandaker, M.U.; Kebaili, I.; Alam, M.M.; Din, I.U.; Muhammad, S.; Razzaq, Z.; Rehman, I.U.; et al. Cytotoxic and photocatalytic studies of hexagonal boron nitride nanotubes: A potential candidate for wastewater and air treatment. *RSC Adv.* **2022**, *12*, 6592. [CrossRef] [PubMed]

biomimetics

MDPI

Article

Polyvinylpyrrolidone Nanofibers Incorporating Mesoporous Bioactive Glass for Bone Tissue Engineering

Ricardo J. R. Matos [1], Jorge C. Silva [2,*], Paula I. P. Soares [1,*] and João Paulo Borges [1]

[1] i3N/CENIMAT, Department of Materials Science, NOVA School of Science and Technology, NOVA University Lisbon, Campus de Caparica, 2829-516 Caparica, Portugal; rj.matos@campus.fct.unl.pt (R.J.R.M.); jpb@fct.unl.pt (J.P.B.)

[2] i3N/CENIMAT, Department of Physics, NOVA School of Science and Technology, NOVA University Lisbon, Campus de Caparica, 2829-516 Caparica, Portugal

* Correspondence: jcs@fct.unl.pt (J.C.S.); pi.soares@fct.unl.pt (P.I.P.S.)

Abstract: Composite biomaterials that combine osteoconductive and osteoinductive properties are a promising approach for bone tissue engineering (BTE) since they stimulate osteogenesis while mimicking extracellular matrix (ECM) morphology. In this context, the aim of the present research was to produce polyvinylpyrrolidone (PVP) nanofibers containing mesoporous bioactive glass (MBG) 80S15 nanoparticles. These composite materials were produced by the electrospinning technique. Design of experiments (DOE) was used to estimate the optimal electrospinning parameters to reduce average fiber diameter. The polymeric matrices were thermally crosslinked under different conditions, and the fibers' morphology was studied using scanning electron microscopy (SEM). Evaluation of the mechanical properties of nanofibrous mats revealed a dependence on thermal crosslinking parameters and on the presence of MBG 80S15 particles inside the polymeric fibers. Degradation tests indicated that the presence of MBG led to a faster degradation of nanofibrous mats and to a higher swelling capacity. The assessment of in vitro bioactivity in simulated body fluid (SBF) was performed using MBG pellets and PVP/MBG (1:1) composites to assess if the bioactive properties of MBG 80S15 were kept when it was incorporated into PVP nanofibers. FTIR and XRD analysis along with SEM–EDS results indicated that a hydroxy-carbonate apatite (HCA) layer formed on the surface of MBG pellets and nanofibrous webs after soaking in SBF over different time periods. In general, the materials revealed no cytotoxic effects on the Saos-2 cell line. The overall results for the materials produced show the potential of the composites to be used in BTE.

Keywords: electrospun polyvinylpyrrolidone; polymeric scaffolds; bone regeneration; mesoporous bioactive glass

Citation: Matos, R.J.R.; Silva, J.C.; Soares, P.I.P.; Borges, J.P. Polyvinylpyrrolidone Nanofibers Incorporating Mesoporous Bioactive Glass for Bone Tissue Engineering. *Biomimetics* **2023**, *8*, 206. https://doi.org/10.3390/biomimetics8020206

Academic Editors: Ryszard Uklejewski and Mariusz Winiecki

Received: 17 April 2023
Revised: 11 May 2023
Accepted: 12 May 2023
Published: 17 May 2023

1. Introduction

Bone transplantation is one of the most common surgical procedures, ranking only behind blood transfusion [1]. Large bone defects result mainly from high-energy traumatic events such as traffic accidents or large bone resection due to different pathologies such as tumors or infections [2]. The body's innate healing mechanisms frequently cannot spontaneously repair these defects. Bone tissue engineering (BTE) has gained more relevance recently since it can offer promising approaches to bone defects treatment by providing stimulus and osteoconductive templates to facilitate the formation of new bone tissue [3]. Additionally, the biomaterials employed in BTE must possess angiogenic properties, promoting the formation of blood vessels within the constructs. This ensures that the newly regenerated tissue obtains sufficient oxygen and nutrients for its survival [4].

In 1969, BioGlass® 45S5 was first introduced by L. Hench and his collaborators [5]. It is well-established in the literature that BioGlass® 45S5 exhibits non-cytotoxicity and can generate a surface hydroxyapatite (HA) layer that closely resembles the composition of natural bone. This discovery paved the way to a wide range of different bioactive glasses

(BGs), thus leading to a wide range of clinical applications in soft and hard tissues (tooth and bone) [6]. Regarding biomedical applications, BG-based materials hold significant importance in dentistry, as BG is capable of creating a chemical bond directly with adjacent tissues and elevating the local pH level to above 5.5; thus, bacteria are deterred from colonizing and thriving at the dental implantation sites, preventing the early stages of caries. Furthermore, the elastic modulus of BG-based materials is comparable to that of the nearby tissues, thus reducing the probability of local bone resorption [7–9].

BGs are mostly applied in BTE since they allow the development of scaffolds that possess both osteogenic and angiogenic properties [10]. BioGlass® 45S5 particles have demonstrated interesting results on potential angiogenic effects, considering the increased secretion of vascular endothelial growth factor (VEGF) in vitro and enhancement of vascularization in vivo [11]. Furthermore, Hench and Polak demonstrated that the dissolution products of BGs can exert genetic control over the osteoblast cell cycle, leading to the fast expression of genes that govern osteogenesis [12]. Whether obtained through melting or sol-gel processing, every bioactive glass has the potential to induce bioreactivity or bioactivity by means of its capability to interact with living tissue and form strong bonds to bone. These strong bonds are formed by the precipitation of a hydroxy-carbonate apatite (HCA) layer on the BG surface upon exposure to appropriate physiological fluids or during in vivo applications [13].

Recently, a novel family of mesoporous bioactive glasses (MBGs) exhibiting well-defined mesoporous structures and greater specific surface area has demonstrated remarkable potential for hard tissue engineering applications. The increased interface effects lead to high reactivity, thus improving bioactivity when compared to standard BGs. The synthesis of highly ordered mesoporous materials depends on the use of surfactants as structure-directing agents, which facilitate the formation of channels within the inorganic component. Upon removal of the surfactant, these channels are preserved, resulting in a well-ordered mesoporous structure [14]. In addition, some studies are focusing on improving the properties of BGs by incorporating doping agents instead of manipulating the ratios of different BG components. The doping process aims to improve various aspects of BGs, including their mechanical properties, bioactivity, and biological outcomes, such as promoting angiogenesis, antibacterial activity, bone cell migration, proliferation, and differentiation, or reducing the patient's inflammatory response. It is possible to incorporate inorganic elements (Ag^{+}, Sr^{2+}, Cu^{2+}, Zn^{2+}, Mg^{2+}, etc.) or organic compounds such as VEGF, bone morphogenetic proteins (BMPs), dexamethasone (DEX), gentamicin, etc. [15]. For example, Bellucci et al. studied the effect of replacing 10 mol% of CaO with MgO or SrO from a silicate BG. The researchers reported that the doped BGs, especially the Mg-doped, further improved the bioactivity [16]. Tang et al. demonstrated that BGs loaded with rhBMP-2 promoted excellent cell attachment, ingrowth, and osteogenesis in vitro and supported complete bone regeneration in vivo using a rabbit radius critical size defect model [17]. Lim et al. produced electrospun nanofiber matrices that incorporated BG nanoparticles loaded with DEX. The research group reported that these scaffolds promoted the odontogenesis of human dental pulp cells (HDPCs) [18].

BGs and MBGs can be produced to feature different morphologies (e.g., powder, granules, monoliths, fiber, coatings, etc.) [19]. In this application area, it is important to take into account both micro-sized and nanoscale powder particles, which can be used in the production of composite material scaffolds, such as the combination of electrospun biodegradable polymers and MBGs, which is the purpose of this work. Several studies have been carried out in order to combine polymers with inorganic phases such as BGs or HA. In this context, Petretta et al. have reported the production of composite scaffolds based on polycaprolactone (PCL) and Mg-containing bioactive glasses [20]. In general, the main benefit of using these composite materials is that they retain the bioactivity of the inorganic bioactive filler while preserving the mechanical properties of the polymer, such as flexibility and the ability to deform under loads. Moreover, the incorporation of MBG powder nanoparticles in a polymeric mat mimics the natural bone structure more accurately as it consists of nanoscale hydroxyapatite crystallites together with the polymeric phase of collagen type I. This natural combination is responsible for the mechanical properties of bone [13]. From this perspective, the size of the BG or MBG particles incorporated into the polymeric matrices also plays an important role in enhancing bioactivity. Research studies suggest that replicating the nanoscale characteristics of bone can enhance the attachment and growth of bone-forming cells [21].

Polyvinylpyrrolidone (PVP) is a synthetic polymer that exhibits high solubility in water as well as numerous organic solvents. PVP has found extensive application in the production of hydrogels, wound dressings, nanofibers/scaffolds, and drug and gene delivery systems. During World War II, PVP gained significant recognition as a blood plasma substitute [22].

The ability of the electrospinning technique to produce nanofibrous scaffolds that mimic ECM while incorporating other materials in particulate form makes it the ideal technique to produce composite scaffolds for bone tissue engineering [23,24]. In this paper, we have successfully prepared MBG 80S15 by using Pluronic F127 as a structure-directing agent through an evaporation-induced self-assembly (EISA) process. Nanosized MBG 80S15 powder was further incorporated into electrospun PVP nanofibers, and its properties were studied to assess the potential of such composites for use in BTE applications. Furthermore, the coupling of PVP with MBG nanoparticles produces composite 2D materials where it is possible to control the degradation and bioactivity rates to fit the natural processes of bone repair.

2. Materials and Methods

Analytical-grade chemical reagents were utilized in this research without the need for additional purification.

2.1. Synthesis of Mesoporous Bioactive Glass MBG 80S15

Mesoporous Bioactive Glass MBG 80S15 is a ternary bioactive glass (SiO_2–CaO–P_2O_5) which is produced by a sol-gel technique adapted from a previously described method by Yan et al. [25]. MBG was synthesized by using the nonionic block copolymer Pluronic F127 ($EO_{106}PO_{70}EO_{106}$) (Sigma-Aldrich, Lisbon, Portugal) as structure-directing agent (EO is poly(ethylene oxide), and PO is poly(propylene oxide)).

In a typical synthesis of MBG 80S15, in order to achieve a molar ratio of Si/Ca/P = 80:15:5, tetraethyl orthosilicate (Sigma-Aldrich) (TEOS, 6.7 g), calcium nitrate-4-hydrate (PanReac AppliChem, Darmstadt, Germany) $Ca(NO_3)_2{\cdot}4H_2O$ (1.4 g), triethyl phosphate (Sigma-Aldrich) (TEP, 0.73 g), and 0.5 *M* HCl (1.0 g) were dissolved in absolute ethanol (Sigma-Aldrich) (60 g). TEOS, TEP and calcium nitrate-4-hydrate were used as precursors of SiO_2, P_2O_5 and CaO, respectively.

The mixture was allowed to react for 30 min at room temperature under magnetic stirring at 250 rpm (2Mag Mix 15 Eco Stirrer, Germany) for the acid hydrolysis of TEOS. Then, 1.82 g of Pluronic F127 was added to the mixture and the resultant solution was also stirred at room temperature for 60 h to undergo an evaporation-induced self-assembly (EISA) process. The gel obtained was heated at 60 °C (Memmert Incubator Modell 100–800, Schwabach, Germany) under atmospheric pressure for 24 h to completely remove the alcohol. Subsequently, the dried gel was calcined in a muffle furnace (Nabertherm GmbH HTCT 01/16, Lilienthal, Germany) at 700 °C for 5 h (ramp of 2 $^{\circ}C{\cdot}min^{-1}$) to obtain the final MBG 80S15 products. The resulting powder was ground in a planetary ball mill (Planetary Mono Mill Pulverisette 6—Fritsch, Idar-Oberstein, Germany) for 6 h at 400 rpm using zirconia balls with a 2 mm diameter. The final average grain size obtained, according to SEM images using ImageJ software, (version 1.52n, National Institutes of Health, Bethesda, MD, USA), was about 174 ± 119 nm (100 measurements).

2.2. Optimization of Electrospinning Process by DOE

A solution of 18% w/v PVP ($M_w \sim 1,300,000$ g/mol) (Sigma-Aldrich) in ethanol (Sigma-Aldrich) was prepared for electrospinning, which was carried out at a temperature of 20 °C with a humidity level of approximately 40%.

In this study, the environmental conditions and the solution parameters were kept constant. Four operating variables (flow rate, tip-collector distance, applied voltage, and needle gauge) were studied by applying a custom Design of Experiments and response surface methodology (RSM) as a statistical method. The custom DOE was performed using JMP software, version 13.0 (S.A.S. Institute Inc., Cary, NC, USA) to investigate the effect of above-mentioned electrospinning process parameters and the interactions between these variables on the average electrospun PVP fiber diameter.

The range of each variable was chosen considering the set of values that allow the PVP electrospinning process based on the literature and previous experiments (Table 1). Within the range defined for each variable, the software selected a set of experiments consisting of a minimum of 27 runs. To determine how to optimize the parameters, several outputs were measured: the samples were analyzed by SEM to observe the overall structure of the collected mats and the average fiber diameter was calculated using ImageJ Software by selecting at least 25 random fibers [26]. Using the RSM, the overall desirable response (the set of variables that minimizes the average fiber diameter) was obtained from the software's calculations within the pre-selected range of each variable. The software also estimated the resulting average PVP fiber diameter for that set of variables, which was about 652 ± 179 nm.

Table 1. Range of four independent variables in a custom design of experiments (JMP software, version 13.0) aiming to minimize average electrospun PVP fiber diameter and respective desirable response using RSM.

Variable	Type	Range	Desirable Response	Estimated Output (nm)
Applied voltage (kV)	Continuous	15–20	18.6	652 ± 179
Flow rate (mL·h^{-1})	Continuous	0.1–0.4	0.1	
Tip-collector distance (cm)	Continuous	15–20	17.3	
Needle gauge	Categorical	G23; G27	G27	

In addition, for all the experiments and subsequent membrane production, a 1 mL syringe was loaded with the polymer solution which was dispensed using a syringe pump (KDS100, KD Scientific, Holliston, MA, USA) at different rates from 0.1 to 0.4 mL·h^{-1}. A high voltage supply (Glassman High Voltage INC EL 30kV, High Bridge, NJ, USA) was connected to the metallic needle (ITEC Iberiana Technical, Braga, Portugal), and a sheet of aluminum foil was used as collector and placed in front of the needle tip (schematically represented in Figure 1) at distances ranging from 15 to 20 cm to collect the fibers.

2.2.1. PVP/MBG Composite Production

To obtain the solution for electrospinning, the MBG powder was first combined with ethanol and subsequently sonicated for 15 min at 320 W for further dispersion. PVP was weighed and dissolved in the previously dispersed solution containing the solvent to obtain a final polymer concentration of 18 w/v %. The final blend was magnetically stirred for at least 30 min, resulting in a well-dispersed MBG powder in the polymeric solution. Different percentages of the MBG powder were added to ethanol for further dispersion: 10; 28; 56; 83; and 100% by weight with respect to PVP content. The quantity of MBG powder utilized in this study must be significant to promote bioactivity and regulate the degradation rate of PVP to some extent [27]. In this context, the composites containing 100% MBG by weight with respect to PVP are of major interest. This composite will be denominated PVP/MBG (1:1) as the mass content of MBG and PVP is the same.

Electrospinning was carried out with the same processing parameters estimated from the DOE to plain PVP, except for the needle gauge. The composites were electrospun with a G23 needle (ITEC Iberiana Technical, Braga, Portugal) to prevent clogging. The membranes produced were kept in a desiccator until the crosslinking treatment due to the water solubility of PVP.

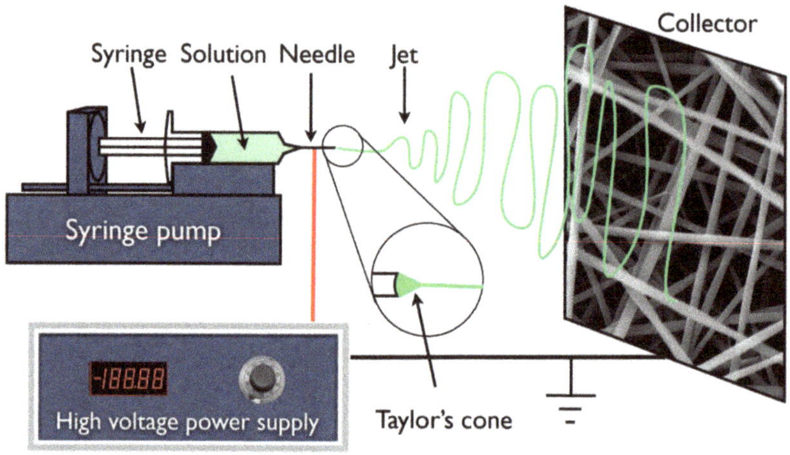

Figure 1. Schematic representation of an electrospinning apparatus in horizontal setup with a static plate collector.

2.2.2. Thermal Crosslinking

The water solubility of PVP restricts the use of electrospun PVP-based nanofibers without any further treatment. Nonetheless, the solubility of PVP can be reduced by heating it in air between 150 and 200 °C [28]. The as-spun PVP and PVP/MBG composites were crosslinked in an incubator (Memmert Incubator Modell 100–800, Schwabach, Germany) at different temperatures within the above-mentioned range. Final temperatures of 150, 165, and 180 °C were studied for different times: 3, 4, 5, 8, 12, and 24 h. To investigate the effect of thermal crosslinking, the membranes were immersed for 24 h in water, dried at 60 °C and analyzed by SEM.

2.3. Characterization

The images captured using transmission electron microscopy (TEM) were acquired using a Hitachi H-8100II (Tokyo, Japan) instrument equipped with thermo-ionic emission LaB6, offering a resolution of 2.7 Å. The analysis of the membranes was conducted on the thinnest membrane that could be technically extracted (around 15–20 min of deposition).

Powder X-ray diffraction (XRD) was employed to confirm the crystalline phases present in the samples. X-ray diffraction patterns were obtained using an X'Pert PRO PANAlytical X-ray diffractometer (Malvern, UK). The measurements were taken over a range of 2θ values from 15° to 80°, using Cu-Kα radiation (λ = 1.54060 Å) with a step size of 0.033.

The Nicolet 6700-Thermo Electron Corporation Attenuated Total Reflectance-Fourier Transform Infrared (ATR-FTIR) spectrometer (Watertown, WI, USA) was employed to obtain the FTIR spectra of the samples. The measurements were taken in the 480–4000 cm^{-1} range with a resolution of 2 cm^{-1}.

Carl Zeiss Auriga (Oberkochen, Germany) scanning electron microscopy (SEM) equipment was used to analyze the fiber morphology. The samples were previously coated with a thin layer of either iridium–palladium or carbon, depending on which elements were being studied by complementary EDS analysis.

N_2 adsorption–desorption isotherms were measured with a Gas Porosimeter Micromeritics (ASAP 2010, Norcross, GA, USA) at −196 °C. Prior to the measurements, the samples underwent a 12 h outgassing process at 300 °C in vacuum. The specific surface area was calculated using the Barrett–Emmett–Teller (BET) method. The pore volume and pore size distribution were determined using the Barrett–Joyner–Halanda (BJH) method based on the adsorption branches of the isotherms. The total pore volume was estimated from the adsorbed amount at a maximum relative pressure.

Thermo-gravimetric analysis (TGA) curves were recorded with the Simultaneous Thermal Analyser (STA 449 F3 Jupiter Netzsch (Selb, Germany)) under an inert nitrogen atmosphere at a heating rate of 10 °C/min from room temperature to 900 °C. Differential TGA (DTGA) was obtained by differentiating the respective TGA curve.

2.4. Swelling and Degradation Assays

The success of thermal crosslinking was studied by testing the dissolution of crosslinked mats immersed in phosphate buffer solution (PBS, pH 7.4) at 37 °C at different time points: 1, 3, 7, 13, 17, 21 and 28 days. At these time points, the degradation of PVP and PVP/MBG composites fibers in physiological solutions was assessed by measuring the membranes' weight loss. The samples were removed from PBS, gently rinsed with Millipore water to remove saline and dried at 37 °C until constant mass was reached The percentage of weight loss (W_L %) was computed as $W_L\ \% = 100 \times \frac{(W_0 - W_r)}{W_0}$, where W_0 and W_r are the initial and the residual weight of the sample, respectively.

The swelling ratio is a measure of the liquid-absorbing capacity of a matrix. The dry weight of each sample was measured (W_d) before soaking in PBS (pH 7.4). The mass of each sample was planned to be at least 10 mg, ensuring the membrane stability over the experiment. At pre-determined intervals of time (5, 15, and 30 min; 1 and 5 h; 1, 3, and 7 days) the specimens were removed from the PBS. The excess liquid on each matrix was removed, and the wet matrices were weighed (W_w). The swelling ratio (Q) was calculated as: $Q = \frac{W_w - W_d}{Wd}$.

In both degradation and swelling experiments, all samples were cut with dimensions of 20 mm × 20 mm and placed in a plastic container (Falcon tube, 50 mL) filled with PBS (pH 7.4). The experiments were performed in an incubator with an orbital shaker (Comecta—Heated Incubator with Shaker S-100D, Barcelona, Spain) at 37 °C. An average of at least three samples was taken for each time point. The samples were weighed with an accuracy of 0.1 mg on an analytical balance (Sartorius Quintix Model 224-1S, Goettingen, Germany).

2.5. Mechanical Response—Tensile Tests

The tensile tests were carried out using a Rheometric Scientific uniaxial tensile testing machine equipped with a 20 N load cell and controlled by Minimat software (Minimat Control Software Version 1.60 February 1994 (c) PL Thermal Science 1984-94 Rheometric Scientific Ltd., New Castle, DE, USA). All experiments were conducted at room temperature and approximately 50% humidity, with a velocity of 1 mm·min^{-1}. At least 10 samples of each membrane type were tested, and each sample was cut into a rectangular shape measuring 10 mm in width and 20 mm in length. The thickness of the samples was measured using a digital micrometer (Mitutoyo 0–25 mm, Aurora, IL, USA). The stress–strain curves were used to determine the ultimate tensile strength, Young's modulus and yield strength.

2.6. Cytotoxicity Assays

The cytotoxicity of the membranes was evaluated using the extract method and the Saos-2 cell line (American Type Culture Collection, HTB-85), following the standard ISO-10993 for the biological evaluation of medical devices, specifically Part 5: tests for in vitro cytotoxicity. To produce the extract, different samples were cut to have an identical mass of 30 mg and pre-sterilized at 120 °C for 2 h. Each sample was placed in 1.5 mL (resulting in an extract concentration of 20 mg·mL^{-1}) of McCoy 5A culture medium supplemented with

2.2 g/L sodium bicarbonate (both from Sigma Aldrich, Portugal), penicillin (100 U/mL) and streptomycin (100 μg/mL) (Gibco Life Technologies, Billings, MT, USA) and 10% FBS (Fetal Bovine Serum, S. America origin, Biowest, Nuaillé, France) at 37 °C for 48 h.

Cells were seeded at a density of 30,000 cell·cm^{-2} in 96-well plates and grown in fully supplemented McCoy followed by incubation at 37 °C in 5% CO_2 for 24 h. The subsequent step was to exchange the medium for the extract to assess whether the cells would survive the next 48 h. Five concentrations of the extract were used: 20, 10, 5, 2.5 and 1.25 mg·mL^{-1} (two-fold serial dilution). Four replicates were used for each concentration. After this period, the culture medium was removed, and a solution of resazurin (Alfa Aesar, MA, USA) containing 10% 0.2 mg·mL^{-1} resazurin solution in PBS and 90% complete culture medium was added to each well. After a 2-h incubation period, the absorbance was measured at 570 nm and 600 nm using a multi-well plate reader (BioTek ELX800 UV, Winooski, VT, USA). The negative control cells were cultured in complete medium while positive control cells were exposed to 10% DMSO to induce cell death. Cell viability was determined as a percentage of the negative control, calculated by [% cell viability = treated cells/control cells × 100].

2.7. Bioactivity Assay

The bioactivity of PVP/MBG composites, related to the surface deposition of hydroxy-carbonate apatite (HCA) layers, was evaluated by immersing the samples in simulated body fluid (SBF) solution. The SBF had a composition and ionic concentration similar to those of human body plasma, and it was prepared by dissolving respective amounts of the reagent chemicals (NaCl (Sigma-Aldrich), $NaHCO_3$ (Sigma-Aldrich), KCl (Sigma-Aldrich), $MgCl_2{\cdot}6H_2O$ (Sigma-Aldrich), HCl (Honeywell Fluka, Lisbon, Portugal), $CaCl_2$ (Carl Roth, Karlsruhe, Germany), Na_2SO_4 (Sigma-Aldrich), $K_2HPO_4{\cdot}3H_2O$ (VWR, Lisbon, Portugal), and $(CH_2OH)_3(CNH)_2$ (Apollo Scientific, Cheshire, UK) as the buffering agent) into deionized water according to the procedure described by Kokubo et al. [29,30]. The prepared SBF was buffered at pH 7.4 with tris (hydroxymethyl aminomethane) and 1 *M* hydrochloric acid at 37 °C. The polymeric matrices were cut into samples of 15 × 15 mm and subsequently immersed in 50 mL SBF in a falcon tube for different time periods (1, 3, 5 and 10 days). This experiment followed international standards ISO/FDIS 23317 [31]; the volume of the SBF solution (V_{SBF}) should have been at least the volume (mL) determined according to the following formula: $V_{SBF} = \frac{S_A}{10}$, where S_A is the surface area of the samples (mm^2). During the experiment, the samples were kept at 37 °C in an incubator with orbital shaker (54 rpm), and the SBF solution was refreshed every two days. After the soaking periods, the samples were removed from SBF, rinsed with Millipore water and dried at room temperature. The bioactivity assay was performed for both the PVP/MBG (1:1) composites and MBG powder. For the latter, it was necessary to produce pellets by dry pressing. Each pellet was produced from 300 mg of powder in a hardened steel mold with an inner diameter of 13 mm. The powder was pressed uniaxially at a load of 5 tonnes for 5 min at room temperature.

Formation of a bone-like apatite layer on the surface of the nanocomposite mats was determined by X-ray diffraction (XRD), Fourier transform infrared spectroscopy (FTIR) analysis and scanning electron microscopy coupled with energy dispersive spectroscopy (SEM–EDS).

3. Results

3.1. Mesoporous Bioactive Glass MBG 80S15

MBG synthesis was performed by the sol-gel method using Pluronic F127 as a structure-directing agent. During the synthesis process, the evaporation of organic solvent progressively increases block copolymer concentration. When the concentration of Pluronic F127 reaches critical micelle concentration, micellar aggregates form through the self-assembly process of silica-surfactant composite micelles, which organize into liquid-crystalline mesophases. Calcination allows the removal of the organic surfactant template and retention of the inorganic framework to obtain highly ordered mesostructured bioactive glass materials [14,32].

Figure 2A shows the nitrogen adsorption–desorption isotherms of the MBG 80S15 calcined powder. The isotherms obtained exhibit a type IV pattern, which is consistent with a mesoporous structure. Additionally, at a relative pressure of approximately 0.40, the sample displays type H1 hysteresis loops, which are indicative of cylindrical pores, in accordance with the p6mm mesostructure. The BET surface area of the MBG calculated from the linear part of the BET plot reached 232 $m^2 \cdot g^{-1}$. The pore size distribution curves were obtained by applying the BJH model to the adsorption and desorption branches. The MBG presents a relatively narrow pore size distribution, and the peak pore size was around 5.9 nm and 6.0 nm for adsorption and desorption branches, respectively. This proximity between pore size results can be an indicator of a perfectly cylindrical channel. The pore volume was also calculated using the BJH model and the result was about 0.25 $cm^3 \cdot g^{-1}$ for both the adsorption and desorption branches.

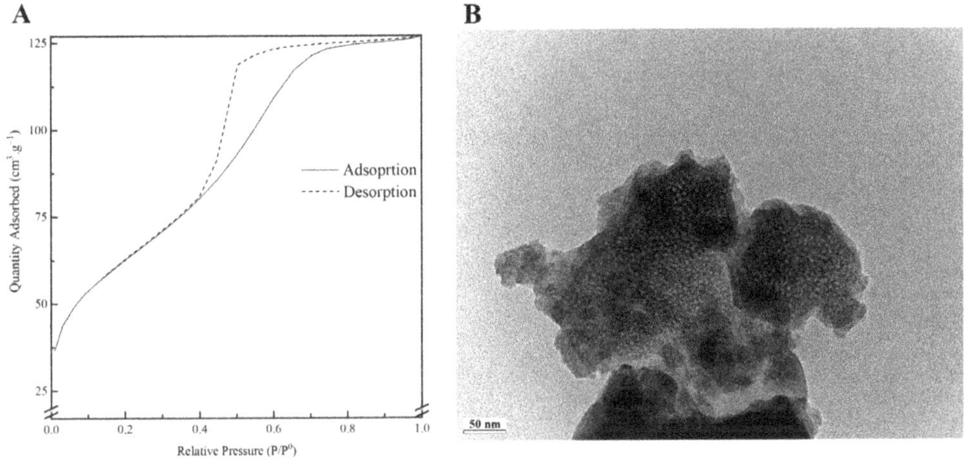

Figure 2. Nitrogen adsorption-desorption isotherms (**A**) and TEM image (**B**) of MBG 80S15 powder.

A TEM image of MBG 80S15 is shown in Figure 2B. Brighter spot areas, which can be an indicator of the expected mesoporous structure, can be observed. However, the cylindrically ordered structure of the mesoporous BG is not evident. Analysis of the sample structure by TEM is hindered by powder aggregation, which increases the thickness and the opacity of the samples. Moreover, the TEM equipment resolution also limits the proper visualization of the MBG mesoporosity.

SEM image of the MBG 80S15 obtained after 6 h of planetary ball-milling is presented in Figure 3A. It can be observed that the size of MBG particles is in the nanoscale range, as shown in the respective particle size distribution graph Figure 3B. The average diameter of MBG particles is around 174 $\pm$ 119 nm. It is worth pointing out the reduced dimensions of these particles, considering they become easier to disperse in an electrospinning solution compared to micro-sized MBG particles.

Figure 4A represents the XRD pattern of the produced MBG 80S15. It can be observed that there are no diffraction peaks appearing on the pattern of the sample, except for a broad reflection at 2θ = 15–35° (characteristic of silicate glasses) which suggests that the MBG 80S15 produced exists in an amorphous state. The temperature assigned to the occurrence of crystallization is about 810 °C [32]. It is important to note that MBG was calcinated at 700 °C to ensure that the mesoporous structure retained its stability. When MBGs are calcined at temperatures higher than 810 °C, the inorganic wall becomes crystalline and the mesostructure collapses.

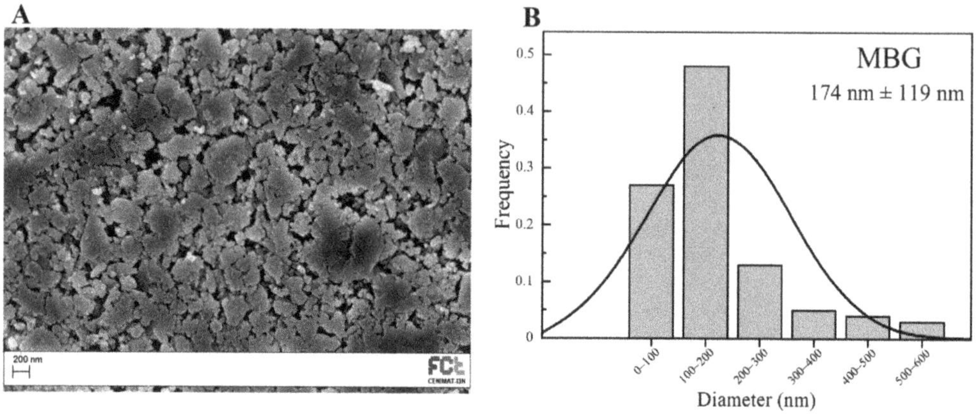

Figure 3. SEM image of MBG 80S15 pellet (**A**) and the respective particle size distribution graph after 6 h of ball-milling (**B**).

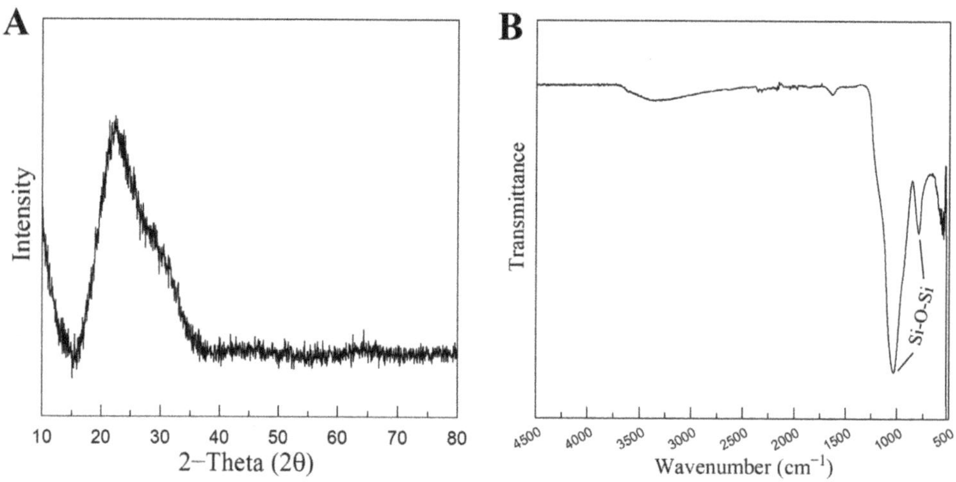

Figure 4. XRD pattern (**A**) and FTIR spectra (**B**) of MBG 80S15 powder calcined at 700 °C for 5 h.

Figure 4B shows the FTIR spectrum of the produced MBG 80S15 powder. It is shown that the sample exhibits the characteristic absorption bands of a silicate glass. The spectrum shows the Si-O-Si asymmetrical stretching vibration at 1040 cm^{-1} and Si-O-Si symmetrical stretching vibration at 806 cm^{-1}. The band associated with Si-O bending (rocking) vibration at 470 cm^{-1} was not detected due to the equipment resolution (the measuring process starts at 500 cm^{-1}) [33,34].

3.2. PVP/MBG 80S15 Composites

Custom Design of Experiments and RSM are statistical methods that allow the definition of a set of electrospinning processing parameters that minimize average PVP fiber diameter as described above (Section 2.2). JMP software (version 13.0) estimated an average resulting PVP fiber diameter of about 652 ± 179 nm for this set of parameters: flow rate of 0.10 mL·h^{-1}; applied voltage of 18.6 kV; tip-collector distance of 17.3 cm; needle gauge G27. The SEM image and respective average size distribution are shown in Figure 5A,B. SEM micrographs revealed that the electrospun PVP nanofiber mats were composed of randomly oriented, uniform and bead-free nanofibers, with an average diameter size of 656 ± 171 nm, which is quite close to the estimated value from DOE.

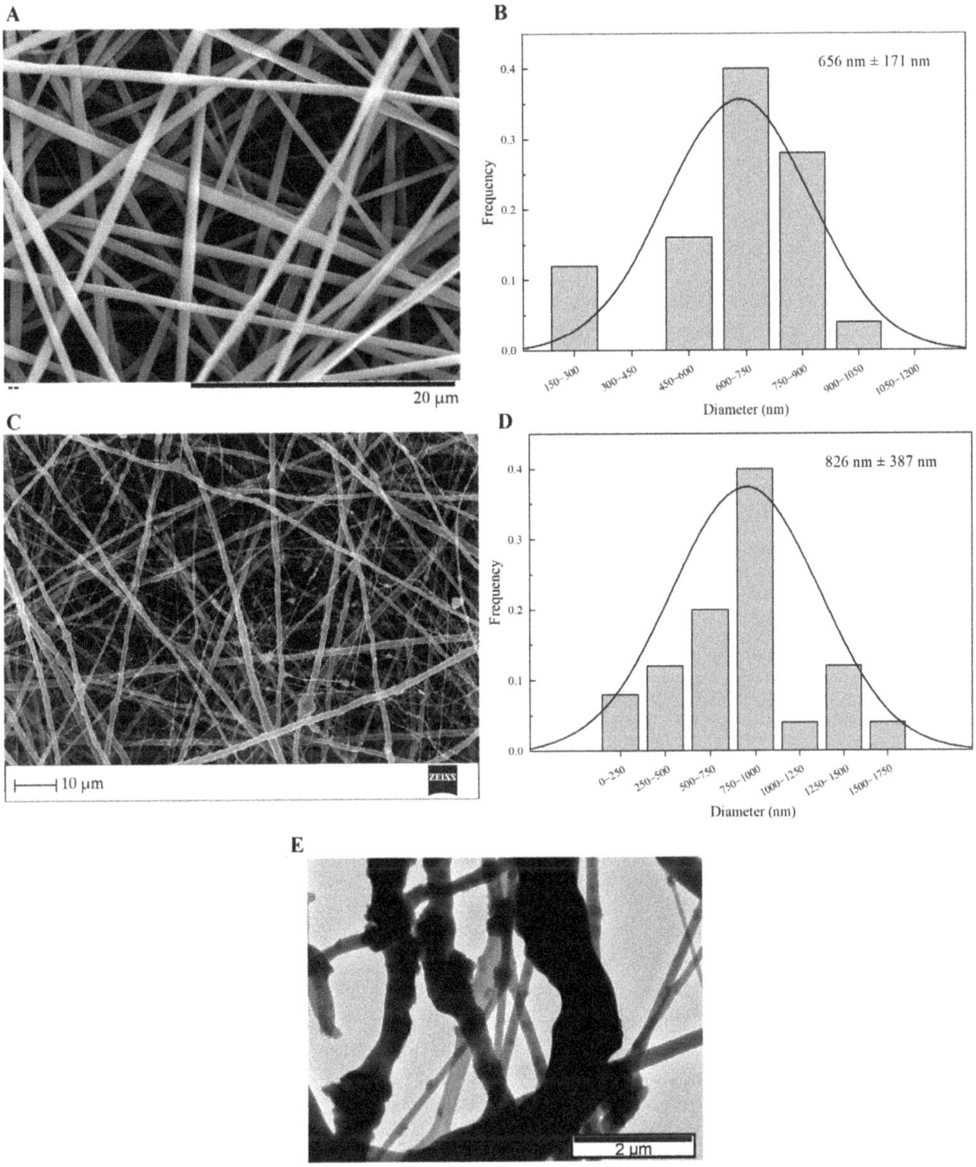

Figure 5. SEM images and the respective fiber diameter distribution graph of plain PVP membrane (**A**,**B**), and PVP/MBG (1:1) composite (**C**,**D**). TEM image of PVP/MBG (1:1) composite (**E**).

The PVP/MBG composites were electrospun in the same conditions as plain PVP, except for the needle gauge (G23) to avoid clogging. The SEM image of PVP/MBG (1:1) composite and the respective fiber size distribution are represented in Figure 5C,D. It can be observed that the incorporation of MBG particles into the PVP fibers changes the fiber's surface. The plain PVP fibers have a smooth surface, which becomes rough and irregular when the MBG particles are incorporated into the fibers, as expected due to the particles' aggregates as shown in the TEM image (Figure 5E). Consequently, the average fiber diameter and standard deviation of PVP/MBG (1:1) composite fibers increase to 826 nm and 387 nm, respectively. However, composite fibers exhibited a relatively homogeneous distribution of embedded MBG 80S15 particles.

TGA analysis was performed on plain PVP, MBG 80S15, and composite electrospun mats PVP/MBG (1:1). The TGA and DTGA curves obtained are reported in Figure 6A,B. All samples show a slight initial mass loss (around 9%) at 125 °C, attributed to the loss of the physically adsorbed water. The thermogram curve for MBG 80S15 previously sintered at 700 °C does not show another considerable mass loss, thus presenting a residual mass of 86.4% at 900 °C. Meanwhile, the thermogram curve of the plain PVP membrane exhibited a steep loss in weight between 400 and 480 °C, and there was no significant residual mass left after heating to 900 °C, indicating complete degradation of the polymer. In contrast, for PVP/MBG (1:1) nanocomposite fiber mats, a weight loss of 58.6% was observed within the tested temperature range. This weight loss is associated with water removal and PVP degradation. The residual mass of 41.4% corroborated the presence of MBG in the nanocomposite mats. The residual mass is consistent with the theoretical amount of MBG content in the composite, considering the initial amount introduced in the electrospinning solution. These results, along with the SEM and TEM analyses, provided confirmation that composite structures were effectively prepared in the present study.

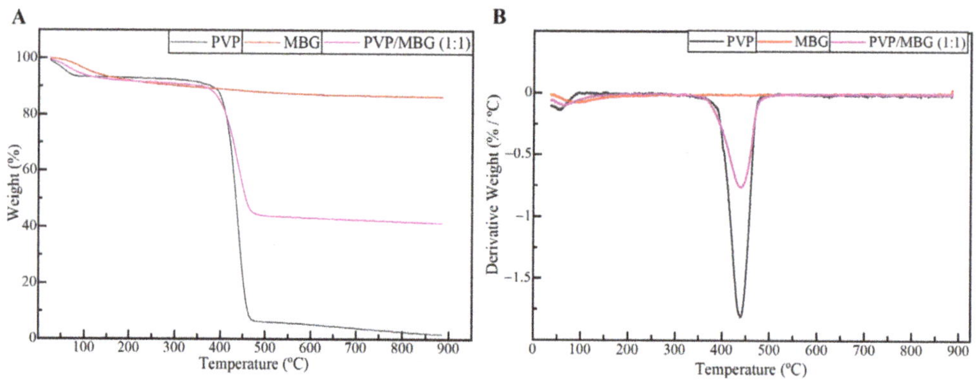

Figure 6. TGA (**A**) and DTGA (**B**) curves of plain PVP (black), PVP/MBG (1:1) composite (purple), and MBG 80S15 calcined at 700 °C (red).

3.3. *Swelling and Degradation Assays*

3.3.1. Thermal Crosslinking

PVP is water soluble, thus the as-spun nanofibers need to be crosslinked to make them water resistant and suitable for bone tissue engineering (BTE). Thermal treatment was found to be useful as a crosslinking agent of polymer chain segments, as it is easy to apply. Several crosslinking conditions were tested depending on the temperature applied (range: 150–200 °C) and time (range: 3–24 h) [28,35]. However, at a temperature of 165 °C, the membranes showed the most promising results in terms of reducing PVP solubility by keeping the overall nanofiber structure without burning. This temperature was kept constant, and different crosslinking times were tested as shown in Figures 7 and 8.

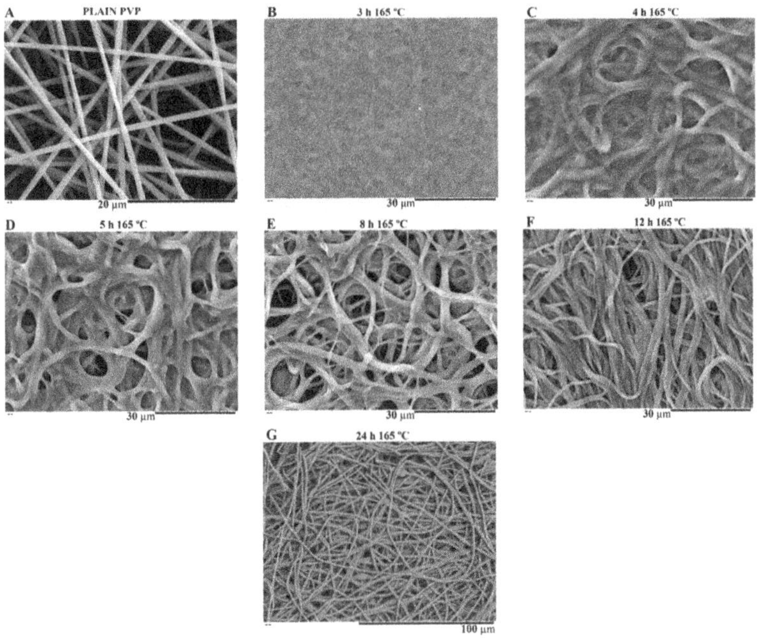

Figure 7. SEM image of electrospun plain PVP before soaking into PBS (**A**), and SEM images of plain PVP thermally crosslinked at 165 °C for 3 h (**B**), 4 h (**C**), 5 h (**D**), 8 h (**E**), 12 h (**F**) and 24 h (**G**) after soaking in PBS for 1 day.

Figure 8. SEM image of electrospun PVP/MBG (1:1) composite before soaking in PBS (**A**) and SEM images of PVP/MBG (1:1) composites thermally crosslinked at 165 °C for 3 h (**B**), 4 h (**C**), 5 h (**D**), 8 h (**E**), 12 h (**F**) and 24 h (**G**) after soaking in PBS for 1 day.

SEM images of Figure 7 show the electrospun plain PVP before and after soaking in PBS for 24 h, depending on previous crosslinking time. Figure 8 shows a similar study for PVP/MBG (1:1) composites, to compare how the MBG particles incorporation affects the thermal crosslinking. Both pure PVP and PVP/MBG (1:1) thermally crosslinked for 3 h appear to lose their fibrous structure and form a kind of rough film without porosity. However, several differences were noticed: the fibrous structure of the plain PVP membranes crosslinked for 4 and 5 h was not completely lost but appears as if the adjacent fibers have fused together and are significantly reduced in porosity, whereas the composites crosslinked for 4 and 5 h completely lose their fiber structure. In the case of the membranes crosslinked for 8 h, the fibrous structure of the membranes is still observed in both plain PVP and PVP/MBG (1:1), but the thickening of the fibers was clearly observed. In general, as crosslinking time increases, most of the morphological characteristics were observed to remain unchanged. The results also show that MBG nanoparticles' incorporation into the PVP fibers also reduces the degree of composite crosslinking for the same crosslinking time and temperature when compared to pure PVP. The MBG incorporation yielded thicker and more heterogeneous fibers; thus, the effect of the temperature on PVP crosslinking can be slightly affected due to the filtering effect [36].

3.3.2. Degradation Assay

The rate at which a scaffold degrades is an essential parameter in bone tissue engineering, as it should match the rate of ECM neogenesis. The duration necessary for bone recovery varies based on the fracture's shape and location, the condition of the adjacent soft tissues and individual factors such as species, age, general health and concurrent illnesses or injuries. In general, bone healing time ranges from 6 weeks to 6 months [37–39]. Therefore, in vitro biodegradation was studied by measuring weight loss of fiber mats in PBS at 37 °C for 4 weeks. Figure 9A represents the degradation ratio of PVP/MBG composites thermally crosslinked at 165 °C for 24 h with different amounts of MBG incorporated into the PVP fibers, ranging from 10 to 100% of the polymer's mass. Pure PVP (also described as plain PVP) was used as a control and revealed no mass loss during the assay, while the composite with a weight percentage of MBG equal to 100% of PVP mass (also mentioned as PVP/MBG (1:1)) revealed a mass loss around 30% after 28 days. Moreover, the data show that scaffold mass loss increases with MBG content for the same soaking time in PBS. This indicated that the inclusion of MBG particles accelerated the degradation rate of the fibrous mats. It was explained as being due to the high affinity of the MBG 80S15 with water, which increased the ability of the composites to absorb and retain water from the medium during the incubation periods and, consequently, raised the hydrolytic degradation of PVP fibers [40]. The MBG dissolution over time also explains the increasing mass loss of composites with higher MBG contents.

Figure 9B,C represent the degradation assay for electrospun plain PVP membranes and PVP/MBG (1:1) composites, respectively. The samples were thermally crosslinked at 165°C for different times: 3, 4, 5, 8, 12 and 24 h. Generally, polymer degradation kinetics is affected by chemical and structural characteristics. In general, both graphics show that the degradation ratio decreases as the crosslinking times increase, except for the PVP/MBG (1:1) composites crosslinked for 8, 12 and 24 h, which presented similar degradation ratios over time. This behavior was expected because thermal crosslinking can enhance the thermal and chemical stabilities of PVP to form a stable network structure [41]. However, under the same thermal crosslinking conditions, the incorporation of MBG accentuated the degradation ratio compared to pure PVP. Furthermore, membranes with a low degree of crosslinking quickly led to the high exposure of MBG particles to the medium. Consequently, the particles could be separated from PVP fibers, thus increasing the degradation ratio even further. The degradation ratio data of PVP/MBG (1:1) thermally crosslinked for 3 h is not shown in the graph because the samples partially lose their physical integrity. It is worth pointing out that all samples studied showed a fast degradation ratio during the first 2 h, then degradation was very slow or reached a plateau.

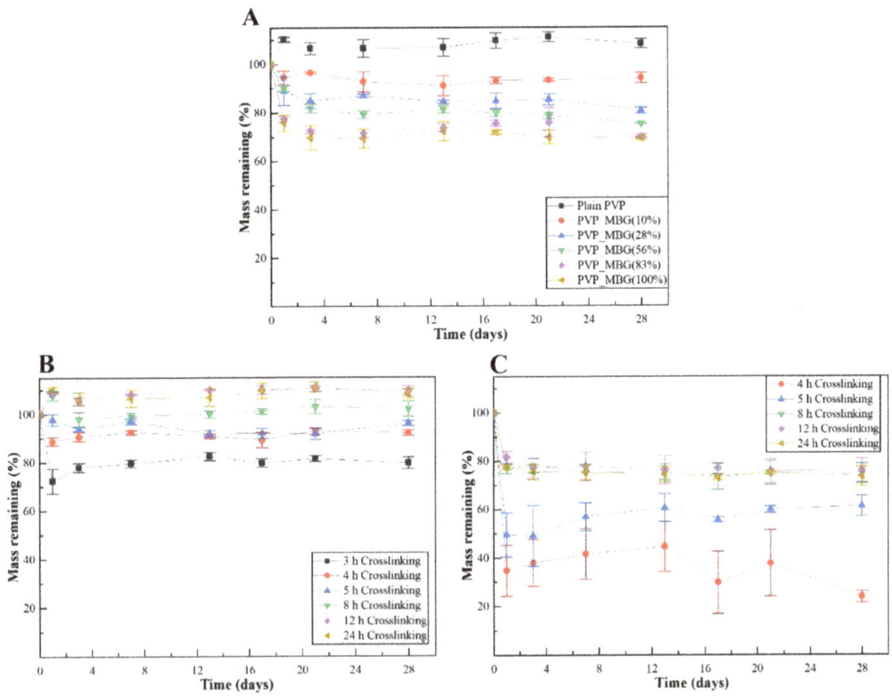

Figure 9. Degradation assay for PVP/MBG composites (thermally crosslinked at 165 °C for 24 h) with different MBG 80S15 content: 0, 10, 28, 56, 83 and 100% of PVP mass (**A**); degradation assay for electrospun PVP membranes (**B**) and PVP/MBG (1:1) composites (**C**) thermally crosslinked at 165 °C for different times: 3, 4, 5, 8, 12 and 24 h.

In general, the degradation ratios studied are in agreement with the overall structural changes of the samples after 24 h of immersion in PBS (Figures 7 and 8).

3.3.3. Swelling Assay

Figure 10A,B show the swelling assay for different PVP/MBG composites depending on MBG content and thermal crosslinking duration, respectively. For each of the membranes studied, the water absorption continues to increase until equilibrium is reached (approximately 24 h of incubation). Plain PVP was used as a control and its swelling ratio was calculated at around 500%, whereas the data show as the amount of incorporated MBG increases, the swelling ratio also increases. The PVP/MBG (1:1) composite had the highest swelling ratio around 750–800%. The hydrophilic character of PVP, the high surface area of the matrices and the presence of MBG nanoparticles inside the fibers are factors that positively affect the swelling capability of the scaffolds since it is dependent on the degree of water uptake by pure polymer or composites. On the other hand, the PVP/MBG (1:1) composites crosslinked for 8, 12, and 24 h showed different swelling capacities (Figure 10B). As discussed above, Figure 8E–G illustrate the respective composites after 24 h of immersion in PBS. It can be noticed that PVP fibers reduce the structural changes after being immersed in PBS as the crosslinking time increases, thus the water uptake from the medium and the capacity to retain such water reaches higher levels. Another study also reported the degree of swelling of polyester elastomers is an important parameter in characterizing their crosslinking degree [42]. Scaffolds with swelling capacity are highly desirable, particularly for drug delivery purposes where the drug can be trapped in the polymeric composite material and then released in a controlled manner.

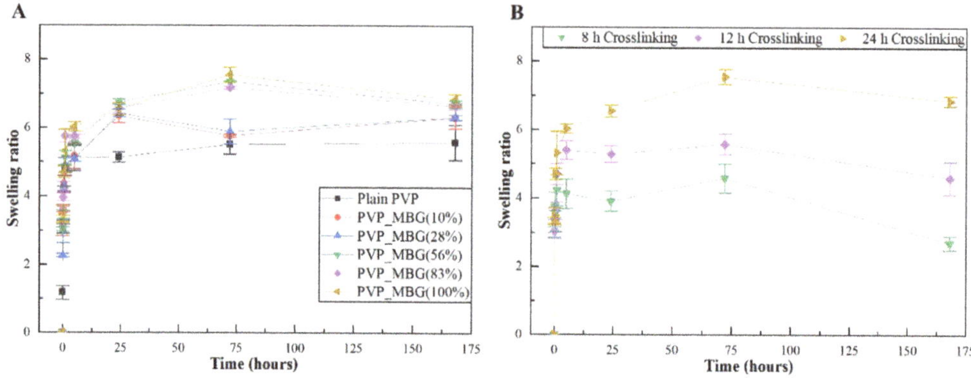

Figure 10. Swelling assay for PVP/MBG composites (thermally crosslinked at 165 °C for 24 h) with different MBG 80S15 content: 0, 10, 28, 56, 83, and 100% of PVP mass (**A**); Swelling assay for PVP/MBG (1:1) composites crosslinked for different times: 8, 12, and 24 h (**B**).

3.4. Mechanical Response—Tensile Tests

The different membranes of PVP and PVP/MBG (1:1) composites thermally crosslinked at 165 °C for 8, 12, and 24 h were submitted to uniaxial stress–strain tests as shown in Figure 11. These results allow us to investigate the effect of thermal crosslinking and MBG nanoparticle incorporation on the mechanical properties of the electrospun matrices.

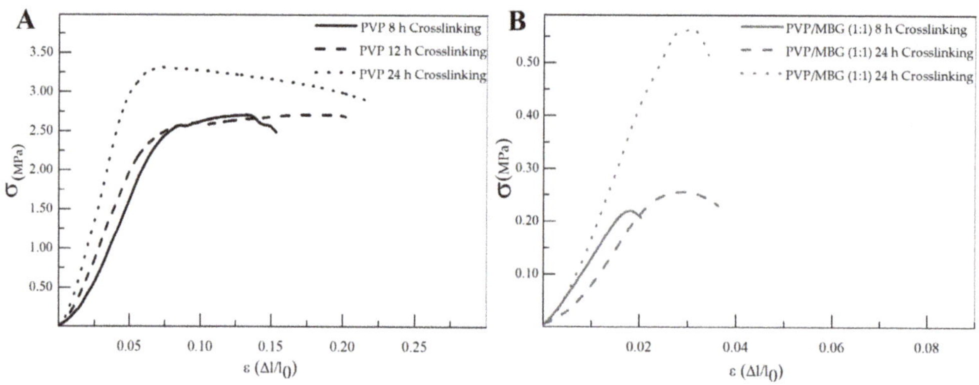

Figure 11. Representative stress–strain curves of plain PVP (**A**) and PVP/MBG (1:1) composite membranes (**B**) crosslinked for different times: 8, 12 and 24 h.

The comparison between plain PVP (Figure 11A) and PVP/MBG (1:1) (Figure 11B) stress–strain curves shows that mechanical properties were significantly modified with the MBG nanoparticles incorporation. With the incorporation of MBG nanoparticles, the composite membranes exhibit minimal plastic deformation as their rupture occurs almost immediately after the elastic region. This behavior is characteristic of brittle materials. The following three mechanical parameters that characterize the membranes were obtained from the stress–strain curves: Young's modulus (E); the yield strength (σ_{YS}); and the ultimate tensile strength (UTS). The results for each membrane are shown in Table 2 and corroborate the changes in the mechanical properties. The presence of MBG in the PVP fibers implies a significant reduction in the three mechanical parameters calculated. The Young's modulus decreases by a factor of around 3 up to 3.5 times with the presence of MBG nanoparticles. This decrease along with the reduction in UTS values could be justified by the increase in the inhomogeneity in average fiber diameter, as seen in the SEM and TEM images of the composites. In fact, the poor dispersion of MBG nanoparticles at high concentrations leads to the formation of clusters, which decreases the homogeneity of PVP fibers. A higher concentration of MBG is favorable for maintaining the bioactivity of the scaffolds, but it may also contribute to the brittle mechanical properties observed in the composite samples. Reducing the amount of MBG particles inside the composite matrices leads to a trade-off between high mechanical properties and bioactivity. However, the high bioactivity of the composite fibers in necessary for BTE applications [43,44]. In fact, previous studies have indicated that there exists a certain threshold concentration of bioactive glass particles beyond which the nanoparticles can cause defects instead of reinforcing the polymer matrices, resulting in increased brittleness of the nanocomposites [40].

Table 2. Calculated mechanical parameters for each membrane type: Young's modulus (E), ultimate tensile strength (UTS) and yield strength (σ_Y).

Sample	E (MPa)	UTS (MPa)	σ_Y (MPa)
PVP 8 h crosslinking	39.2 ± 10.9	2.45 ± 0.50	2.77 ± 056
PVP 12 h crosslinking	45.7 ± 14.1	2.27 ± 0.35	2.75 ± 0.44
PVP 24 h crosslinking	70.0 ± 16.2	3.01 ± 0.28	3.39 ± 0.38
PVP/MBG (1:1) 8 h crosslinking	11.9 ± 3.4	0.22 ± 0.07	0.23 ± 0.05
PVP/MBG (1:1) 12 h crosslinking	13.2 ± 2.5	0.23 ± 0.03	0.26 ± 0.05
PVP/MBG (1:1) 24 h crosslinking	22.9 ± 3.9	0.58 ± 0.09	0.59 ± 0.07

The time of thermal crosslinking also affected the mechanical properties of the samples. In general, the changes were very similar for plain PVP and PVP/MBG (1:1) composites. For instance, the plain PVP samples crosslinked for 8 and 12 h exhibited similar mechanical behavior under tensile forces, which also occurred for PVP/MBG (1:1) composites crosslinked for 8 and 12 h. Furthermore, when the samples were crosslinked for 24 h, the mechanical parameters reached higher values. The Young's modulus increases in both PVP and PVP/MBG (1:1) composites by a factor of around 1.8 up to 1.9 when the crosslinking time increases from 8 to 24 h. In general, it is expected that crosslinking processes affect the mechanical properties of polymers since new interactions between polymer chains are being created. These results show that longer crosslinking time is responsible for enhancing the mechanical strength of the samples. Thus, for periods of 8 and 12 h at 165 °C, the crosslinking process is not yet complete, while 24 h of thermal crosslinking means higher bonding levels between the PVP chains.

3.5. Cytotoxicity Assays

Cytotoxicity assays were performed to evaluate the cytotoxic effect of MBG powder, electrospun PVP and PVP/MBG (1:1) composites using the extract method. Figure 12 shows cell viability after incubation with 20 mg·mL^{-1} of each type of membrane studied and the respective two-fold dilutions. In general, the results show an absence of cytotoxic effects in Saos-2 cells, except for plain PVP membrane thermally crosslinked at 165 °C for 8 h and an extract concentration of 20 mg·mL^{-1}. This may conflict with some researchers who have reported an increasing interest in PVP towards its use in biomedical applications due to its inertness, chemical stability and biocompatibility [22]. To understand these results, it is worth pointing out that PVP was previously thermally crosslinked. At elevated temperatures, the stability of PVP decreases and leads to the opening of pyrrolidone ring fractions. This phenomenon results in the transfer of hydrogen atoms from one radical to another, which leads to the formation of double bonds by hydrogen donor and saturation of acceptor by the crosslinking process [45]. Furthermore, it can be found in the literature that the biocompatibility of PVP-based materials depends on PVP content and on changes in pyrrolidone rings, which may lead to the consequent loss of biocompatibility [46]. There are few reports about the thermal crosslinking process of PVP, and the effect of MBG nanoparticle incorporation on the thermal behavior of electrospun PVP fibers. However, the results presented here may lead to the hypothesis that electrospun PVP matrices were not fully crosslinked at 165 °C for 8 h (in concordance with tensile tests), which may indicate the presence of open pyrrolidone rings, consequently decreasing its biocompatibility. However, after a two-fold dilution of the extract concentration to 10 mg·mL^{-1}, cell viability rises to 100%. This is in concordance with the cell viability of the PVP/MBG (1:1) composite crosslinked in the same conditions of the above-mentioned PVP membrane (165 °C for 8 h). The concentration of 20 mg·mL^{-1} for this composite means an individual concentration equal to 10 mg·mL^{-1} of both PVP and MBG 80S15. For this concentration, both individual materials revealed no cytotoxic effects; thus the PVP/MBG (1:1) composite also revealed the absence of cytotoxicity.

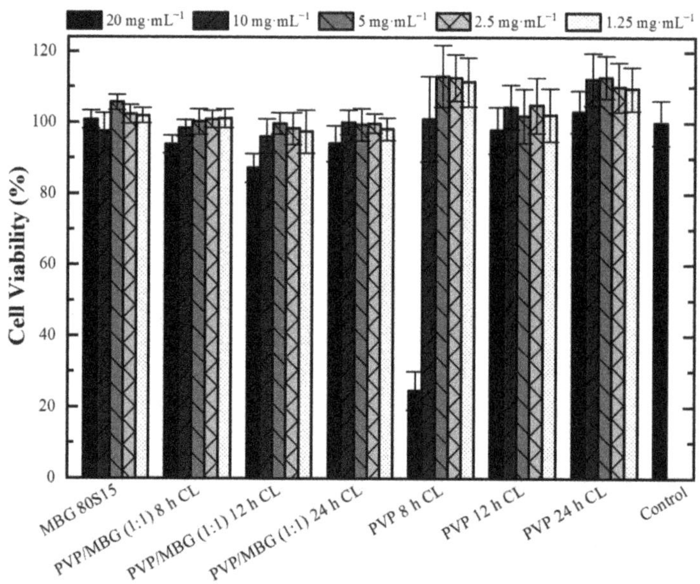

Figure 12. Osteoblast cell line viability after 48 h exposure to diverse types of PVP membranes through indirect method (CL stands for crosslinking). Data is expressed as average ± standard deviation for at least five independent experiments.

3.6. Bioactivity Assay

The bioactivity of MBG 80S15 pellets and PVP/MBG (1:1) composites was investigated in vitro by soaking the samples in SBF to detect the formation of HA on the surface. The samples were examined by SEM, XRD and FTIR.

Figure 13 shows SEM images of MBG 80S15 before and after different immersion intervals in SBF. Before soaking in SBF solution (Figure 13A), the MBG particle's surface was smooth. After soaking for 24 h, the surfaces of the MBG pellets showed important changes. The surface appeared to be completely covered by a layer consisting of crystallite structures with needle-like shapes. EDS analysis confirmed that the chemical composition can be assigned to an HCA phase, given a Ca/P ratio slightly lower than 1.67 (Table 3). Furthermore, the thickness of the HCA layer continued to grow over time. After 5 days of immersion, cauliflower-like clusters with needlelike crystallites formed spontaneously to reduce surface energy. These clusters grew over time, so that after MBG soaking in SBF for 10 days the surface was completely covered in cauliflower-like clusters.

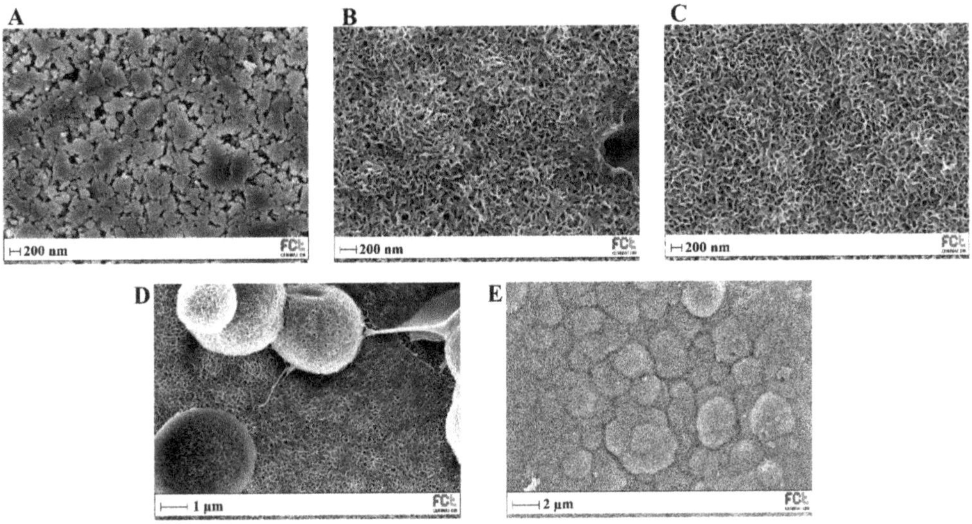

Figure 13. SEM images of MBG 80S15 before (**A**) and after soaking in SBF for 1 day (**B**), 3 days (**C**), 5 days (**D**) and 10 days (**E**).

Table 3. Atomic concentration (At. %) of silicon (Si) and respective Ca/P ratio during the bioactivity assay of the different types of membrane studied and MBG 80S15 pellets before (control) and after soaking in SBF for 1, 3, 5 and 10 days. These results were obtained from EDS analysis (CL stands for crosslinking).

EDS (At. %)	Control		1 Day		3 Days		5 Days		10 Days	
Sample	Ca/P	Si	Ca/P	Si	Ca/P	Si	Ca/P	Si	Ca/P	Si
MBG 80S15 pellet	1.16	75.39	1.60	69.76	1.59	9.01	1.52	8.46	1.58	2.24
PVP/MBG (1:1) 8 h CL	1.49	73.59	1.48	89.89	0.75	85.96	1.23	34.48	1.64	0.60
PVP/MBG (1:1) 12 h CL	1.49	73.59	1.05	92.32	1.13	36.11	1.58	31.73	1.83	5.75
PVP/MBG (1:1) 24 h CL	1.49	73.59	1.43	85.69	1.47	77.99	3.72	85.66	1.22	66.24

The XRD results of the MBG 80S15 pellets soaked in SBF for 1, 3, 5 and 10 days are shown in Figure 14A. As mentioned before, MBG showed the common broad band of amorphous materials at about 15–35° (2θ) before soaking in SBF. After 1 day of immersion in SBF, the XRD pattern of MBG 80S15 showed a major peak located at 32° (2θ) and another peak located at 26° (2θ), which are assigned to the (211) and (002) reflection of an apatite-like phase [14]. After soaking for 3 days, the four new diffraction peaks at 39, 46, 49 and 53° (2θ) were assigned to the (310), (222), (213) and (004) reflections of HCA, respectively. As the immersion and mineralization time increased to 5 days, the diffraction peaks became more intense. The above-mentioned newly formed diffraction peaks are in accordance with the standard card of HCA (JCPD 24-0033). The XRD results demonstrate that MBG 80S15 has good bioactivity in vitro and its mineralization product is HCA. Furthermore, the crystallinity of HCA became higher as immersion time increased [1].

The growth of HCA on MBG 80S15 surface was further investigated by FTIR spectroscopy analysis. Figure 14B illustrates the IR spectra for MBG 80S15C samples after the same time periods soaked in SBF that were studied by SEM. Before immersion in SBF, the adsorption bands at 800 and 1040 cm^{-1} can be indexed to Si-O-Si bonds. However, when soaking in SBF for 3 days, three P–O crystalline vibrational bands near to 568, 602 and 963 cm^{-1} were detected (the noise near 500 cm^{-1} hindered the accurate analysis of the 568 cm^{-1}

band). Furthermore, three other new vibrational bands were detected near to 1456, 1414 and 870 cm^{-1}, which are assigned with C-O bond indicating the growth of the HCA layer on the surface of MBG 80S15 [47–49].

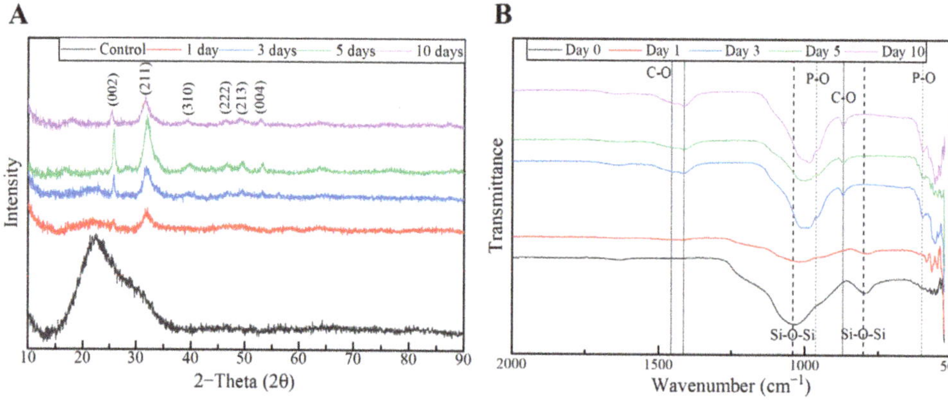

Figure 14. XRD patterns (**A**) and FTIR spectra (**B**) of MBG 80S15 after soaking in SBF for different periods of time.

The above SEM, XRD and FTIR results indicate that the MBGs 80S15 can induce the formation of an HCA layer on their surface even for short SBF soaking periods, which corroborates the superior in vitro bone forming bioactivity, pointing out the interesting properties of the MBG 80S15 as a material for bone tissue engineering.

The ability of a biomaterial to form an HCA layer on its surface when exposed to SBF is often used to estimate its bone-bonding potential. In order to study the formation of this layer, the PVP/MBG (1:1) composite scaffolds were also analyzed by SEM, XRD and FTIR before and after being soaked in SBF. Furthermore, the bioactive behavior of the composite scaffolds is also dependent on the degradation ratio of the polymeric matrices, which in turn is dependent on the crosslinking conditions. To study the best crosslinking conditions considering the effect on enhancing the bioactivity of the scaffolds, the bioactivity assay was carried out on scaffolds crosslinked at 165 °C for 8 h (Figure 15), 12 h (Figure 16) and 24 h (Figure 17). SEM analysis reveals the surface morphology of the scaffolds before and after soaking in SBF for 1, 3, 5 and 10 days. After immersion for 1 day, the surface of the samples crosslinked for 8, 12 and 24 h becomes rougher. The samples' porosity decreases when compared with the samples before immersion in SBF: this effect is more accentuated for the samples crosslinked for 8 and 12 h. Several precipitates were clearly observed on the nanocomposite surface over the immersion period. However, this effect is delayed in samples crosslinked for 24 h, probably due to the slower degradation of the PVP fibers hindering contact between the MBG glass particles and the medium. The bioactivity of the scaffolds can be clearly seen in the samples crosslinked for 8 and 12 h with the micrographs showing the spherical structures associated with the typical cauliflower-like structure. However, there are PVP fibers that have not completely degraded yet. Figure 15E and the respective inset micrograph show that HCA growth stretches the PVP fibers. This phenomenon can accelerate the degradation rate of the scaffolds when compared with the degradation rate studied for immersion in PBS. It is worth pointing out that the above-mentioned makes it impossible to evaluate the bioactivity of composites crosslinked for 3, 4 or 5 h since the samples did not keep their physical integrity when soaked in SBF.

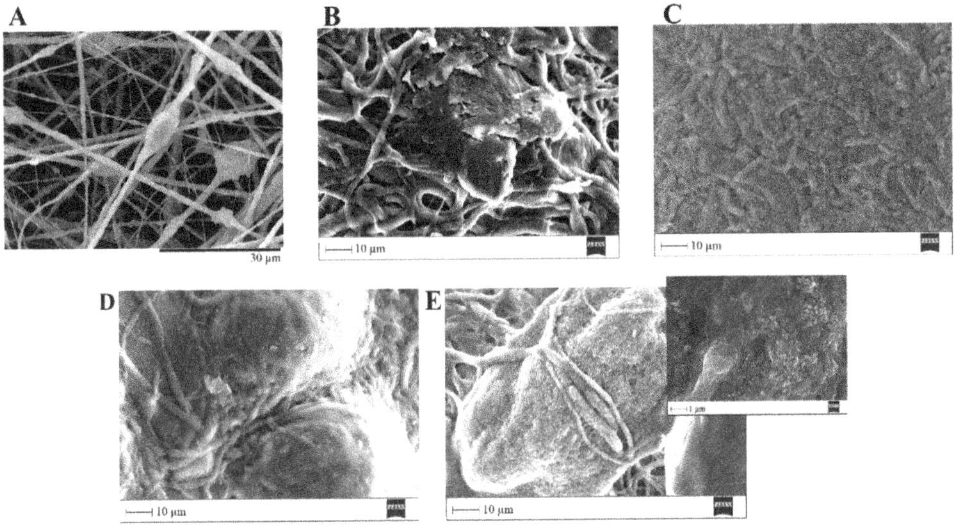

Figure 15. SEM images of PVP/MBG (1:1) composite crosslinked for 8 h before (**A**) and after soaking in SBF for 1 day (**B**), 3 days (**C**), 5 days (**D**) and 10 days (**E**) and respective inset micrograph at higher magnification.

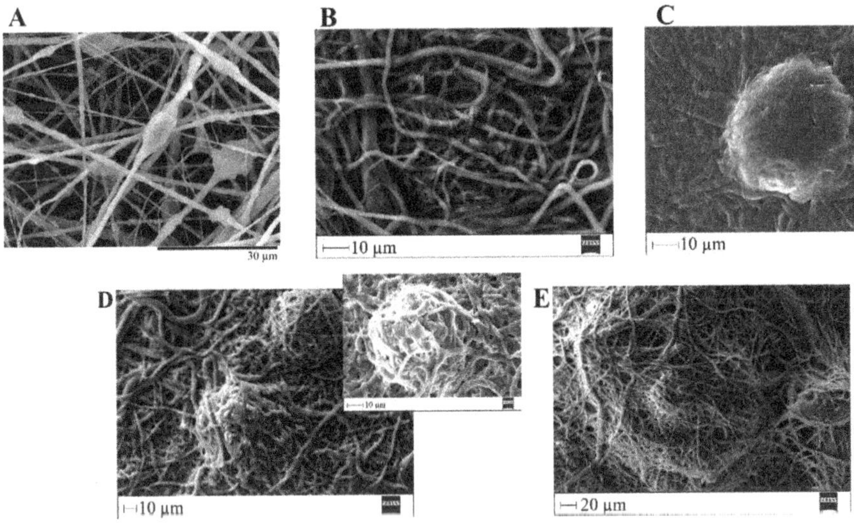

Figure 16. SEM images of PVP/MBG (1:1) composite crosslinked for 12 h before (**A**) and after soaking in SBF for 1 day (**B**), 3 days (**C**), 5 days (**D**) and respective inset micrograph and 10 days (**E**).

EDS analysis (Table 3) of the nanofibrous mats surface before and after soaking in SBF allowed the determination of the atomic concentration of silicon, calcium and phosphorus, respectively. In general, there was a decline in silicon concentration while calcium and phosphorus concentrations increased over time for all the samples. Moreover, the earliest apatite precipitate is usually "Ca-deficient", exhibiting a Ca/P ratio below 1.67. Throughout the HCA layer formation, the HA fraction theoretically increases, and the Ca/P ratio also increases asymptotically towards Ca/P $\approx$ 1.67 (for the times studied this observation is not obvious through EDS analysis) [50].

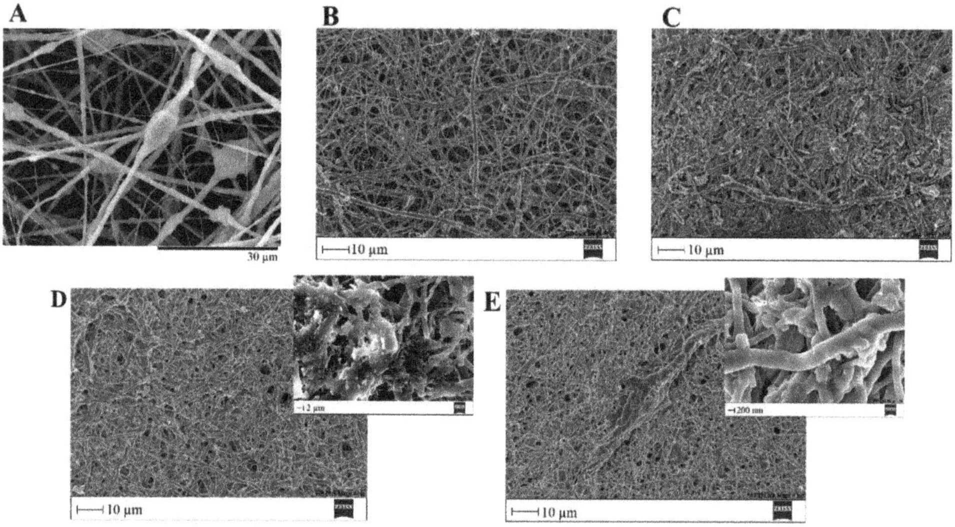

Figure 17. SEM images of PVP/MBG (1:1) composite crosslinked for 24 h before (**A**) and after soaking in SBF for 1 day (**B**), 3 days (**C**), 5 days (**D**) and respective inset micrograph and 10 days (**E**) and respective inset micrograph.

In general, the SEM analysis of the bioactivity assay of PVP/MBG (1:1) composites revealed quite similar results considering the HCA precipitation for samples crosslinked for 8 and 12 h, while the HCA precipitation on fibers' surface seems to be delayed for the samples crosslinked for 24 h.

To confirm the formation of the HCA layer on fiber's surface, the PVP/MBG (1:1) composites thermally crosslinked for 8, 12 and 24 h were also analyzed by XRD and FTIR before and after being soaked in SBF. XRD patterns are illustrated in Figure 18A,C,E. All samples exhibited no diffraction peaks before soaking in SBF, as expected due to the amorphous structure of both MBG and the PVP. However, significant changes in the XRD patterns occur after immersion in SBF for the three types of composite membranes. The samples crosslinked for 8 h revealed different diffraction peaks as the immersion time increased at $2\theta = 26°, 32°, 40°, 46°, 49°, 53°$ and $57°$, which correspond to the reflection of crystalline planes with the Miller indices (002), (211), (310), (222), (213), (004) and (322), respectively. In the case of composites crosslinked for 12 h, as the SBF immersion time increased, the XRD patterns show diffraction peaks at $2\theta = 32°, 46°, 53°$ and $57°$, which correspond to the reflection of crystalline planes with the Miller indices (211), (222), (004) and (322), respectively. Lastly, in the above-mentioned conditions, the XRD patterns of the composites crosslinked for 24 h show diffraction peaks at $2\theta = 26°, 32°, 46°$ and $57°$, which correspond to the reflection of crystalline planes with the Miller indices (002), (211), (222) and (322), respectively. As previously discussed for MBG 80S15 bioactivity assay, the newly formed diffraction peaks for the three types of polymeric composites studied are in concordance with the standard card of HCA (JCPD 24-0033).

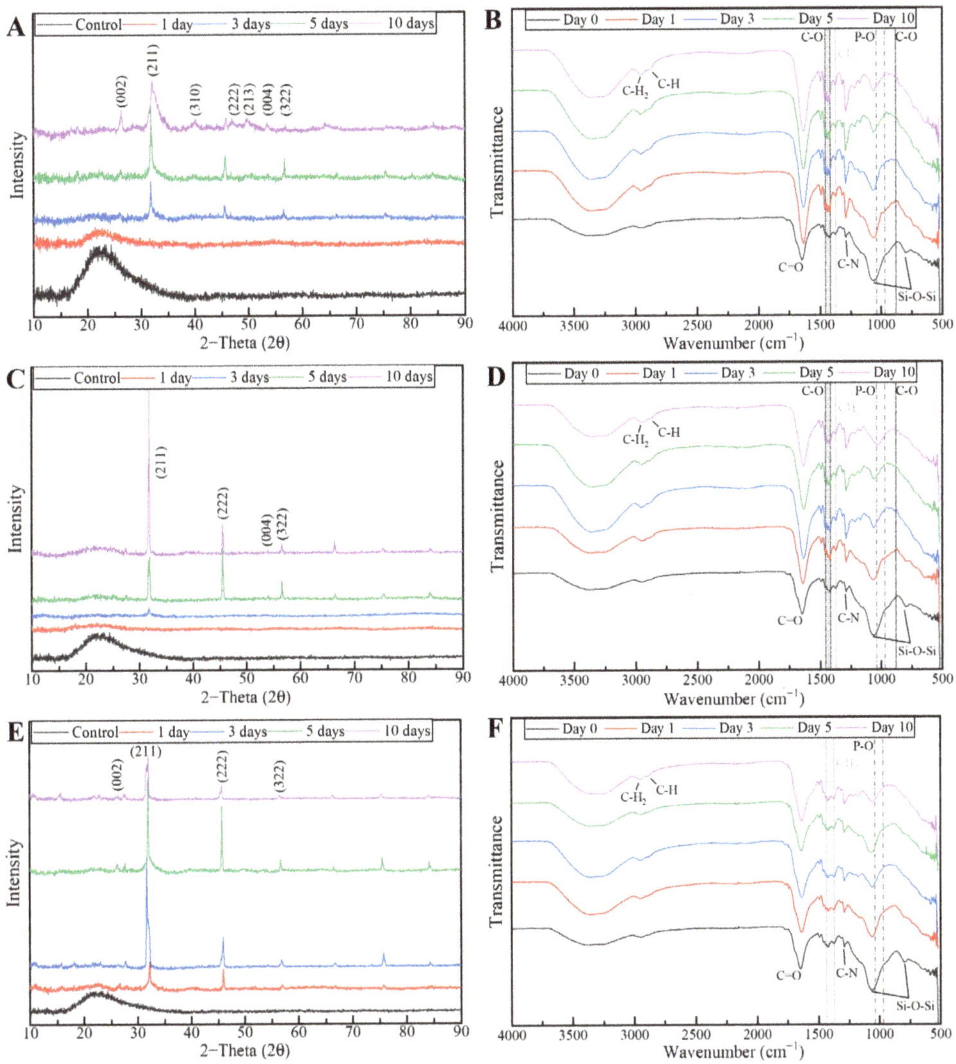

Figure 18. XRD patterns and FTIR spectra of PVP/MBG (1:1) composites crosslinked for 8 h (**A**,**B**), 12 h (**C**,**D**) and 24 h (**E**,**F**) after soaking in SBF for different periods of time: 1, 3, 5 and 10 days.

These XRD results are in accordance with FTIR analysis (Figure 18B,D,F), revealing the formation of the HCA structure. Starting with the FTIR spectrum of PVP/MBG (1:1) before immersion in SBF, the following absorption bands are assigned to PVP: the band located around 1648 cm^{-1} can be ascribed to the stretching vibration of the C=O in the pyrrolidone group; the broad band near to 2850–3000 cm^{-1} is associated with CH stretching modes, which can be assigned to five overlapping signals: asymmetric CH_2 stretching (polymer chain: 2983 cm^{-1}, ring: 2954 cm^{-1}), symmetric CH_2 stretching (polymer chain: 2919 cm^{-1}; ring: 2885 cm^{-1}) and ternary CH (2852 cm^{-1}); the bands at 1427 cm^{-1} and 1372 cm^{-1} also correspond to the CH deformation modes from the CH_2 group; the absorption bands at 1288 cm^{-1} are related to C–N bending vibration from the pyrrolidone structure [51]. The absorption bands at 470 (not seen in the graphs due to noise), 802 and 1058 cm^{-1} can be assigned to Si-O-Si bonds of the MBG 80S15. After immersion in SBF, the membranes crosslinked for 8 and 12 h showed similar FTIR spectra over time: the P–O symmetric stretching vibrational bands near 970 cm^{-1} were detected (the noise near 500 cm^{-1} hindered the accurate analysis of the other two characteristics bands of the P–O bond: 568 and 600 cm^{-1}, associated with bending modes); after 5 days of immersion in SBF, the vibrational band near 970 cm^{-1} was no longer detected due to the shift of the 1058 cm^{-1} band assigned to Si-O-Si band. This shift was related to the overlap of a new band at 1036 cm^{-1}, which is associated with the asymmetric stretching of P–O bond [52]; three vibrational bands near 1458, 1417 and 880 cm^{-1} were also detected and assigned to the C-O bond. However, it was noticed that the bands near 1458 and 1417 cm^{-1} increased in intensity over immersion time, while the band at 880 cm^{-1} revealed only a slight change in intensity.

The above-mentioned changes in the FTIR spectra were not detected for the samples crosslinked for 24 h, except for the shift associated with the new band at 1036 cm^{-1} after 10 days after immersion in SBF. The FTIR analyses are in accordance with the SEM results, indicating the growth of the HCA layer over time, which seems to be delayed for the scaffolds crosslinked for 24 h.

Hench has reported a detailed analysis of the reactions involved in the five-stage mechanism of apatite formation upon contact of BGs with SBF. The five stages include: (1) rapid ion exchange of alkali ions with hydrogen ions from the liquid medium, (2) dissolution of the glass network, (3) polymerization of silica gel and (4 and 5) chemisorption and crystallization of the carbonated hydroxyapatite layer [53]. Stages (1) and (2) explain the fading of the band at 802 cm^{-1} (Si-O-Si), which was noticed in all the scaffolds studied. Furthermore, stages 1 and 2 strongly depend on PVP degradation rate, thus allowing us to state that crosslinking directly affects the bioactivity of the polymeric composite scaffolds studied. On the other hand, as shown in the swelling assay, the high wettability of PVP may favor SBF interaction with MBG nanoparticles, allowing apatite mineralization inside PVP fibers and subsequent growth in the radial direction [54].

The findings of this study are consistent with previous research that has demonstrated that the inclusion of BG particles in polymeric matrices enhances the in vitro formation of hydroxyapatite on the surface of nanocomposites when exposed to an SBF medium. For instance, Lin et al., [55], Allo et al., [56] and Liverani et al. [43] performed different studies and achieved similar conclusions. The researchers found that the incorporation of bioactive glasses into PCL [55,56] and PCL/chitosan [43] nanofibrous matrices significantly enhanced their apatite-formation ability in SBF compared with individual polymeric membranes. In addition, according to Han et al. [57], composite nanofibers demonstrated superior bioactivity compared to pure PAN-based carbon nanofibers. Additionally, Yang et al. [58] observed that the addition of BG nanoparticles to carbon nanofiber composites increased the rate of heterogeneous apatite nucleation [54].

4. Discussion and Conclusions

In general, the main goals proposed here have been achieved, confirming the hypothesis put forward at the beginning of this study. The composites produced can be used as bone scaffolds considering their inherent properties.

The characterization of the MBG 80S15 produced revealed the desired properties for BTE, i.e., non-toxic effects on the Saos-2 cell line and enhanced bioactivity considering the rapid HCA layer formation within the first 24 h after immersion in SBF. This is directly related to the high surface area of the MBG 80S15 powder due to its mesoporosity, thereby increasing MBG reactivity through the accelerated ion exchange with the medium.

The MBG 80S15 powder was successfully incorporated into the PVP-fibrous mats by means of electrospinning. The resulting composite materials were thermally crosslinked to reduce PVP water solubility, aiming to tailor the degradation ratio of PVP and PVP/MBG (1:1) mats. The literature reports the formation of some radicals in the temperature range of 150 to 200 °C, which can be attributed to the dissociation of methylene of the pyrrole ring or the methine of the main polymer chain. As a result, crosslinks in the polymer chains are formed [59,60].

In general, all PVP/MBG (1:1) composites thermally crosslinked at 165 °C for 8, 12 and 24 h showed HCA-forming ability. However, the impact of thermal crosslinking duration on scaffolds' bioactivity should be considered as previously discussed. When the scaffolds were thermally crosslinked at 165 °C for 8 and 12 h, the bioactivity displayed similar results, whereas the scaffolds crosslinked for 24 h showed slower reactivity. Future experiments should be carried out to study the relation between thermal crosslinking conditions and ion release from the composites to the medium, in order to compare the HCA layer formation.

These polymeric scaffolds also revealed the absence of cytotoxic effects on Saos-2 cells as intended for BTE applications. Moreover, the degradation ratio and swelling capacity of the resulting composites can be tailored by varying the amount of MBG powder incorporated or the crosslinking conditions. It is worth pointing out the importance of thermal crosslinking conditions due to the direct relationship between swelling capacity, degradation ratio and MBGs' ion exchange with the medium. Consequently, the bioactivity of the composites depends on crosslinking conditions.

The overall results indicated that the described electrospun nanocomposite fiber mats are promising bone scaffolds, as they combine the high bioactivity of MBG 80S15 nanosized powder and the benefit of the flexibility of electrospun PVP mats. The major hurdle to face is to combine bioactivity and mechanical strength, since the results showed a trade-off between bioactivity and mechanical strength in PVP/MBG (1:1) scaffolds. This effect resulted from the inhomogeneity of PVP nanofibers' diameter with the incorporation of large amounts of MBG powder, which tends to agglomerate and form clusters.

For future work, it would be interesting to study the properties of these composites using in vivo models, in order to assess the ability of the scaffolds to induce osteogenesis and angiogenesis. Moreover, it would be of great interest to encapsulate other biomaterials within PVP fibers, such as magnetic nanoparticles or drugs, in order to perform local treatment of bone disease through magnetic hyperthermia or drug delivery and simultaneously induce bone regeneration.

Considering the results presented, the scaffolds thermally crosslinked for 12 h showed the most promising results. These scaffolds showed rapid HCA layer formation, which was comparable to that obtained for scaffolds thermally crosslinked for 8 h. However, the composites crosslinked for 12 h are easier to handle, e.g., they keep their physical properties for longer periods when immersed in SBF solution. They also revealed high biocompatibility and suitable swelling capacity for further drug delivery applications combined with BTE.

Author Contributions: Conceptualization, R.J.R.M., P.I.P.S. and J.P.B.; methodology, R.J.R.M.; validation, P.I.P.S. and J.P.B.; formal analysis, R.J.R.M.; investigation, R.J.R.M.; writing—original draft preparation, R.J.R.M.; writing—review and editing, P.I.P.S., J.P.B. and J.C.S.; supervision, P.I.P.S., J.P.B. and J.C.S.; funding acquisition, P.I.P.S., J.P.B. and J.C.S. All authors have read and agreed to the published version of the manuscript.

Funding: The authors acknowledge the support of FCT, in the scope of the projects LA/P/0037/2020, UIDB/50025/2020 and UIDP/50025/2020 of the Associate Laboratory Institute of Nanostructures, Nanomodelling and Nanofabrication-i3N. Ricardo Matos also acknowledges FCT for the PhD grant with reference SFRH/BD/140090/2018.

Institutional Review Board Statement: Not applicable.

Informed Consent Statement: Not applicable.

Data Availability Statement: The data presented in this study are available on request from the corresponding author.

Conflicts of Interest: The authors declare no conflict of interest.

References

1. Chen, J.; Zeng, L.; Chen, X.; Liao, T.; Zheng, J. Preparation and characterization of bioactive glass tablets and evaluation of bioactivity and cytotoxicity in vitro. *Bioact. Mater.* **2018**, *3*, 315–321. [CrossRef] [PubMed]
2. Calori, G.M.; Mazza, E.; Colombo, M.; Ripamonti, C. The use of bone-graft substitutes in large bone defects: Any specific needs? *Injury* **2011**, *42*, 2. [CrossRef] [PubMed]
3. El-Rashidy, A.A.; Roether, J.A.; Harhaus, L.; Kneser, U.; Boccaccini, A.R. Regenerating bone with bioactive glass scaffolds: A review of in vivo studies in bone defect models. *Acta. Biomater.* **2017**, *62*, 1–28. [CrossRef] [PubMed]
4. Amini, A.R.; Laurencin, C.T.; Nukavarapu, S.P. Bone Tissue Engineering: Recent Advances and Challenges. *Crit. Rev. Biomed. Eng.* **2012**, *40*, 363–408. [CrossRef] [PubMed]
5. Hench, L.L.; Splinter, R.J.; Allen, W.C.; Greenlee, T.K. Bonding Mechanisms at the Interface of Ceramic Prosthetic Materials. *J. Biomed. Mater. Res.* **1971**, *5*, 117–141. [CrossRef]
6. Ibrahim, M.Z.; Sarhan, A.A.D.; Yusuf, F.; Hamdi, M. Biomedical materials and techniques to improve the tribological, mechanical and biomedical properties of orthopedic implants—A review article. *J. Alloy. Compd.* **2017**, *714*, 636–667. [CrossRef]
7. Raszewski, Z.; Chojnacka, K.; Kulbacka, J.; Mikulewicz, M. Mechanical Properties and Biocompatibility of 3D Printing Acrylic Material with Bioactive Components. *J. Funct. Biomater.* **2023**, *14*, 13. [CrossRef]
8. Rizwan, M.; Hamdi, M.; Basirun, W.J. Bioglass® 45S5-based composites for bone tissue engineering and functional applications. *J. Biomed. Mater. Res. Part A* **2017**, *105*, 3197–3223. [CrossRef]
9. Al-eesa, N.A.; Fernandes, S.D.; Hill, R.G.; Wong, F.S.L.; Jargalsaikhan, U.; Shahid, S. Remineralising fluorine containing bioactive glass composites. *Dent. Mater.* **2021**, *37*, 672–681. [CrossRef]
10. Chen, Q.Z.; Thompson, I.D.; Boccaccini, A.R. 45S5 Bioglass®-derived glass-ceramic scaffolds for bone tissue engineering. *Biomaterials* **2006**, *27*, 2414–2425. [CrossRef]
11. Gorustovich, A.A.; Roether, J.A.; Boccaccini, A.R. Effect of Bioactive Glasses on Angiogenesis: A Review of In Vitro and In Vivo Evidences. *Tissue. Eng. Part. B Rev.* **2010**, *16*, 199–207. [CrossRef]
12. Hench, L.L.; Polak, J.M. Third-Generation Biomedical Materials. *Bodybuild. Bionic. Hum.* **2002**, *295*, 1014–1017. [CrossRef]
13. Boccaccini, A.R.; Erol, M.; Stark, W.J.; Mohn, D.; Hong, Z.; Mano, J.F. Polymer/bioactive glass nanocomposites for biomedical applications: A review. *Compos. Sci. Technol.* **2010**, *70*, 1764–1776. [CrossRef]
14. Cao, Y.; Zhang, L.; Xu, Y.; Wang, X.; Liu, B. Preparation and bioactivity of mesoporous bioactive glasses with different phosphorus contents. *Adv. Mater. Res.* **2011**, *287–290*, 1997–2002. [CrossRef]
15. Granel, H.; Bossard, C.; Nucke, L.; Wauquier, F.; Rochefort, G.Y.; Guicheux, J.; Jallot, E.; Lao, J.; Wittrant, Y. Optimized Bioactive Glass: The Quest for the Bony Graft. *Adv. Healthc. Mater.* **2019**, *8*, 1801542. [CrossRef]
16. Bellucci, D.; Sola, A.; Salvatori, R.; Anesi, A.; Chiarini, L.; Cannillo, V. Role of magnesium oxide and strontium oxide as modifiers in silicate-based bioactive glasses: Effects on thermal behaviour, mechanical properties and in-vitro bioactivity. *Mater. Sci. Eng. C.* **2017**, *72*, 566–575. [CrossRef]
17. Tang, W.; Lin, D.; Yu, Y.; Niu, H.; Guo, H.; Yuan, Y.; Liu, C. Bioinspired trimodal macro/micro/nano-porous scaffolds loading rhBMP-2 for complete regeneration of critical size bone defect. *Acta. Biomater.* **2016**, *32*, 309–323. [CrossRef]
18. Lim, H.C.; Nam, O.H.; Kim, M.J.; El-Fiqi, A.; Yun, H.M.; Lee, Y.M.; Jin, G.Z.; Lee, H.H.; Kim, H.W.; Kim, E.C. Delivery of dexamethasone from bioactive nanofiber matrices stimulates odontogenesis of human dental pulp cells through integrin/BMP/mTOR signaling pathways. *Int. J. Nanomed.* **2016**, *11*, 2557–2567. [CrossRef]
19. Kim, H.W.; Lee, H.H.; Chun, G.S. Bioactivity and osteoblast responses of novel biomedical nanocomposites of bioactive glass nanofiber filled poly(lactic acid). *J. Biomed. Mater. Res. A* **2008**, *85*, 651–663. [CrossRef]
20. Petretta, M.; Gambardella, A.; Boi, M.; Berni, M.; Cavallo, C.; Marchiori, G.; Maltarello, M.C.; Bellucci, D.; Fini, M.; Baldini, N.; et al. Composite scaffolds for bone tissue regeneration based on pcl and mg-containing bioactive glasses. *Biology.* **2021**, *10*, 398. [CrossRef]
21. Palin, E.; Liu, H.; Webster, T.J. Mimicking the nanofeatures of bone increases bone-forming cell adhesion and proliferation. *Nanotechnology* **2005**, *16*, 1828–1835. [CrossRef]
22. Kurakula, M.; Rao, G.S.N.K. Moving polyvinyl pyrrolidone electrospun nanofibers and bioprinted scaffolds toward multidisciplinary biomedical applications. *Eur. Polym. J.* **2020**, *136*, 109919. [CrossRef]
23. Querido, D.; Vieira, T.; Ferreira, J.L.; Henriques, C.; Borges, J.P.; Silva, J.C. Study on the Incorporation of Chitosan Flakes in Electrospun Polycaprolactone Scaffolds. *Polymers.* **2022**, *14*, 1496. [CrossRef] [PubMed]
24. Gonçalves, A.; Simões, B.T.; Almeida, F.V.; Fernandes, S.N.; Valente, M.; Vieira, T.; Henriques, C.; Borges, J.P.; Soares, P.I. Engineering dual-stimuli responsive poly(vinyl alcohol) nanofibrous membranes for cancer treatment by magnetic hyperthermia. *Biomater. Adv.* **2023**, *145*, 213275. [CrossRef]
25. Yan, X.; Yu, C.; Zhou, X.; Tang, J.; Zhao, D. Highly ordered mesoporous bioactive glasses with superior in vitro bone-forming bioactivities. *Angew. Chem. Int. Ed.* **2004**, *43*, 5980–5984. [CrossRef]
26. Schneider, C.A.; Rasband, W.S.; Eliceiri, K.W. NIH Image to ImageJ: 25 years of image analysis. *Nat. Methods* **2012**, *9*, 671–675. [CrossRef]
27. Tamjid, E.; Bagheri, R.; Vossoughi, M.; Simchi, A. Effect of particle size on the in vitro bioactivity, hydrophilicity and mechanical properties of bioactive glass-reinforced polycaprolactone composites. *Mater. Sci. Eng. C* **2011**, *31*, 1526–1533. [CrossRef]

28. Newsome, T.E.; Olesik, S.V. Electrospinning silica/polyvinylpyrrolidone composite nanofibers. *J. Appl. Polym. Sci* **2014**, *131*, 40966. [CrossRef]
29. Kokubo, T.; Takadama, H. How useful is SBF in predicting in vivo bone bioactivity? *Biomaterials* **2006**, *27*, 2907–2915. [CrossRef]
30. Kokubo, T.; Kushitani, H.; Sakka, S.; Kitsugi, T.; Yamamum, T. Solutions able to reproduce in vivo surface-structure changes in bioactive glass-ceramic A-W3. *J. Biomed. Mater. Res.* **1990**, *24*, 721–734. [CrossRef]
31. ISO/FDIS 23317 Implants for Surgery-In Vitro Evaluation for Apatite-Forming Ability of Implant Materials Implants Chirurgicaux-Évaluation In Vitro de la Capacité de Formation D'apatite des Matériaux D'implants. 2007. Available online: www.iso.org (accessed on 11 April 2022).
32. Yan, X.X.; Deng, H.X.; Huang, X.H.; Lu, G.Q.; Qiao, S.Z.; Zhao, D.Y.; Yu, C.Z. Mesoporous bioactive glasses. I. Synthesis and structural characterization. *J. Non. Cryst. Solids* **2005**, *42*, 3209–3217. [CrossRef]
33. Yi, J.; Wei, G.; Huang, X.; Zhao, L.; Zhang, Q.; Yu, C. Sol-gel derived mesoporous bioactive glass fibers as tissue-engineering scaffolds. *J. Solgel. Sci. Technol* **2008**, *45*, 115–119. [CrossRef]
34. Luz, G.M.; Mano, J.F. Preparation and characterization of bioactive glass nanoparticles prepared by sol-gel for biomedical applications. *Nanotechnology* **2011**, *22*, 494014. [CrossRef]
35. Hatch, K.M.; Hlavatá, J.; Paulett, K.; Liavitskaya, T.; Vyazovkin, S.; Stanishevsky, A.V. Nanocrystalline cellulose/polyvinylpyrrolidone fibrous composites prepared by electrospinning and thermal crosslinking. *Int. J. Polym. Sci.* **2019**, *2019*, 7103936. [CrossRef]
36. Fogaça, R.; Catalani, L.H. PVP hydrogel membranes produced by electrospinning for protein release devices. *Soft Mater* **2013**, *11*, 61–68. [CrossRef]
37. Kumar, K.; Gill, R.S.; Batra, U. Challenges and opportunities for biodegradable magnesium alloy implants. *Mater. Technol.* **2018**, *33*, 153–172. [CrossRef]
38. Bellucci, D.; Sola, A.; Cannillo, V. Hydroxyapatite and tricalcium phosphate composites with bioactive glass as second phase: State of the art and current applications. *J. Biomed. Mater. Res. Part. A* **2016**, *104*, 1030–1056. [CrossRef]
39. Bigham-Sadegh, A.; Oryan, A. Basic concepts regarding fracture healing and the current options and future directions in managing bone fractures. *Int. Wound. J.* **2015**, *12*, 238–247. [CrossRef]
40. Kouhi, M.; Morshed, M.; Varshosaz, J.; Fathi, M.H. Poly (ε-caprolactone) incorporated bioactive glass nanoparticles and simvastatin nanocomposite nanofibers: Preparation, characterization and in vitro drug release for bone regeneration applications. *Chem. Eng. J.* **2013**, *228*, 1057–1065. [CrossRef]
41. Jin, X.; Li, L.; Xu, R.; Liu, Q.; Ding, L.; Pan, Y.; Wang, C.; Hung, W.; Lee, K.; Wang, T. Effects of thermal cross-linking on the structure and property of asymmetric membrane prepared from the polyacrylonitrile. *Polymers.* **2018**, *10*, 539. [CrossRef]
42. Indira, R.; Tamizharuvi, T.; Rajendran, T.V.; Jaisankar, V. Synthesis and Characterisation of Certain Novel Polyester Elastomers. *Indian J. Adv. Chem. Sci.* **2013**, *1*, 250–255. Available online: www.ijacskros.com (accessed on 12 April 2023).
43. Liverani, L.; Lacina, J.; Roether, J.A.; Boccardi, E.; Killian, M.S.; Schmuki, P.; Schubert, D.W.; Boccaccini, A.R. Incorporation of bioactive glass nanoparticles in electrospun PCL/chitosan fibers by using benign solvents. *Bioact. Mater.* **2018**, *3*, 55–63. [CrossRef] [PubMed]
44. Serio, F.; Miola, M.; Vernè, E.; Pisignano, D.; Boccaccini, A.R.; Liverani, L. Electrospun filaments embedding bioactive glass particles with ion release and enhanced mineralization. *Nanomaterials* **2019**, *9*, 9020182. [CrossRef] [PubMed]
45. Milosavljevic, V.; Jelinkova, P.; Jimenez Jimenez, A.M.; Moulick, A.; Haddad, Y.; Buchtelova, H.; Krizkova, S.; Heger, Z.; Kalina, L.; Richtera, L.; et al. Alternative Synthesis Route of Biocompatible Polyvinylpyrrolidone Nanoparticles and Their Effect on Pathogenic Microorganisms. *Mol. Pharm* **2017**, *14*, 221–233. [CrossRef]
46. Liu, X.; Xu, Y.; Wu, Z.; Chen, H. Poly(N-vinylpyrrolidone)-Modified surfaces for biomedical applications. *Macromol. Biosci.* **2013**, *13*, 147–154. [CrossRef]
47. Wu, Z.; Zhou, X.; Zhang, Y.; Luan, J.; Yang, X.; Wu, Z.; Wang, B. Synthesis and characterization of mesoporous bioactive glasses with highly ordered structures and high surface areas by a self-assembly process. *J. Non. Cryst. Solids* **2019**, *517*, 1–8. [CrossRef]
48. Zhu, Y.; Wu, C.; Ramaswamy, Y.; Kockrick, E.; Simon, P.; Kaskel, S.; Zreiqat, H. Preparation, characterization and in vitro bioactivity of mesoporous bioactive glasses (MBGs) scaffolds for bone tissue engineering. *Microporous Mesoporous Mater.* **2008**, *3*, 494–503. [CrossRef]
49. Yan, X.; Huang, X.; Yu, C.; Deng, H.; Wang, Y.; Zhang, Z.; Qiao, S.; Lu, G.; Zhao, D. The in-vitro bioactivity of mesoporous bioactive glasses. *Biomaterials* **2006**, *27*, 3396–3403. [CrossRef]
50. Edén, M. Structure and formation of amorphous calcium phosphate and its role as surface layer of nanocrystalline apatite: Implications for bone mineralization. *Materialia* **2021**, *17*, 101107. [CrossRef]
51. Safo, I.A.; Werheid, M.; Dosche, C.; Oezaslan, M. The role of polyvinylpyrrolidone (PVP) as a capping and structure-directing agent in the formation of Pt nanocubes. *Nanoscale. Adv.* **2019**, *1*, 3095–3106. [CrossRef]
52. Gandou, Z.; Nounah, A.; Belhorma, B.; Yahyaoui, A. Nanosized calcium-deficient carbonated hydroxyapatite synthesized by microwave activation. *J. Mater. Environ. Sci.* **2015**, *6*, 983–988. Available online: https://www.researchgate.net/publication/281800019 (accessed on 12 April 2023).
53. Hench, L.L. Bioceramics: From Concept to Clinic. *J. Am. Ceram. Soc.* **1991**, *74*, 1487–1510. [CrossRef]
54. Gönen, S.Ö.; Taygun, M.E.; Küçükbayrak, S. Fabrication of bioactive glass containing nanocomposite fiber mats for bone tissue engineering applications. *Compos. Struct.* **2016**, *138*, 96–106. [CrossRef]

55. Lin, H.M.; Lin, Y.H.; Hsu, F.Y. Preparation and characterization of mesoporous bioactive glass/polycaprolactone nanofibrous matrix for bone tissues engineering. *J. Mater. Sci. Mater. Med.* **2012**, *23*, 2619–2630. [CrossRef]
56. Allo, B.A.; Lin, S.; Mequanint, K.; Rizkalla, A.S. Role of bioactive 3D hybrid fibrous scaffolds on mechanical behavior and spatiotemporal osteoblast gene expression. *ACS Appl. Mater. Interfaces* **2013**, *5*, 7574–7583. [CrossRef]
57. Han, B.; Zhang, X.; Liu, H.; Deng, X.; Cai, Q.; Jia, X.; Yang, X.; Wei, Y.; Li, G. Improved bioactivity of PAN-based carbon nanofibers decorated with bioglass nanoparticles. *J. Biomater. Sci. Polym. Ed.* **2014**, *25*, 341–353. [CrossRef]
58. Yang, Q.; Sui, G.; Shi, Y.Z.; Duan, S.; Bao, J.Q.; Cai, Q.; Yang, X.P. Osteocompatibility characterization of polyacrylonitrile carbon nanofibers containing bioactive glass nanoparticles. *Carbon NY* **2013**, *56*, 288–295. [CrossRef]
59. Tan, L.; Zheng, X.; Chen, L.; Wang, Y. Quality testing of human albumin by capillary electrophoresis using thermally cross-linked poly(vinyl pyrrolidone)-coated fused-silica capillary. *J. Sep. Sci.* **2014**, *37*, 2974–2982. [CrossRef]
60. Bogatyrev, V.M.; Borisenko, N.V.; Pokrovskii, V.A. Thermal Degradation of Polyvinylpyrrolidone on the Surface of Pyrogenic Silica. *Orig. Russ. Appl. Chem.* **2001**, *74*, 814–819.

biomimetics

MDPI

Article

TCP Doped with Metal Ions Reinforced with Tetragonal and Cubic Zirconia

Vanessa M. Ferro [1,2], Beatriz C. Silva [1], Duarte F. Macedo [1], Natanael F. Fernandes [2] and Abílio P. Silva [1,*]

[1] C-MAST—Centre for Mechanical and Aerospace Science and Technologies, Universidade da Beira Interior, Rua Marquês d'Ávila e Bolama, 6201-001 Covilhã, Portugal; vanessa.ferro@ubi.pt (V.M.F.)

[2] CICS-UBI—Centro de Investigação em Ciências da Saúde, Universidade da Beira Interior, Av. Infante D. Henrique, 6200-506 Covilhã, Portugal

* Correspondence: abilio@ubi.pt; Tel.: +351-275242055

Abstract: Ceramic biocomposites based on bioactive tricalcium phosphate doped with metal ions are a strategy for obtaining good biomimetics for human bone composition. Manufacturing with PMMA porogen also induces bone-like porosity morphology. The poor strength of tricalcium phosphate can be overcome by designing ceramic composites reinforced with tetragonal and cubic zirconia. In this work, five different bioceramic composites were manufactured without and with induced porosity and their physical, mechanical, microstructural, and biological properties were studied. With the addition of tetragonal and cubic zirconia, an improvement in strength of 22% and 55%, respectively, was obtained, corresponding to up to 20.7 MPa. PMMA was suitable for adding porosity, up to 30%, with interconnectivity while an excellent hOB cellular viability was achieved for all biocomposites.

Keywords: bioceramic composites; TCP; ZrO_2; metal ions; bone regeneration; biocompatibility

Citation: Ferro, V.M.; Silva, B.C.; Macedo, D.F.; Fernandes, N.F.; Silva, A.P. TCP Doped with Metal Ions Reinforced with Tetragonal and Cubic Zirconia. *Biomimetics* **2023**, *8*, 599. https://doi.org/10.3390/biomimetics8080599

Academic Editors: Hermann Ehrlich and Bo Su

Received: 17 October 2023
Revised: 19 November 2023
Accepted: 7 December 2023
Published: 12 December 2023

1. Introduction

The average life expectancy has increased, causing the aging of the world population. With this, health complications related to joints and bone tissue, such as osteoporosis and osteoarthritis, can arise [1,2]. Therefore, finding alternatives that provide quality of life to the patient is essential. Inert bioceramics (e.g., zirconia, ZrO_2), do not form biochemical bonds with surrounding tissues, i.e., do not react with the body [3]. Physical, chemical, and mechanical properties such as high compressive strength, wear and corrosion resistance, hardness, elastic modulus similar to that of steel, high fracture toughness, and stability in a physiological environment, make ZrO_2 a material of the highest interest for the manufacture of orthopedic and dental prostheses [2,4].

At room temperature and in pure form, ZrO_2 has a stable monoclinic crystalline structure (m-ZrO_2) up to 1170 °C. With increasing temperature, it becomes metastable: tetragonal (t-ZrO_2) stable up to 2370 °C and above this temperature, cubic (c-ZrO_2) [4–6]. The m-ZrO_2 is not suitable for high temperature applications due to the volume expansion associated with the transformation of t-ZrO_2 to m-ZrO_2 [7,8], known as martensitic transformation [5,9]. In this transformation there is an increase in volume of approximately 4.5% during cooling [6,8] which is detrimental to the mechanical behavior of ZrO_2 because the stress induced during this transformation leads to the formation of cracks [5]. In this sense, it can be concluded that the oral environment is a strongly predisposing factor for uncontrolled martensitic transformation [6].

Stabilizers such as yttrium oxide (Y_2O_3) can be added to ZrO_2 to inhibit the transformation of the t-phase to m-phase. Briefly, this addition results in the part of the Zr^{+4} atoms being replaced by Y^{+3} atoms, stabilizing the polymorphic modifications of ZrO_2 when subjected to the sintering process. This avoids the volume variations caused by phase transformations [6,10] that form yttrium-stabilized zirconia (YSZ), where 3YSZ is mostly tetragonal and 8YSZ is mostly the cubic crystalline phase [11]. However, several studies describe that the use of Y_2O_3 has disadvantages mainly in its degradation at low temperature; this phenomenon is known as "Aging Phenomenon" while its use is limited by unpredictable behavior in the body [1,6,9,12].

On the other hand, active bioceramics can establish connections with the surrounding tissue and suffer progressive degradation as the new tissues form [13]. TCP can be used in cement and bone implants for its bioactivity, osteoinduction, and bioresorbability [14,15]. By its reabsorbable nature, part of the material dissolves, providing raw material for the growth and maintenance of the new tissue, facilitating osteoconduction [13,16]. Chemically, it is $Ca_3(PO_4)_2$ and characterized by a Ca/P ratio 1.5 [17]. As a function of the temperature at which it is sintering, it can present three polymorphic forms: β-TCP (~<1150 °C), α-TCP (~1150–1460 °C), and α'-TCP (~>1460 °C) [18,19].

Metal ions such as magnesium (Mg^{2+}), manganese (Mn^{2+}), zinc (Zn^{2+}), iron (Fe^{3+}), and strontium (Sr^{2+}), among others, can be added to form biomimetic bone composition and also improve several properties of these bioceramics [14,20]. Considering the cations in the human body, magnesium is the fourth in abundance and the second intracellular cation in tissues [21,22]. The human body contains about 30 g of magnesium, 50% stored in bones [23]. Magnesium plays a vital role in many enzymatic reactions, among which are the transmission of nerve impulses and the synthesis of fatty acids and proteins. Also, Mg^{2+} has an elemental role in biology because ATP needs to be bound to a magnesium ion to be biologically active as well as in the formation of the transition state where ATP is synthesized from ADP and inorganic phosphate. Mg^{2+} ions can replace Ca^{2+} ions by increasing the transformation temperature of β to α, stabilizing β-TCP, increasing mechanical strength, and improving osteoblast function and biocompatibility. Moreover, Mn^{2+} ions improve mechanical properties and corrosion resistance, while Zn^{2+} improves osteoblast function, corrosion resistance, and has antibacterial action. Other cations, such as Fe^{3+} preserve the structural stability of bioceramics and increase osseointegration, while Sr^{2+} inhibits bone resorption and stimulates the proliferation of osteoblasts and bone formation [14,20,24–29].

However, besides the essential effect of metal ions for human life, higher doses can be toxic [23]. For instance, levels of Mg^{2+} above 1.1 mM are generally considered hypermagnesemic. This effect promotes patients clinically to suffer from nausea, vomiting, lethargy, headaches, and/or flushing. When Mg^{2+} levels rise above 3.0 mM, it can cause serious heart defects, characterized by brachycardia, hypotension, and in extreme hypermagnesemia can result in coma, asystole, and death from cardiac arrest [22].

Manganese (Mn^{2+}) ions accumulate in the bones, liver, pancreas and mainly in the brain. Pathologies, as polycythemia, dystonia, hepatic cirrhosis, have been related to excess of manganese. In particular, symptoms as in Parkinsonism have been reported to overexposure of manganese. The molecular mechanisms involved include oxidative stress, protein misfolding, apoptosis, mitochondrial dysfunctions, and interference in the homeostasis of other metal essential ions. The Mn^{2+} amount should not exceed 5 mg/m^3, even for short periods; however, its toxicity depends on its chemical form: -Mn^{2+}, the common form, is not dangerous but MnO_4^- is very toxic [23].

High concentration of Zinc (Zn^{2+}) can be found in vesicles in the brain, in bones, and in muscles. It is reported that the recommended dose is around 15 mg/day [23]. Poisoning with zinc phosphide, a rodenticide, causes cardiovascular, respiratory, renal, and hepatobiliary failure, among other complications [30].

In vitro studies have shown that high doses of iron could lead to osteoblast apoptosis via caspase 3. Iron overload also decreases the formation of mineralization nodes and inhibits growth of hydroxyapatite crystals, altering their crystallinity [30]. The toxicity generated by the excess of iron can cause cirrhosis, liver carcinoma, heart failure, diabetes mellitus, and osteoporosis. Therefore, in the presence of molecular oxygen, loosely bound iron is able to undergo a redox cycle (Fe^{3+}/Fe^{2+}) generating poisonous oxygen-derived free radicals [23].

For successful applications, bone structures must mimic the porosity of native bone and allow it to grow through the interconnectivity of the structure [31]. Polymethylmethacrylate (PMMA) has been considered for its particular characteristics, such as mechanical strength, moldability to fill complex defects, low cost, having approval by the FDA and with clean and easy thermal elimination at high temperatures [32–34].

The aim of this work was to produce dense ceramic biocomposites and PMMA induced porous ceramic biocomposites, both compositions based in TCP, doped with metal ions of magnesium (Mg^{2+}), manganese (Mn^{2+}), zinc (Zn^{2+}), and iron (Fe^{3+}). The two types of manufactured samples were reinforced with 10 wt% and 20 wt% of tetragonal zirconia (3YSZ) and cubic zirconia (8YSZ). A detailed evaluation of their physical, mechanical, microstructural, and biological properties was performed.

2. Materials and Methods

2.1. Materials and Manufacture Methods

Following the procedure described in the literature [14], TCP was doped with 10 mol% of Mg^{2+} and 5 mol% of a mixture of Zn^{2+}, Mn^{2+}, and Fe^{3+}, totaling 15 mol% of metal ions. Being similar to the maximum amount of Ca^{2+} ions that can be replaced in the TCP structure, substituted TCP (sTCP) was obtained in this way [24]. For this purpose, calcium carbonate ($CaCO_3$, PanReac AppliChem, Barcelona, Spain), ammonium phosphate dibasic ($(NH_4)_3HPO_4$ Acros organics, Geel, Belgium, 98+%, CAS: 423375000), magnesium oxide (MgO, Alfa Aesar, Ward Hill, MA, USA, 96% min, CAS: 1309-48-4), manganese (II) oxide (MnO, 99%, Alfa Aesar, Ward Hill, MA, USA, CAS 011870-36), iron (III) oxide (Fe_2O_3, 99.9%, Alfa Aesar, Ward Hill, MA, USA, CAS: 1309-37-1), and zinc oxide (ZnO, 99.0%, PanReac AppliChem, Barcelona, Spain, CAS: 1314-13-2) were used. Also, ZrO_2 stabilized with 3 mol% of yttrium (3YSZ, TZ-3Y, t-ZrO_2, Tosoh, Tokyo, Japan, Lot: Z308782P), ZrO_2 stabilized with 8 mol% of yttrium (8YSZ, TZ-8Y, c-ZrO_2, Tosoh, Tokyo, Japan, Lot: Z807724P), and PMMA microspheres (Acros Organics, Geel, Belgium, CAS: 9011-14-7), with average size of 100 μm, were also added to induce porosity in some samples. Calcium carbonate, ammonium phosphate dibasic, 10 mol% of Mg, and 1.67 mol% of Zn^{2+}, 1.67 mol% of Mn^{2+} and 1.67 mol% of Fe^{3+} in total amount of 50 g was mixed with 100 g of isopropyl alcohol (Labchem, Laborspirit, Santo Antão do Tojal, Portugal) in a high energy ball mill (Fritsch, Pulverisette 6, Idar-Oberstein, Germany) at 500 rpm for 150 min, divided into cycles of 30 min. To promote this mixing and simultaneous milling process, 50 g of YSZ spheres (Fritsch, Idar-Oberstein, Germany) were used. The mixture was dried in a stove (Carbolite, NR200-F, Derbyshire, UK) at 60 °C for 48 h. After this process, the mixture was sieved (Retsch, AS200, Düsseldorf, Germany) up to 63 μm to remove the YSZ spheres and some agglomerates. The laser diffraction method (Beckman Coulter, LS200, Brea, CA, USA) was used to analyzed the particle size distribution of the powders after milling. Then, the powder was placed in crucibles and calcined in an electric furnace (Termolab, MLR, Águeda, Portugal) in air for 10 h at a temperature of 1000 °C. After calcination of the sTCP reference material, the other four ceramic composites with 3YSZ and 8YSZ were mixed according to the compositions and nomenclatures summarized in Table 1.

To each of the compositions with YSZ, 3YSZ and 8YSZ were added according to the corresponding mass fraction. Then, 15 g of the doped or substituted TCP (sTCP) was added according to the mass fraction, and 15 g of YSZ balls and 30 g of isopropyl alcohol were mixed in a high energy ball mill [35]. The milling, drying, and sieving processes occurred under the previous conditions. The particle size distribution (PSD) of the powders of the reference material (sTCP, which give rise to the material designated 10T) and the other compositions with the mixture between 80% of sTCP, 20% of cubic zirconia (8T2cZ), and 20% of tetragonal zirconia (8T2tZ) are shown in Figure 1.

Table 1. Nomenclature and composition of the materials studied in molar fraction (x_i), volumetric fraction (v_i) and mass fraction (w_i).

Materials	Molar Fraction x_i			Volumetric Fraction v_i			Mass Fraction w_i		
	sTCP	3YSZ	8YSZ	sTCP	3YSZ	8YSZ	sTCP	3YSZ	8YSZ
10T	1	0	0	1	0	0	1	0	0
9T1tZ	0.7814	0.2186	0	0.9473	0.0527	0	0.9	0.1	0
8T2tZ	0.6138	0.3862	0	0.8888	0.1112	0	0.8	0.2	0
9T1cZ	0.7814	0	0.2186	0.9468	0	0.0532	0.9	0	0.1
8T2cZ	0.6138	0	0.3862	0.8877	0	0.1123	0.8	0	0.2

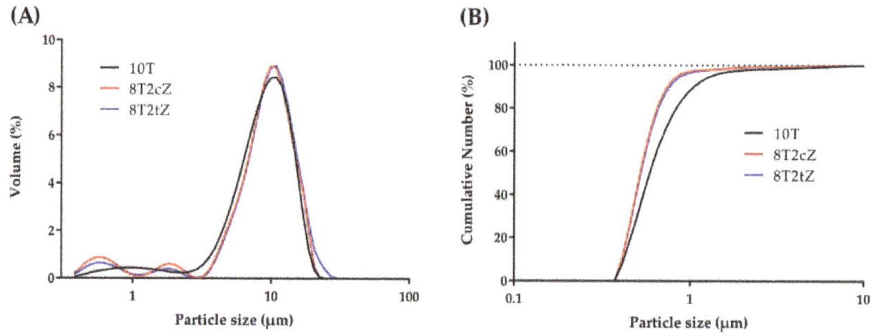

Figure 1. Particle size distribution of powders of the mixtures 10T (sTCP), 8T2cZ, and 8T2tZ: (**A**) differential volume; (**B**) cumulative number.

The PSD of 10T powders is bimodal: with a major peak at 10 μm and another minor peak at 1 μm. Both compositions have zirconia (8T2cZ and 8T2tZ) present, instead of one, with two small minor peaks of 0.5 and 1.8 microns, respectively.

To induce porosity, PMMA spheres were added to each of the five compositions in the mass fraction 60:40 (each of the five compositions: PMMA, respectively). In order to obtain a homogeneous mixture, the mixture was placed in the Fritsch mill for 20 s at 500 rpm.

Cylindric pellets of bioceramic composites with mass of ~0.65 g, ~13 mm diameter and 3 mm thickness were made using a universal electromechanical testing machine (Shimadzu, AGS-X, Kyoto, Japan) with a 13 mm diameter stainless steel matrix in which uniaxial pressure of 50 MPa was applied for 10 s. Then, the samples were sintered in air at a temperature of 1300 °C for 120 min with a heating rate of 5 °C/min.

2.2. Microstructural and Mechanical Characterization

To confirm if the doping reaction occurred, the XRD test was performed in the dense biocomposites 10T, 8T2tZ, and 8T2cZ, since 8T2tZ and 8T2cZ have the highest concentration of each ZrO_2 studied and 10T the same concentration of each ion without the addition of 3YSZ and 8YSZ. An X-ray diffractometer (DMAX III/C, Rigaku, Tokyo, Japan) with the Bragg-Brentano (θ/2θ) horizontal geometry was used. The X-ray tube of copper (wavelength of 1.5405 Å) operated at 40 kV at 30 mA using CuKα radiation. The intensity of diffracted radiation as function of the 2θ diffraction angle was obtained between 5° and 90°. The diffractograms obtained were compared with the theoretical cards available in the ICDD database of the MDI/JADE, version 6 analysis software. The contents of each crystalline phase of the compositions in %vol were quantified through Rietveld refinements using the FullProf software, version May2021 [36]. Scanning electron microscopy (SEM) (Hitachi S-2700, Tokyo, Japan), was performed for microstructure imaging by applying the SE mode at an accelerating voltage of 20 kV. Chemical analysis was performed using SEM with energy dispersive X-ray probe (EDX, Brucker Quantax 400, Elk Grove Village, IL,

USA). This analysis was carried out using the average of three different areas, where the gold peaks were not considered and all other peaks corresponding to Ca, P, and O and also the metals such as Zr, Mg, Mn, Fe, and Zn were quantified.

Fourier transform infrared spectroscopy (FTIR) was performed using a Nicolet™ iS 10 FTIR spectrometer (Thermo Scientific Inc., Waltham, MA, USA). Infrared spectra were recorded in the range of 525–4000 cm^{-1} at ambient temperature and with a resolution of 4 cm^{-1} (32 scans).

Apparent porosity and bulk density were measured according to ASTM C20-00 [37]. In this procedure three different weights, in grams, were calculated: dry sample weight (D), saturated weight (W), after boiling for 2 h and resting 12 h entirely covered with water, and suspended immersed in water weight (S). Considering the density of the water equal to 1 g/cm^3, the apparent porosity, P, in %, expressing the relationship of the volume of the open pores in the specimen to its exterior volume, is calculated using Equation (1):

$$P = \frac{W - D}{W - S} \times 100 \tag{1}$$

The bulk density, BD, in g/cm^3, is the quotient of its dry weight divided by the exterior volume, including pores, and is calculated by Equation (2):

$$BD = \frac{D}{W - S} \tag{2}$$

The diametral compression test was conducted with a universal electromechanical testing machine (Shimadzu, AGS-X, Japan) with a load cell of 10,000 N [38] and a displacement rate of 0.5 mm/min [39]. The tensile strength, σ_x, was determined by Equation (3) where F corresponds to the maximum force applied, d, the diameter, and e, the thickness of the cylindric sample:

$$\sigma_x = \frac{2 \times F}{\pi \times d \times e} \tag{3}$$

In Figure 2, the illustrations of dense and porous ceramic samples are observed after the fracture occurs during the diametral compression test.

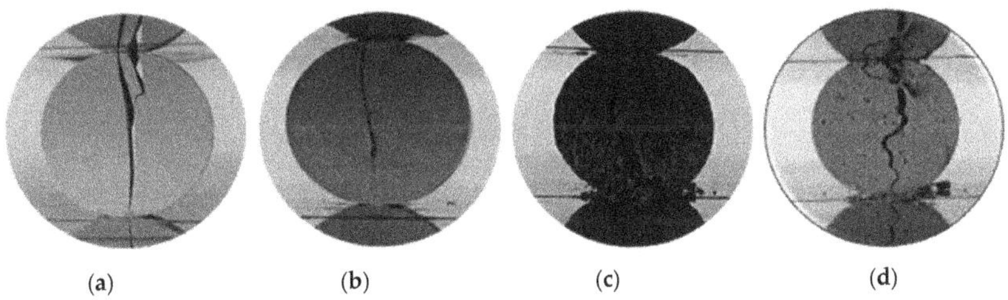

Figure 2. Illustration of a valid diametral compression test: (**a**) 10T and (**b**) 8T2tZ dense samples; (**c**) 10T, and (**d**) 8T2tZ porous samples.

2.3. Biological Characterization

The resazurin assay [40] was used to evaluate the cytocompatibility of bioceramic composites in human osteoblasts (hOB). Thus, a cell line of hOB (Cell Applications, Inc., San Diego, CA, USA), t-flasks of 75 cm^3 cell culture (Orange Scientific, Braine-l'Alleud, Belgium), Dulbecco's Modified Eagle Medium/Nutrient F-12 (DMEM-F12, Sigma-Aldrich, Sintra, Portugal), sodium bicarbonate ($NaHCO_3$, Labchem, Laborspirit, Santo Antão do Tojal, Portugal, CAS: 144-55-8), double deionized water (ultrapure water, obtained using an ultrapure water purification system Milli-Q Advantage A10, filtered at 0.22 µm and 18.2 MΩ cm at 25 °C), bovine fetal serum (FBS, Biochrom AG, Berlin, Germany, CAS: 9014-81-7), trypsin (Sigma-Aldrich, Sintra, Portugal, CAS: 9002-07-7), resazurin (Sigma-Aldrich, Sintra, Portugal, CAS: 62758-13-8), spectrofluorimeter (SpectraMax Gemini EM Molecular

Devices, San José, CA, USA), scanning electron microscope (Hitachi, S-3400 N, Tokyo, Japan) and a turbomolecular pumped coater (Quorum Technologies, Q150R ES, Lewes, UK) were used.

Discs similar to those used in the mechanical characterization (Figure 2) of ~13 mm in diameter and 3 mm in thickness were broken into six identical parts, and before starting the biological tests, were sterilized by ultraviolet irradiation (UV) for 1 h [1]. The cells were seeded at a density of 15,000, 10,000, and 2500 cells/well in three 48-well plates for 24 h at 37 °C. Then, the medium was removed, and the cells were incubated with 10% of material in relation to the well area [41] and 300 μL of DMEM-F12 in all wells. After 1, 3, and 7 days of incubation, the medium and the material were removed and the hOB were incubated with 220 μL resazurin 10% (*v*/*v*) [40,42]. Cell viability was determined by measuring resorufin fluorescence at λ_{ex} = 545 nm and λ_{em} = 590 nm [40,42]. The plate corresponding to day 1 of incubation with material contained a density of 15,000 cells/well, day 3, 10,000 cells/well, and day 7 plate, 2500 cells/well. Negative control (K^-) cells were incubated only with culture medium and positive control (K^+) cells were incubated with bleach.

The cellular attachment to porous biocomposites was assessed by SEM analysis in BSE 3D mode (Backscattered electrons) with an acceleration voltage of 20 kV. The cells were fixed with 500 μL of glutaraldehyde at 2.5% (*v*/*v*) for 1 h. After, the samples were dehydrated with increasing ethanol concentrations (50, 70, and 99%) for 5 min each, frozen at −80 °C for 1 h and freeze-dried for 3 h. Then, the samples were coated with gold using a turbomolecular pumped coater.

3. Results and Discussion

3.1. Microstructural Properties

Representative SEM micrographs of the fracture surfaces of the sintered ceramic composites are shown in Figure 3. In all cases there are mixed fracture surfaces, that is, transgranular (examples illustrated by letter "T" in Figure 3) and intergranular fractures (examples illustrated by letter "I" in Figure 3). Regarding porosity, it is evident that an interconnected porosity network exists even in dense biocomposites. In addition to perfectly spherical pores, that is, pores caused by PMMA spheres, more elongated pores are also observed.

The macropores caused by the PMMA spheres, visible in the image at 200× magnification, present an average size of 100 μm, between 120–150 μm and between 110–140 μm for the 10T, 8T2tZ, and 8T2cZ biocomposites, respectively. In this way, bone growth and cell colonization have good conditions to occur.

XRD assays were performed to confirm the presence of ZrO_2 crystalline phases in the dense biocomposites 8T2tZ and 8T2cZ. Figure 4 shows the X-ray diffraction spectra of the analyzed biocomposites. In the 10T biocomposites, in addition to other small, identified peaks, four main peaks of higher intensity were identified for 2θ = 27°, 32°, 34°, and 53° that correspond to β-TCP and are coincident with their theoretical card. The predominant structure is β-TCP (~78%vol according to Table 2). Also, in 10T samples, HA is the second most present phase, and α-TCP, with a very low value, can be neglected. The existence of α-TCP was not expected, because of the doping of TCP with 10% MgO. According to the literature [24], with the addition of Mg^{2+} to TCP, the transformation temperature of β-TCP to α-TCP is expected to increase. This is also promoted by the relative high temperature of sintering. In biocomposites 8T2tZ and 8T2cZ, four main peaks were identified for 2θ = 30.2°, 31.1°, 50.4°, and 59.7° corresponding to t-ZrO_2, β-TCP, t-ZrO_2, and also t-ZrO_2, respectively, for 8T2tZ; in 8T2cZ main peaks for 2θ = 30°, 31.1°, 50.1° and 59.7° were identified, corresponding to c-ZrO_2, β-TCP, c-ZrO_2, and also c-ZrO_2, respectively. For both cases, the peaks coincide with the respective theoretical cards.

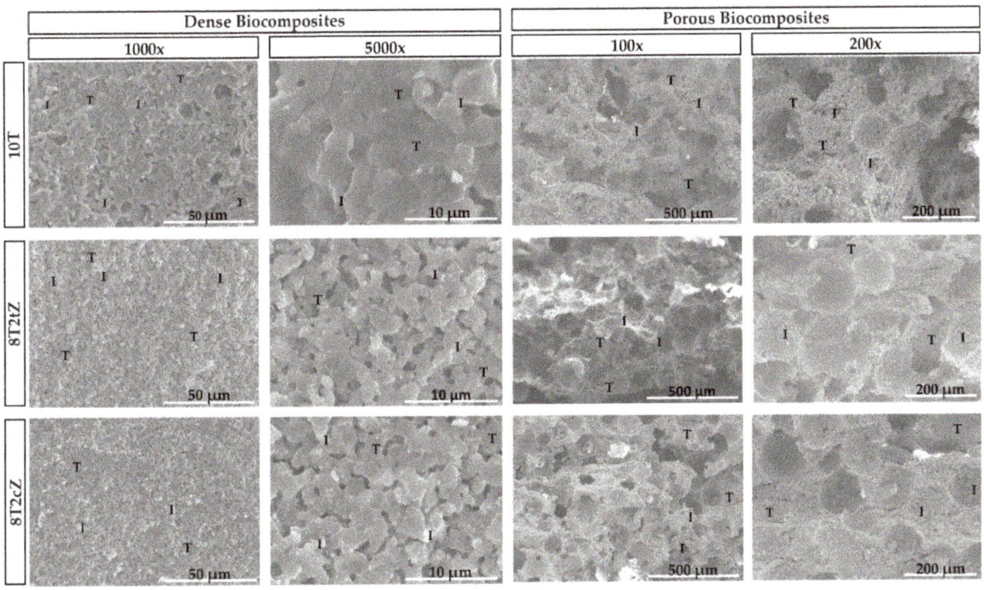

Figure 3. SEM characteristic fracture surfaces of the ceramic biocomposites. Dense biocomposites with a magnification of 1000× and 5000× and porous biocomposites with a magnification of 100× and 200×. Examples of transgranular and intergranular fracture surfaces are illustrated by letters "T" and "I", respectively.

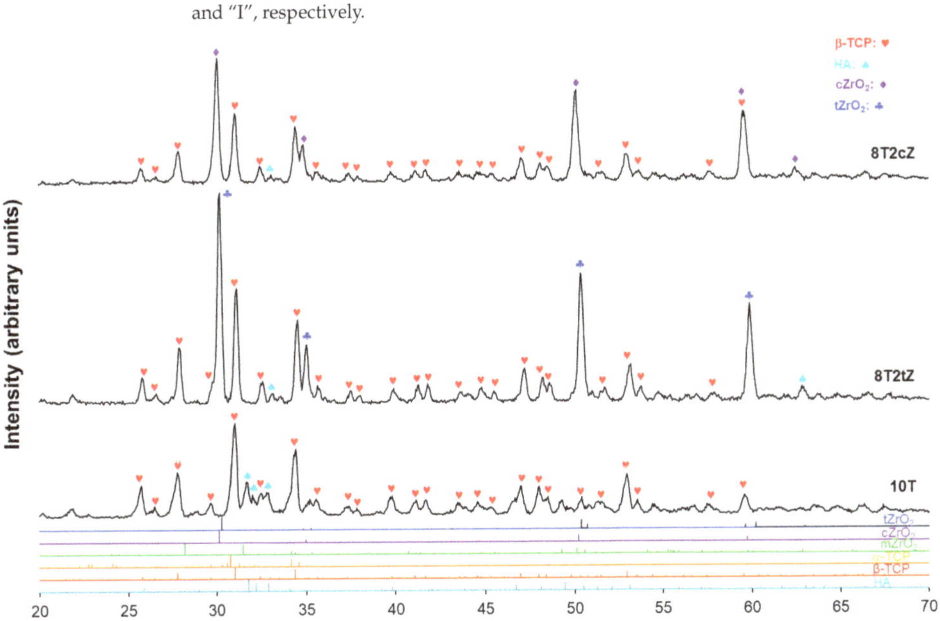

Figure 4. X-ray diffraction spectra of the biocomposites 10T, 8T2tZ, and 8T2cZ and X-ray diffraction spectra of the theoretical cards #37-1484, #50-1089, #49-1642, #09-0432, #09-0169, and #09-0348, corresponding to m-ZrO_2, t-ZrO_2 (♣), c-ZrO_2 (♦), HA (♠), β-TCP (♥), and α-TCP, respectively.

Table 2. Quantitative analysis (%vol) of XRD for bioceramic composites 10T, 8T2tZ, and 8T2cZ, by Rietveld refinement.

Phases (%vol)	10T	8T2tZ	8T2cZ
β-TCP	77.60 ± 0.00	71.09 ± 1.00	77.50 ± 0.00
HA	21.9 ± 0.00	3.05 ± 0.02	-
α-TCP	0.72 ± 0.00	-	-
t-ZrO_2	-	25.86 ± 0.54	1.48 ± 0.00
c-ZrO_2	-	-	21.02 ± 0.00

The main phases of the biocomposites analyzed were quantified by Rietveld refinement, as shown on Table 2. According to Table 2, the predominant phase is β-TCP, presenting values of 77.60, 71.09, and 77.50%vol for biocomposites 10T, 8T2tZ, and 8T2cZ, respectively. HA was only detected in 8T2tZ and 10T, with values of 3.05 and 21.69%vol, respectively. In biocomposites with 8YSZ, mostly this phase corresponds to ZrO_2, c-ZrO_2, whose value is 21.02%vol, and a small percentage of t-ZrO_2, 1.48%vol, which indicates that during the manufacturing process the c-ZrO_2 was converted into t-ZrO_2, probably due to high temperatures. In the biocomposites with 3YSZ, only t-ZrO_2, was quantified, 25.86%vol.

No metal oxides were detected, including MgO, which indicates that during the calcination process, the Ca^{2+} ions were replaced by the metal ions of Mg^{2+}, Mn^{2+}, Zn^{2+}, and Fe^{3+} into sites of the crystalline structure of β-TCP.

In order to understand the presence of several peaks corresponding to HA, the XRD was carried out for sTCP after the calcination process and compared with the XRD after the sintering process (Figure S1). The main peaks correspond, in both spectra, to the TCP beta phase. The presence of TCP alpha crystalline phase peaks is unclear. In the calcined sTCP sample, the peaks for $2\theta = 29°$, 31.8°, 32.2°, 32.9°, 34.0°, 39.2°, 46.7°, 49.5°, and 63.0° correspond to the hydroxyapatite phase. In the sintered phase at 1300 °C, the number of HA peaks decreases, essentially leaving the relevant peaks at 31.8°, 32.2°, and 32.9°.

The formation of hydroxyapatite in this case most likely occurs through what in the literature is called mechanochemical synthesis [43,44]. These mechanochemical reactions resulted in the formation of a defective phase of calcium-deficient HA, which, when calcined at up to 720 °C, leads to the formation of HA and β-TCP [43]. The principle of this dynamic synthesis during grinding is related to the energy during grinding (the impact that the mill balls have on the powdered grains). In other words, a reaction and an interdiffusion mechanism are promoted between different grains or with an intimate chemical reaction of the molecules between them [44]. In this sense, similar to the present work, after 5 h of mechanical activation, Yeong et al. [43] obtained a 2θ of 31.8° as the most prominent peak, corresponding to the crystalline plane of HA (211).

In fact, the growth of the HA phase, which is directly related to the improvement of biocompatibility and osteoinduction, is important in bone reconstruction [1,18,45].

From the elementary chemical analysis (EDX), Table 3, the presence of the elements Ca, P, Zr, and the remaining added metal ions were verified.

The elements Ca and P come from the chemical formula of TCP, ($Ca_3(PO_4)_2$), and the Zr from the doping with the two types (tetragonal and cubic) of ZrO_2 used in the composition of the ceramic biocomposites. The Ca/P ratio, in wt%, of 2.68, 3.53, and 3.37, and in mol% of the 2.07, 2.73, and 2.60 for the ceramic biocomposites 10T, 8T2tZ, and 8T2cZ, respectively is slightly high when compared to the works reported in the literature, namely, Wu et al. [46], which refers to bone minerals with Ca/P ratio between 1.37 and 1.87 mol%. With the addition of the 3YSZ and 8YSZ, this Ca/P ratio, tends to increase, with its highest value being in ceramic biocomposites with 20% of 3YSZ.

Figure 5 shows the characteristic FTIR spectra of the biocomposites.

Table 3. Chemistry analysis (EDX) of the ceramic biocomposites 10T, 8T2tZ, and 8T2cZ. The error is the standard deviation (sd).

Element	10T			8T2tZ			8T2cZ		
	wt%	mol%	sd wt%	wt%	mol%	sd wt%	wt%	mol%	sd wt%
Ca	46.7	30.02	2.2	36.8	24.61	2.01	37.93	28.65	2.03
P	17.43	14.5	1.12	10.42	9.02	0.79	11.24	10.95	0.84
O	33.51	53.97	9.09	36.33	60.87	9.76	27.59	52.04	7.85
Mg	0.78	0.82	0.15	0.65	0.72	0.14	0.62	0.78	0.13
Mn	0.47	0.22	0.11	0.35	0.17	0.09	0.25	0.14	0.09
Fe	0.48	0.22	0.11	0.14	0.07	0.08	0.31	0.17	0.09
Zn	0.64	0.25	0.13	0.37	0.15	0.11	0.55	0.25	0.12
Zr	-	-	-	14.95	4.39	1.1	21.5	7.11	1.52

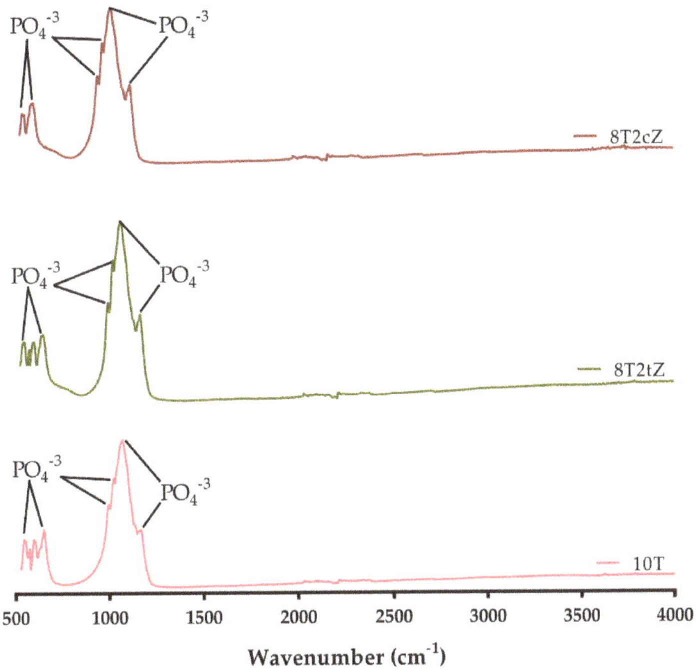

Figure 5. FTIR spectra of different biocomposites 10T, 8T2tZ, and 8T2cZ.

From the FTIR analysis, the characteristic covalent bonds of the different ionic groups are observed. The ceramic compound TCP is formed by Ca^{2+} and PO^{3-} ions. The peak at 943.19 cm^{-1} and 972.12 cm^{-1} is related to the presence of pure β-TCP. Thus, the bands between the range 900–1200 cm^{-1} represent the stretching mode of the $PO_4{}^{-3}$ group [47,48]. The results show the characteristic peaks of the covalent bonds present in $PO_4{}^{-3}$. Hydroxyapatite (HA) has the OH^- ion. The characteristic peaks at 630 cm^{-1} and 3571 cm^{-1} were attributed to the stretching mode of the hydroxyl group (OH^-) [47]; however, these peaks were not clearly detected or were too small compared to others. The band at 465–627 cm^{-1} and the band at 900–1000 cm^{-1} were expected due to Zr-O, which indicates the formation of cubic ZrO_2 and tetragonal ZrO_2 crystalline phases, respectively [49,50]. In this sense, it is believed that the peaks with the greater width of the two samples with zirconia (~600 cm^{-1}), in comparison with the sharp peaks of the 10T composition, and the more pronounced drop (~820 cm^{-1}), are due to the presence of Zr-O bonds, thus being almost undetectable due to their low content and the presence of high PO_4^{3-} peaks in this region.

3.2. Mechanical Properties

The apparent porosity of the 10T biocomposite was 12.9%. The porosity increased to 24.6% and 21.9% for the dense biocomposites reinforced with 3YSZ, i.e., 9T1tZ and 8T2tZ, respectively. The similar samples manufactured with PMMA show higher porosity, namely, 26.2%, 29.0%, and 29.3% for 10T, 9T1tZ, and 8T2tZ, respectively. The dense biocomposites of 9T1cZ and 8T2cZ showed lower apparent porosity with reinforcement of 10 wt% of 8YSZ (14.9%) and similar porosity for 20 wt% (20.7%). The samples manufactured with PMMA present similar porosity of 28.9% and 27.6%, for 10 wt% and 20 wt%, respectively. In both manufacturing conditions, the addition of 3YSZ or 8YSZ increased the apparent porosity in relation to the standard, 10T, see Figures 2, 3 and S2.

The bulk density of the sTCP doped with metal ions (reference material, 10T) presented a value of 2.80 g/cm^3. The density increased for 3.19 g/cm^3 and 3.36 g/cm^3 and for 3.18 g/cm^3 and 3.35 g/cm^3 for the dense biocomposites reinforced with 3YSZ and 8YSZ, respectively. The samples manufactured with PMMA, with 10 wt% and 20 wt% of 3YSZ, present a density of 2.30 g/cm^3 and 2.62 g/cm^3, respectively. While the porous samples reinforced with 10 wt% and 20 wt% of 8YSZ, show 2.73 g/cm^3 and 2.60 g/cm^3, respectively. The addition of 3YSZ and 8YSZ in dense biocomposites increased the density related to initial doped sTCP. However, in the samples manufactured with PMMA the density decreased or presented similar values (Figures 3 and S3).

These results are in agreement with the literature [1,18,31,45]. There, it was proven that, with the addition of 10 and 20% ZrO_2, the apparent porosity increased while the bulk density decreased, and this is what happened with the porous biocomposites analyzed in this work. With the addition of the 3YSZ and 8YSZ, porosity and bulk density increased, which can be explained by the effect of the sintering temperature (1300 °C). Considering the particle size distribution of the mixture up to 10 μm, with both zirconias up to 2 μm, the sintering conditions by pressure-less, unassisted sintering where, in general, densification is accompanied by (an undesirable) grain coarsening, the success of avoiding the grain growth is related to the control of the competition between densification and grain growth. That is extremely difficult because the driving forces for both are proportional to the reciprocal grain size and hence comparable in magnitude [51]. Thus, for the composition, particle size, and single step sintering in air the temperature used is not sufficient to promote the grain boundary atomic migration of zirconia. In these conditions, only the approximation of zirconia particles promotes the formation of zirconia agglomerates and partial grain growing with poor densification and superior porosity [52]. Nonetheless, the addition of ions like Mg^{2+} promotes the densification of materials [18,45]. Another explanation for the apparent porosity results in the use of the biocomposites manufactured with PMMA (40 wt%). Lee et al. [31] showed that, with different percentages of PMMA, starting with 40%, the porosity tends to increase gradually, and even with 40% the value is already quite considerable.

The diametrical compression tests revealed that the dense biocomposites with 3YSZ and 8YSZ presented mechanical strength values between 12.85 and 16.40 MPa and between 13.36 and 20.74 MPa, respectively (Table 4).

These results revealed that, in comparison with sTCP doped (13.4 MPa), the dense biocomposites with 20 wt% of 3YSZ and 8YSZ are higher. The samples 8T2cZ, are those that have higher mechanical strength, with a value of 20.7 MPa (55% higher than reference doped sTCP). As expected in induced porous biocomposites, the mechanical strength values were greatly decreased. From the initial 1.28 MPa (TCP doped) the addition of the 20 wt% of ZrO_2 decreases to 0.1 and 0.14 MPa for 3YSZ and 8YSZ, respectively (Figure S4).

Table 4. Diametral compression for dense and porous bioceramic composites.

sTCP	sTCP + t-ZrO_2		sTCP + c-ZrO_2	
10T	**9T1tZ**	**8T2tZ**	**9T1cZ**	**8T2cZ**
		Dense biocomposites		
13.36 ± 1.6	12.85 ± 1.6	16.40 ± 1.4	15.22 ± 1.5	20.74 ± 1.6
		Porous biocomposites		
1.28 ± 0.3	0.16 ± 0.1	0.10 ± 0.0	0.20 ± 0.0	0.14 ± 0.1

The addition of ZrO_2 to the ions doped with sTCP creates a ceramic microstructure in which the sTCP base matrix is reinforced with micro- and nanoparticles of zirconia (3YSZ and 8YSZ) which is due to its superior mechanical resistance (strength, toughness, and hardness). Furthermore, it is evident that the superior mechanical properties of the cubic phase (8YSZ) contribute more effectively to the increase in resistance than the tetragonal phase (3YSZ) [53].

Similar to previous studies, biocomposites with reinforcement of ZrO_2 have higher mechanical strength than "pure" biocomposites, i.e., biocomposites without the presence of ZrO_2 [1,54,55]. In this study the most resistant biocomposites have 20% of 3YSZ or 8YSZ. On the other hand, these results can also be explained by the addition of Mg^{2+}, that replaces the Ca^{2+} ions in its sites (doping effect), which have already been shown to increase the mechanical strength of the biocomposites [45]. Regarding biocomposites manufactured with PMMA, low mechanical strength was already reported by Lee et al. [31].

3.3. Biological Properties

The biocompatibility of dense ceramic biocomposites, both with and without porosity induced by PMMA reinforcement with 3YSZ and 8YSZ, was evaluated using a resazurin assay, with the corresponding results displayed in Figure 6. hOB were chosen as the cell model due to their pivotal role in bone matrix production and remodeling [56], a crucial aspect of the osseointegration process. Notably, the data obtained from the resazurin assay, as shown in Figure 6A,B, demonstrated that even after 7 days of incubation, the hOB cells remained highly metabolically active when in contact with both dense and porous biocomposites.

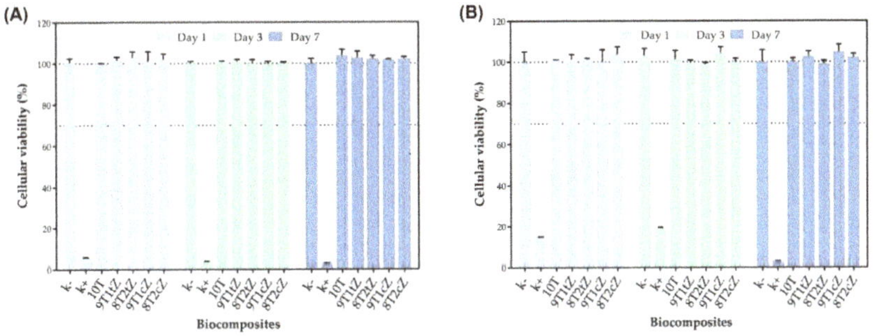

Figure 6. Evaluation of the cytocompatibility of biocomposites: (**A**) dense samples reinforced with 3YSZ and 8YSZ; (**B**) porous samples reinforced with 3YSZ and 8YSZ.

In fact, their cell viability consistently exceeded the 70% threshold, indicating strong biocompatibility. It is worth noting that the porous biocomposites were made from PMMA, a structure porosity that, according to the literature [57], could potentially hinder biocompatibility and bioactivity. However, the results of this study are consistent with previous research investigating the in vitro cytotoxicity of biocomposites containing β-TCP and

ZrO_2 [1,45,58]. Also, as in other reported studies [57,58], the use of PMMA during manufacturing did not leave residues that could harm the cytotoxicity results.

As in previous studies [45], no relevant differences were detected in cell viability and proliferation between the pure TCP composition and the TCP compositions doped with combinations of the four metal ions, while the bioceramic composites of TCP with zirconia did not show high cell viability. This cell viability can be attributed to the range of molar concentrations of ions incorporated into the crystalline structure of β-TCP, which closely mimics the composition of natural human bone [14,45,59].

SEM was used to visualize cell attachment and growth on the PMMA-induced porous biocomposites. Figure 7 shows the results at 3000× and 5000× magnification after 3 days of incubation. In addition, SEM images of the biocomposites without hOB cells, magnified to 1200×, are included for comparison.

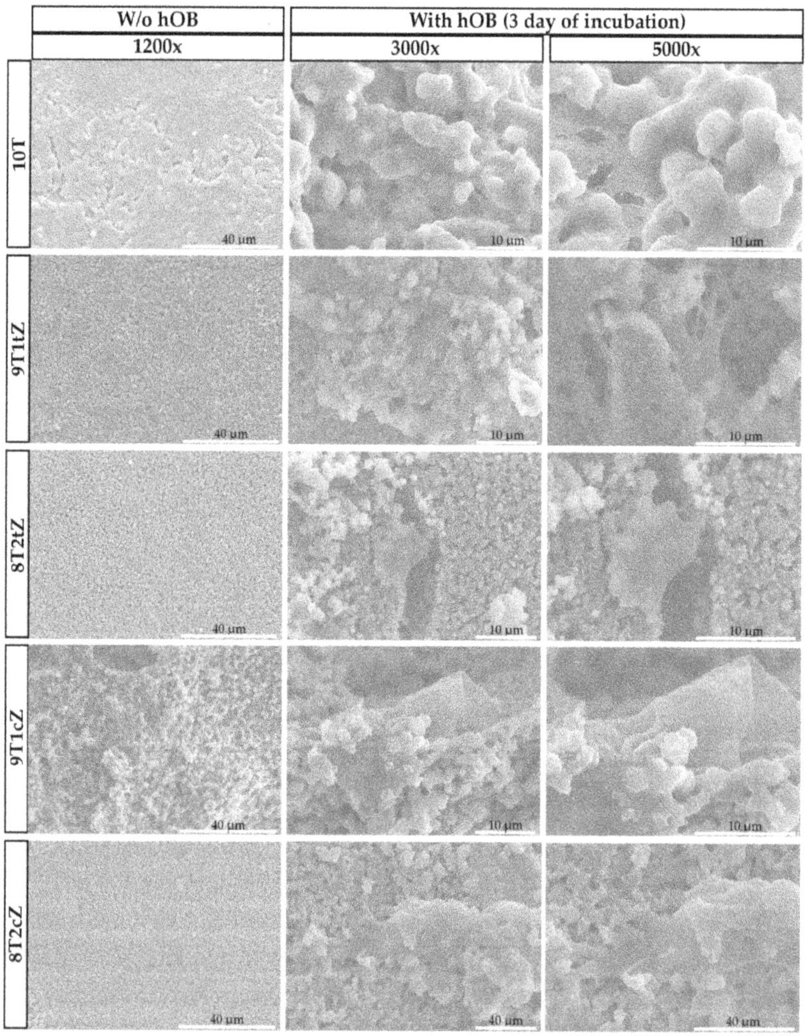

Figure 7. SEM images of cellular attachment to porous biocomposites after 3 days and without cellular attachment, with magnifications of 3000×, 5000× and 1200×, respectively.

These images clearly show that hOB cells not only adhered to the surface but also infiltrated the interior of the biocomposites over time. This observation is consistent with previous research using β-TCP [60]. These results may be associated with the porosity present in the materials, whose macropores have an ideal average size between 100–200 μm for cell growth, and according to the literature [61–63] this directly influences cell proliferation, promoting cell adhesion and growth.

4. Conclusions

Ceramic biocomposites with different percentage of metal ion doped sTCP, 3YSZ, and 8YSZ, in dense form and with porosity induced by PMMA, were manufactured. The ceramic microstructure and the composition obtained have a good similarity with human bone biomimicry. The effect of doping with 15 mol% metal ions of Mg^{2+}, Zn^{2+}, Mn^{2+}, and Fe^{3+} and simultaneously strengthening with two ZrO_2 (tetragonal and cubic crystalline phases) was assessed through the physical, mechanical, microstructural, and biological properties.

XRD, EDX, and FTIR analyses show that metal ions are present in the composition and replace the calcium in the crystal structure of the biocomposites. The β-TCP crystalline phase is predominant; however, hydroxyapatite is also present in the reference material (sTCP), most likely due to mechanochemical synthesis during the mixing/grinding process. As expected, the tetragonal and cubic phases are also present in the biocomposites reinforced by the respective ZrO_2. The images of the fracture surfaces show the porosity, interconnectivity, and fracture modes, while in the materials with induced porosity, voids generated by the PMMA spheres of approximately 100 microns are observed.

Moreover, from the analysis of the results, the following summary can be given:

(i) The addition of 20 wt% of 3YSZ increases the mechanical strength to 16.4 MPa, 22% higher than sTCP; and 20 wt% of 8YSZ increases the strength to 20.7 MPa, 55% highest than reference doped sTCP. In this way, as expected, cubic zirconia promotes greater resistance than tetragonal zirconia.

(ii) The apparent porosity increases 65%, from 12.6 to 24.6% with the addition of 10 wt% of 3YSZ and 60%, to 20.6% with the addition of 20 wt% of 8YSZ. This is justified by the sintering conditions of pressure-less, unassisted sintering, where the temperature is not sufficient to promote atomic migration along the zirconia grain boundary, promoting the formation of zirconia agglomerates and partial grain growing with poor densification.

(iii) The use of PMMA increases the apparent porosity for all ceramic biocomposites; however, this effect is more visible for sTCP, where the value obtained of 26.2%, matches an increase of two times. In this case the mechanical resistance is too low for structural applications.

(iv) The resazurin assay revealed that the two types of biocomposites produced did not affect the viability of hOB, presenting a cellular viability, in most cases, of 100%. These results highlight the ideal properties of these biocomposites as bone substitutes, especially the dense ones doped with 8YSZ, which showed a more constant behavior for the analyzed properties.

Supplementary Materials: The following supporting information can be downloaded at: https://www.mdpi.com/article/10.3390/biomimetics8080599/s1, Figure S1: X-ray diffraction spectra of the biocomposites 10T (sTCP) after calcination and after sintering and X-ray diffraction spectra of the theoretical cards #09-0432, #09-0169, and #09-0348, corresponding to HA, β-TCP and α-TCP, respectively; Figure S2: Apparent porosity of the composites: (a) dense reinforced with 3YSZ; (b) manufactured with PMMA and reinforced with 3YSZ; (c) dense reinforced with 8YSZ; (d) manufactured with PMMA and reinforced with 8YSZ; Figure S3: Bulk density of the composites: (a) dense reinforced with 3YSZ; (b) manufactured with PMMA and reinforced with 3YSZ; (c) dense reinforced with 8YSZ; (d) manufactured with PMMA and reinforced with 8YSZ; Figure S4: Mechanical strength of the composites: (a) dense reinforced with 3YSZ; (b) manufactured with PMMA and reinforced with 3YSZ; (c) dense reinforced with 8YSZ; (d) manufactured with PMMA and reinforced with 8YSZ.

Author Contributions: Conceptualization, A.P.S.; methodology, A.P.S., D.F.M. and N.F.F.; software, D.F.M.; formal analysis, V.M.F., D.F.M. and N.F.F.; investigation, V.M.F. and B.C.S.; resources, A.P.S.; writing—original draft preparation, V.M.F.; writing—review and editing, A.P.S., D.F.M. and N.F.F.; supervision, A.P.S.; project administration, A.P.S.; funding acquisition, A.P.S. All authors have read and agreed to the published version of the manuscript.

Funding: The research work was supported by FCT-Fundação para a Ciência e a Tecnologia, I.P./MCTES, through national funds (PIDDAC), within the scope of the Unit I&D: C-MAST, Projects UIDB/00151/2020 and UIDP/00151/2020; CICS-UBI, Projects UIDB/00709/2020 and UIDP/00709/2020.

Institutional Review Board Statement: Not applicable.

Data Availability Statement: The authors declare that the data supporting the findings of this study are available within the article.

Acknowledgments: The research work was supported by FCT-Fundação para a Ciência e a Tecnologia, I.P./MCTES, through national funds (PIDDAC), within the scope of the Unit I&D: C-MAST (Centre for Mechanical and Aerospace Science and Technology), Projects UIDB/00151/2020 and UIDP/00151/2020; CICS-UBI (Centro de Investigação em Ciências da Saúde), Projects UIDB/00709/2020 and UIDP/00709/2020. N.F.F. acknowledges his PhD fellowship from FCT (2022.14608.BD). D.F.M. also acknowledges his PhD fellowship from FCT (2022.14478.BD).

Conflicts of Interest: The authors declare no conflict of interest.

References

1. Vassal, M.F.; Nunes-Pereira, J.; Miguel, S.P.; Correia, I.J.; Silva, A.P. Microstructural, mechanical and biological properties of hydroxyapatite—$CaZrO_3$ biocomposites. *Ceram. Int.* **2019**, *45*, 8195–8203. [CrossRef]
2. Mantripragada, V.P.; Lecka-Czernik, B.; Ebraheim, N.A.; Jayasuriya, A.C. An overview of recent advances in designing orthopedic and craniofacial implants. *J. Biomed. Mater. Res.* **2013**, *101*, 3349–3364. [CrossRef] [PubMed]
3. Dubok, V.A. Bioceramics—Yesterday, today, tomorrow. *Powder Metall. Metal. Ceram.* **2000**, *39*, 381–394. [CrossRef]
4. Hisbergues, M.; Vendeville, S.; Vendeville, P. Review zirconia: Established facts and perspectives for a biomaterial in dental implantology. *J. Biomed. Mater. Res.* **2009**, *88*, 519–529. [CrossRef] [PubMed]
5. Christel, P.; Meunier, A.; Heller, M.; Torre, J.P.; Peille, C.N. Mechanical properties and short-term in vivo evaluation of yttrium-oxide-partially-stabilized zirconia. *J. Biomed. Mater. Res.* **1989**, *23*, 45–61. [CrossRef]
6. Nakonieczny, D.; Walke, W.; Majewska, J.; Paszenda, Z. Characterization of magnesia-doped yttria-stabilized zirconia powders for dental technology applications. *Acta Bioeng. Biomech.* **2014**, *16*, 97–104. [CrossRef]
7. Boobalan, K.; Vijayaraghavan, R.; Chidambaram, K.; Mudali, U.M.K.; Raj, B. Preparation and characterization of nanocrystalline zirconia powders by the glowing combustion method. *J. Am. Ceram. Soc.* **2010**, *93*, 3651–3656. [CrossRef]
8. Kelly, J.R.; Denry, I. Stabilized zirconia as a structural ceramic: An overview. *Dent. Mater.* **2008**, *24*, 289–298. [CrossRef]
9. Nakonieczny, D.S.; Antonowicz, M.; Paszenda, Z.K.; Radko, T.; Drewniak, S.; Bogacz, W.; Krawcyk, C. Experimental investigation of particle size distribution and morphology of alumina-yttria-ceria-zirconia powders obtained via sol–gel route. *Biocybern. Biomed. Eng.* **2018**, *38*, 535–543. [CrossRef]
10. Vasile, B.S.; Andronescu, E.; Ghitulica, C.; Vasile, O.R.; Curechiu, L.; Scurtu, R.; Vasile, E.; Trusca, R.; Pall, L.; Aldica, V. Microstructure and electrical properties of zirconia and composite nanostructured ceramics sintered by different methods. *Ceram. Int.* **2013**, *39*, 2535–2543. [CrossRef]
11. Green, D.J.; Hannink, R.H.J.; Swain, M.V. *Transformation Toughenung of Ceramics*; CRC Press, Inc.: Boca Raton, FL, USA, 1989; Chapter 2; pp. 17–55.
12. Chevalier, J.; Cales, B.; Drouin, J.M. Low-Temperature Aging of Y-TZP Ceramics. *J. Am. Ceram. Soc.* **2004**, *82*, 2150–2154. [CrossRef]
13. Delfino, C.S.; Ribeiro, C.; Vieira, G.F.; Bressiani, A.H.A.; Turbino, M.L. The use of new materials for pulp capping (hydroxyapatite–HAp and tricalcium phosphate-β-TCP). *Cerâmica* **2010**, *56*, 381–388. [CrossRef]
14. Macedo, D.F.; Silva, A.P. Analysis of the effect of doped metal ions in the tricalcium phosphate bioceramic. *J. Phys. Conf. Ser.* **2021**, *1960*, 012014. [CrossRef]
15. Kwon, S.H.; Jun, Y.K.; Hong, S.H.; Kim, H.E. Synthesis and dissolution behavior of β-TCP and HA/β-TCP composite powders. *J. Eur. Ceram. Soc.* **2003**, *23*, 1039–1045. [CrossRef]
16. Alcaide, M.; Serrano, M.C.; Pagani, R.; Sánchez-Salcedo, S.; Vallet-Regí, M.; Portolés, M.T. Biocompatibility markers for the study of interactions between osteoblasts and composite biomaterials. *Biomaterials* **2009**, *30*, 45–51. [CrossRef]
17. Thamaraiselvi, T.V.; Rajeswari, S. Biological Evaluation of Bioceramic Materials-A Review. *Trends Biomater. Artif. Organs* **2004**, *18*, 9–17.

18. Frasnelli, M.; Sglavo, V.M. Effect of Mg2+ doping on beta-alpha phase transition in tricalcium phosphate (TCP) bioceramics. *Acta Biomater.* **2016**, *33*, 283–289. [CrossRef]
19. Yashima, M.; Sakai, A. High-temperature neutron powder diffraction study of the structural phase transition between α and α' phases in tricalcium phosphate $Ca_3(PO_4)_2$. *Chem. Phys. Lett.* **2003**, *372*, 779–783. [CrossRef]
20. Ran, J.; Jiang, P.; Sun, G.; Ma, Z.; Hu, J.; Shen, X.; Tong, H. Comparisons among Mg, Zn, Sr, and Si doped nano-hydroxyapatite/chitosan composites for load-bearing bone tissue engineering applications. *Mater. Chem. Front.* **2017**, *1*, 900–910. [CrossRef]
21. Baker, W.L. Treating arrhythmias with adjunctive magnesium: Identifying future research directions. *Eur. Heart J. Cardiovasc. Pharmacother.* **2017**, *3*, 108–117. [CrossRef]
22. de Baaij, J.H.F.; Hoenderop, J.G.J.; Bindels, R.J.M. Magnesium in Man: Implications for Health and Disease. *Physiol. Rev.* **2015**, *95*, 1–46. [CrossRef] [PubMed]
23. Zoroddu, M.A.; Aaseth, J.; Crisponi, G.; Medici, S.; Peana, M.; Nurchi, V.M. The essential metals for humans: A brief overview. *J. Inorg. Biochem.* **2019**, *195*, 120–129. [CrossRef] [PubMed]
24. Enderle, R.; Götz-Neunhoeffer, F.; Göbbels, M.; Müller, F.A.; Greil, P. Influence of magnesium doping on the phase transformation temperature of β-TCP ceramics examined by Rietveld refinement. *Biomaterials* **2005**, *26*, 3379–3384. [CrossRef]
25. Arcos, D.; Vallet-Regí, M. Substituted hydroxyapatite coatings of bone implants. *J. Mater. Chem. B* **2020**, *8*, 1781–1800. [CrossRef] [PubMed]
26. Banerjee, S.S.; Tarafder, S.; Davies, N.M.; Bandyopadhyay, A.; Bose, S. Understanding the influence of MgO and SrO binary doping on the mechanical and biological properties of β-TCP ceramics. *Acta Biomater.* **2010**, *6*, 4167–4174. [CrossRef]
27. Bose, S.; Tarafder, S.; Banerjee, S.S.; Davies, N.M.; Bandyopadhyay, A. Understanding in vivo response and mechanical property variation in MgO, SrO and SiO_2 doped β-TCP. *Bone* **2011**, *48*, 1282–1290. [CrossRef]
28. Singh, R.K.; Srivastava, M.; Prasad, N.K.; Awasthi, S.; Dhayalan, A.; Kannan, S. Iron doped β-Tricalcium phosphate: Synthesis, characterization, hyperthermia effect, biocompatibility and mechanical evaluation. *Mater. Sci. Eng. C* **2017**, *78*, 715–726. [CrossRef]
29. Tarafder, S.; Davies, N.M.; Bandyopadhyay, A.; Bose, S. 3D printed tricalcium phosphate bone tissue engineering scaffolds: Effect of SrO and MgO doping on in vivo osteogenesis in a rat distal femoral defect model. *Biomater. Sci.* **2013**, *1*, 1250–1259. [CrossRef]
30. Rodríguez, J.; Mandalunis, P.M. A review of metal exposure and its effects on bone health. *J. Toxicol.* **2018**, *2018*, 4854152. [CrossRef]
31. Lee, Y.H.; Lee, J.W.; Yang, S.Y.; Lee, H.; Koh, Y.H.; Kim, H.E. Dual-scale porous biphasic calcium phosphate gyroid scaffolds using ceramic suspensions containing polymer microsphere porogen for digital light processing. *Ceram. Int.* **2021**, *47*, 11285–11293. [CrossRef]
32. Sa, Y.; Yang, F.; De Wijn, J.R.; Wang, Y.; Wolke, J.G.C.; Jansen, J.A. Physicochemical properties and mineralization assessment of porous polymethylmethacrylate cement loaded with hydroxyapatite in simulated body fluid. *Mater. Sci. Eng. C* **2016**, *61*, 190–198. [CrossRef]
33. Shi, M.; Kretlow, J.D.; Nguyen, A.; Young, S.; Baggett, L.S.; Wong, M.E.; Kasper, F.K.; Mikos, A.G. Antibiotic-releasing porous polymethylmethacrylate constructs for osseous space maintenance and infection control. *Biomaterials* **2010**, *31*, 4146–4156. [CrossRef]
34. Descamps, M.; Duhoo, T.; Monchau, F.; Lu, J.; Hardouin, P.; Hornez, J.C.; Leriche, A. Manufacture of macroporous β-tricalcium phosphate bioceramics. *J. Eur. Ceram. Soc.* **2008**, *28*, 149–157. [CrossRef]
35. Nunes-Pereira, J.; Carneiro, P.M.C.; Maceiras, A.; Baudín, C.; Silva, A.P. Modelling of elastic modulus of $CaZrO_3$-MgO composites using isotropic elastic and anisotropic models. *J. Eur. Ceram. Soc.* **2020**, *40*, 5882–5890. [CrossRef]
36. Rodriguez-Carvajal, J. Recent developments of the program FULLPROF, commission on powder diffraction. *Comm. Powder Diffr. (IUCr)* **2001**, *26*, 12–19.
37. *ASTM C179-04*; ASTM Standard Test Method for Drying and Firing Linear Change of Refractory Plastic and Ramming Mix Specimens. ASTM: West Conshohocken, PA, USA, 2004; pp. 1–2. Available online: www.astm.org (accessed on 2 May 2023).
38. Jonsén, P.; Häggblad, H.Å.; Sommer, K. Tensile strength and fracture energy of pressed metal powder by diametral compression test. *Powder Technol.* **2007**, *176*, 148–155. [CrossRef]
39. García-Páez, I.H.; Carrodeguas, R.G.; De Aza, A.H.; Baudín, C.; Pena, P. Effect of Mg and Si co-substitution on microstructure and strength of tricalcium phosphate ceramics. *J. Mech. Behav. Biomed. Mater.* **2014**, *30*, 1–15. [CrossRef] [PubMed]
40. O'Brien, J.; Wilson, I.; Orton, T.; Pognan, F. Investigation of the Alamar Blue (resazurin) fluorescent dye for the assessment of mammalian cell cytotoxicity. *Eur. J. Biochem.* **2000**, *267*, 5421–5426. [CrossRef]
41. Sequeira, R.S.; Miguel, S.P.; Cabral, C.S.D.; Moreira, A.F.; Ferreira, P.; Correia, I.J. Development of a poly(vinyl alcohol)/lysine electrospun membrane-based drug delivery system for improved skin regeneration. *Int. J. Pharm.* **2019**, *570*, 118640. [CrossRef]
42. Fernandes, N.; Rodrigues, C.F.; de Melo-Diogo, D.; Correia, I.J.; Moreira, A.F. Optimization of the gsh-mediated formation of mesoporous silica-coated gold nanoclusters for nir light-triggered photothermal applications. *Nanomaterials* **2021**, *11*, 1946. [CrossRef] [PubMed]
43. Yeong, B.; Junmin, X.; Wang, J. Mechanochemical synthesis of hydroxyapatite from calcium oxide and brushite. *J. Am. Ceram. Soc.* **2001**, *84*, 465–467. [CrossRef]
44. Toriyama, M.; Ravaglioli, A.; Krajewski, A.; Celotti, G.; Piancastelli, A. Synthesis of Hydroxyapatite-Based Powders by Mechano-Chemical Method and their Sintering. *J. Eur. Ceram. Soc.* **1996**, *16*, 429–436. [CrossRef]

45. Macedo, D.F.; Cunha, A.F.; Mano, J.F.; Oliveira, M.B.; Silva, A.P. Tricalcium phosphate doped with Mg2+ and combinations of Mn2+, Zn2+ and Fe3+: A DoE study on sintering, mechanical, microstructural and biological properties. *Ceram. Int.* **2022**, *48*, 20467–20477. [CrossRef]
46. Wu, S.; Liu, X.; Yeung, K.W.K.; Liu, C.; Yang, X. Biomimetic porous scaffolds for bone tissue engineering. *Mater. Sci. Eng. R Rep.* **2014**, *80*, 1–36. [CrossRef]
47. Xidaki, D.; Agrafioti, P.; Diomatari, D.; Kaminari, A.; Tsalavoutas-Psarras, E.; Alexiou, P.; Psycharis, V.; Tsilibary, E.C.; Silvestros, S.; Sagnou, M. Synthesis of hydroxyapatite, β-Tricalcium phosphate and biphasic calcium phosphate particles to act as local delivery carriers of curcumin: Loading, release and in vitro studies. *Materials* **2018**, *11*, 595. [CrossRef] [PubMed]
48. Nahar, U.K.; Shovon, B.; Chandra, D.R.; Chandra, P.S.; Shukanta, B.; Muhammed, Y.M.; Islam, M.D.S. Characterization of Beta-Tricalcium Phosphate (β- TCP) Produced at Different Process Conditions. *J. Bioeng. Biomed. Sci.* **2017**, *7*, 2. [CrossRef]
49. Dwivedi, R.; Maurya, A.; Verma, A.; Prasad, R.; Bartwal, K.S. Microwave assisted sol-gel synthesis of tetragonal zirconia nanoparticles. *J. Alloys Compd.* **2011**, *509*, 6848–6851. [CrossRef]
50. Manjunatha, S.; Dharmaprakash, M.S. Microwave assisted synthesis of cubic Zirconia nanoparticles and study of optical and photoluminescence properties. *J. Lumin.* **2016**, *180*, 20–24. [CrossRef]
51. Hotza, D.; García, D.E.; Castro, R.H.R. Obtaining highly dense YSZ nanoceramics by pressureless, unassisted sintering. *Int. Mater. Rev.* **2015**, *60*, 353–375. [CrossRef]
52. Booth, F.; Garrido, L.; Aglietti, E.; Silva, A.; Pena, P.; Baudín, C. CaZrO3-MgO structural ceramics obtained by reaction sintering of dolomite-zirconia mixtures. *J. Eur. Ceram. Soc.* **2016**, *36*, 2611–2626. [CrossRef]
53. Silva, A.P.; Booth, F.; Garrido, L.; Aglietti, E.; Pena, P.; Baudín, C. Young's modulus and hardness of multiphase $CaZrO_3$-MgO ceramics by micro and nanoindentation. *J. Eur. Ceram. Soc.* **2018**, *38*, 2194–2201. [CrossRef]
54. Ayoub, G.; Veljovic, D.; Zebic, M.L.; Miletic, V.; Palcevskis, E.; Petrovic, R.; Janackovic, D. Composite nanostructured hydroxyapatite/yttrium stabilized zirconia dental inserts—The processing and application as dentin substitutes. *Ceram. Int.* **2018**, *44*, 18200–18208. [CrossRef]
55. Castkova, K.; Hadraba, H.; Matousek, A.; Roupcova, P.; Chlup, Z.; Novotna, L.; Cihlar, J. Synthesis of Ca,Y-zirconia/hydroxyapatite nanoparticles and composites. *J. Eur. Ceram. Soc.* **2016**, *36*, 2903–2912. [CrossRef]
56. Albrektsson, T.; Johansson, C. Osteoinduction, osteoconduction and osseointegration. *Eur. Spine J.* **2001**, *10*, S96–S101. [CrossRef]
57. Sugino, A.; Miyazaki, T.; Kawachi, G.; Kikuta, K.; Ohtsuki, C. Relationship between apatite-forming ability and mechanical properties of bioactive PMMA-based bone cement modified with calcium salts and alkoxysilane. *J. Mater. Sci. Mater. Med.* **2008**, *19*, 1399–1405. [CrossRef]
58. Jäger, M.; Wilke, A. Comprehensive biocompatibility testing of a new PMMA-HA bone cement versus conventional PMMA cement in vitro. *J. Biomater. Sci. Polym. Ed.* **2003**, *14*, 1283–1298. [CrossRef] [PubMed]
59. Hurle, K.; Oliveira, J.M.; Reis, R.L.; Pina, S.; Goetz-Neunhoeffer, F. Ion-doped Brushite Cements for Bone Regeneration. *Acta Biomater.* **2021**, *123*, 51–71. [CrossRef]
60. Lu, J.X.; Flautre, B.; Anselme, K.; Hardouin, P.; Gallur, A.; Descamps, M.; Thierry, B. Role of interconnections in porous bioceramics on bone recolonization in vitro and in vivo. *J. Mater. Sci. Mater. Med.* **1999**, *10*, 111–120. [CrossRef]
61. Volkmer, T.M.; Santos, L.A.D. Influence of the induction time on the properties of porous hydroxyapatite obtained by gelcasting foams. *Cerâmica* **2007**, *53*, 429–435. [CrossRef]
62. Flautre, B.; Descamps, M.; Delecourt, C.; Blary, M.C.; Hardouin, P. Porous HA ceramic for bone replacement: Role of the pores and interconnections—Experimental study in the rabbit. *J. Mater. Sci. Mater. Med.* **2001**, *12*, 679–682. [CrossRef]
63. Klawitter, J.J.; Bagwell, J.G.; Weinstein, A.M.; Sauer, B.W. An Evaluation of Bone Growth into Porous High Density Polyethylene. *J. Biomed. Mater. Res.* **1976**, *10*, 311–323. [CrossRef] [PubMed]

biomimetics

MDPI

Article

Bioactive Hydroxyapatite Aerogels with Piezoelectric Particles

Catarina Tavares [1], Tânia Vieira [2], Jorge C. Silva [2,*], João P. M. R. Borges [1,*] and M. Carmo Lança [1]

1 CENIMAT | i3N, Department of Materials Science, School of Science and Technology, NOVA University Lisbon, 2829-516 Caparica, Portugal; cdc.tavares@campus.fct.unl.pt (C.T.); mcl@fct.unl.pt (M.C.L.)

2 CENIMAT | i3N, Department of Physics, School of Science and Technology, NOVA University Lisbon, 2829-516 Caparica, Portugal; ts.vieira@fct.unl.pt

* Correspondence: jcs@fct.unl.pt (J.C.S.); jpb@fct.unl.pt (J.P.M.R.B.)

Abstract: Open-cell foams based on hydroxyapatite (HAp) can mimic the extracellular matrix (ECM) to better replace damaged hard tissues and assist in their regeneration processes. Aerogels of HAp nanowires (NW) with barium titanate (BT) particles were produced and characterized regarding their physical and chemical properties, bioactivity, and in vitro cytotoxicity. Considering the role of piezoelectricity (mainly due to collagen) and surface charges in bone remodeling, all BT particles, of size 280 nm and 2 and 3 μm, contained $BaTiO_3$ in their piezoelectric tetragonal phase. The synthesized nanowires were verified to be AB-type carbonated hydroxyapatite. The aerogels showed high porosity and relatively homogeneous distribution of the BT particles. Barium titanate proved to be non-cytotoxic while all the aerogels produced were cytotoxic for an extract concentration of 1 mg/mL but became non-cytotoxic at concentrations of 0.5 mg/mL and below. It is possible that these results were affected by the higher surface area and quicker dissolution rate of the aerogels. In the bioactivity assays, SEM/EDS, it was not easy to differentiate between the apatite deposition and the surface of the HAp wires. However, a quantitative EDS analysis shows a possible CaP deposition/dissolution cycle taking place.

Keywords: piezoelectricity; aerogel; hydroxyapatite; barium titanate; solvothermal synthesis; bioactivity

Citation: Tavares, C.; Vieira, T.; Silva, J.C.; Borges, J.P.M.R.; Lança, M.C. Bioactive Hydroxyapatite Aerogels with Piezoelectric Particles. *Biomimetics* **2024**, *9*, 143. https://doi.org/10.3390/biomimetics9030143

Academic Editor: Jing Sun

Received: 21 November 2023
Revised: 16 February 2024
Accepted: 19 February 2024
Published: 27 February 2024

1. Introduction

Bone is a dynamic tissue capable of regenerating, growing, and remodeling to preserve its structural integrity and mineral homeostasis. While the natural remodeling ability of bone tissue is remarkable, there exists a threshold limit that dictates the speed and extent to which regeneration can occur. For instance, fractures due to traumatic injuries, age degenerative processes, or even surgical removal of tumors that result in the absence or large defects of bone, lead to the need for new features of biomaterials and medical devices to improve the natural healing process of bone [1,2]. The increasing importance of orthopedic biomaterials is reflected in the growth of their global market. In 2022 this market reached around USD 17,000 million and it is expected to grow by about 10% in 2023 [3]. From these numbers, it is evident that there is a constant demand for innovative biomaterials in this field so that patients' healing ability is improved and the time of recovery can be reduced. Materials that enhance osteointegration and osteogenesis, particularly in the early stages of bone regeneration, will help to reduce the period of patients' immobilization.

One material of interest is hydroxyapatite (HAp), represented by the chemical formula $Ca_{10}(PO_4)_6(OH)_2$. This ceramic belongs to the family of calcium phosphates (CaP), also named apatites. It is a highly biocompatible ceramic, chemically similar to the bone's mineral phase, and natural HAp that has a Ca/P atomic ratio between 1.5 and 1.67 and demonstrates osteoconductive, osteoinductive, and osteointegrative properties [4,5]. The main mechanism for bioactivity is the release of ionic products, such as calcium and phosphate ions, by partial dissolution upon implantation. This process, in turn, leads to the precipitation of biological apatite on the surface of the ceramics in a process called bone mineralization. Consequently, this contributes to the attachment of the implant to the host's bone, improving implant fixation to the surrounding tissues [4,6]. Since HAp bioceramics are brittle in nature and have low fracture toughness compared to natural bone, their applications in orthopedics are usually as non-load-bearing implants, the filling of bone defects, or in load-bearing implants as coatings, either in dense or porous forms [6].

In addition to the chemical composition of the scaffolds, their morphology also plays an important role in the process of bone remodeling and the osteointegration of the implant [7]. For instance, the porosity of a scaffold, controlled by its synthesis method, has been shown to improve implant performance due to better bio-integration and mechanical stability of the implant. The higher surface area and open pore structure improve the transport of oxygen and nutrients, along with higher surface area enhancing the migration of cells responsible for bone tissue deposition [5,8].

Among various HAp morphologies, hydroxyapatite nanowires (HAp NWs) seem to be highly promising in several fields [9–11]. Huang et al. explored the use of highly porous aerogels (~99.7% porosity) composed of HAp NWs as scaffolds in bone regeneration and neovascularization [11]. Compared to dense hydroxyapatite, the HAp NWs aerogels promote adhesion and migration deeply into the pores of osteoblasts and other cells responsible for bone regeneration. Moreover, these aerogels exhibit better mechanical properties, such as high elasticity and high fracture resistance due to the flexibility and 3D network of the HAp nanowires [9,10].

Barium titanate ($BaTiO_3$, BT) is a perovskite ceramic with interesting applications in the biomedical field. In its tetragonal phase, it demonstrates piezoelectric properties and is also biocompatible [12]. As early as 1957, Eiichi Fukada et al., developed research that proved the existence of piezoelectricity in bone [13,14]. Currently, it is now known that piezoelectric effects dominate in dry bone, while in wet bone, regeneration and remodeling electric cues are mainly related to stress generated potentials (SGPs), with a smaller contribution of the direct piezoelectric effect [15,16]. Therefore, surfaces that present stable electrical charges, either by dipole orientation or space charge trapping, can accelerate the regenerative processes of hard tissue [17]. Implementing this surface modification in biomaterials, using materials with permanent dipoles, such as tetragonal BT, and mimicking the natural behavior of bone tissue seems quite helpful as a stimulus for bone remodeling, possibly enhancing the osteoinduction effects of the implant. As highlighted in the recent literature review by Zhang et al. [18], there are an increasing number of studies to enhance the bone regenerative response using composites of HAp and BT, including developments presenting open porous morphology. Also, recently, interest has increased in highly structured HAp materials to overcome the brittleness typical of ceramics [19]. In these are included the aerogels of nanowires of HAp as prospective biomaterials in hard tissue engineering [20,21]. To the best of the authors' knowledge, aerogels of HAp nanowires/BT micro- and nanoparticles are a novelty that attempts to combine the advantages of a highly porous structure as a cell scaffold (better osteoconductitvity) and to accelerate osteogenesis and osteoinduction by the presence of surface charges induced by the piezoelectric BT.

The main purpose of this work was the synthesis and physical, chemical, and in vitro biological characterization of a hydroxyapatite aerogel along with an innovative composite aerogel of HAp embedded with barium titanate particles with significant potential for biomedical applications. All the BT particles and the aerogels were characterized by X-ray diffraction (XRD), Fourier-transform infrared (FTIR), and Raman spectroscopies. The HAp aerogels were analyzed by SEM to make sure that the pore size and interconnectivity of pores were suitable for their expected applications. In addition, the dispersion of the BT particles was observed on the composite aerogels. The biocompatibility of all materials was evaluated by cytotoxicity assays. Finally, the bioactivity of the aerogels with and without the piezoelectric particles was studied through simulated body fluid (SBF) immersion.

2. Materials and Methods

2.1. Barium Titanate

Commercial tetragonal BT particles with sizes of 280 nm (LT#NG04MO0503—Nanografi, Jena, Germany), 2 μm (LT#MKBB0111V—Sigma-Aldrich, Saint Louis, MO, USA), and 3 μm (LT#MFCD00003447—Sigma-Aldrich) were selected because of their piezoelectricity. Additionally, the selection aimed to test the effect of different particle sizes on bioactivity.

2.2. Hydroxypatite-Based Aerogels

The HAp NWs were synthesized based on previous research by Zhi-Chao Xiong et al. [9]. Firstly, a solution of 9.36 g of oleic acid (31997—Alfa Aesar, Kandel, Germany) in 13.50 g of deionized water and 4.75 g of methanol (L13255—Alfa Aesar) was mixed under mechanical stirring. A second solution consisting of 1.05 g of NaOH (reagent grade, 97%—655104—Sigma-Aldrich) dissolved in 15 g of deionized water was poured into the first solution and stirred for 30 min, while two other solutions of 0.33 g of $CaCl_2$ (anhydrous, 93%—12316—Alfa Aesar) dissolved in 12 g of deionized water and 0.94 g of $NaH_2PO_4{\cdot}2H_2O$ (purum p.a., ≥99.0%—71500—Sigma-Aldrich) dissolved in 18 g of deionized water were separately poured into the previous mixture and stirred for 10 min each. Finally, 5 mL of the reaction system was poured into a 100 mL Teflon-lined stainless-steel autoclave for a solvothermal reaction. This reaction was tested for different times such as 5, 7, 18, and 24 h at temperatures of 120, 165, and 180 °C. The resulting solution was stirred for 30 min at 300 rpm. To separate the HAp NWs from the impurities, the solution was centrifuged for 10 min at 3000 rpm and washed with progressively diluted methanol/deionized water ratios. The HAp NWs were stored in deionized water.

Following this, the best solvothermal conditions for the formation of HAp NWs were chosen by observing the previously synthesized and dried HAp NWs under SEM and by XRD characterization. The synthesis parameters selected, 180 °C and 18 h, yielded more differentiated and longer nanowires (see Section 3.2).

Next, four different slurries were made from the solution of HAp NWs synthesized at 180 °C for 18 h and stored in deionized water, where one contained only the HAp NWs and three corresponded to 20% BT/80% HAp (ratio of *w*%/*w*%), using tetragonal BT particles with sizes, respectively, of 280 nm and 2 and 3 μm, dispersed through sonication for 15 min. To produce the aerogel structures, each slurry was put into a mold and frozen at –5 °C for 12 h, followed by lyophilization at −40 °C and 0.001 mbar for 24 h.

The HAp aerogel without BT particles was denominated HAp, and the three HAp aerogels with BT were named, respectively, HAp/BT280 for an aerogel embedded with 280 nm sized particles, HAp/BT2 in the case of 2 μm, and HAp/BT3 for the aerogel with the 3 μm particles.

2.3. Materials Characterization

2.3.1. Structural, Chemical, and Morphologic Analysis

The analysis of the crystallographic structures was conducted in an X'Pert PRO (PANAnalytical, Malvern, UK) X-ray diffractometer with the use of CuKα radiation generated at 45 kV and 40 mA, in the range of $15° < 2θ < 60°$, with a step size of 0.08° for the Hap NWs and in the range of $20° < 2θ < 80°$, with a step size of 0.08° for the $BaTiO_3$, particles. An additional analysis of the BT, in order to better observe the phase composition of the sample, was conducted in the range of $44° < 2θ < 47°$, with a step size of 0.002°.

Additionally, for BT and HAp, a FTIR analysis with the use of the spectrophotometer FT-IR Thermo Nicolet 6700, was carried out in the wavenumber range from 4000 to 400 cm^{-1} to identify the functional groups of BT and HAp. To evaluate the molecular structure of the samples and to confirm the presence of the tetragonal phase in $BaTiO_3$, a Raman spectroscopy analysis was conducted using the Renishaw (Wotton-under-Edge, UK) inVia Qontor micro-Raman spectrometer. Five scans were made, each with an integration time of 1 s, using an incident 632.81 nm laser with an intensity of 0.32 mW and a frequency range from 150 to 1000 cm^{-1}.

To ensure the existence of HAp nanowires, the aerogel's morphology, and the dispersion of the BT particles within them, the samples were observed under a scanning electron microscope/Energy Dispersive X-ray Spectrometry (SEM/EDS), model Hitachi (Tokyo, Japan) TM 3030 Plus. All samples were sputter-coated with either gold or titanium before SEM observation.

2.3.2. Biocompatibility Assays

The colorimetric cytotoxicity assays, with the use of resazurin, were performed on all the $BaTiO_3$ particles and HAp aerogels, with and without BT, in conformity with the standard ISO-10993-5 "Biological evaluation of medical devices, Part 5: Tests for in vitro cytotoxicity" test protocol [22]. All samples subjected to these tests were previously sterilized by heating at 150 °C for 5 h. Extracts of the BT particles were prepared by leaving them in contact with a complete culture medium (McCoy 5A from Sigma Aldrich supplemented with 10% fetal bovine serum from Biowest and 1% penicillin/streptomycin from Gibco, Waltham, MA, USA) at a concentration of 60 mg/mL at 37 °C for 48 h, while the extracts of aerogels were prepared in complete culture medium at a concentration of 1 mg/mL at 37 °C for 48 h. Additionally, SaOs2 cells (provided by the American Type Culture Collection, ATCC, Manassas, VA, USA HTB-85) were trypsinized using TrypLE Express (from Gibco) for 5 min at 37 °C and the cells were counted using a hemocytometer. The cells were seeded at a density of 30,000 cells/cm^2 in 96-well plates and incubated at 37 °C and 5% CO_2 for 24 h. After that period, the culture medium was replaced by the previously mentioned extracts (initial concentration C0) and 3 dilutions (C0/2, C0/4, and C0/8). A positive control (cytotoxic), where cells were treated with 10% dimethyl sulfoxide (DMS), and a negative control (non-cytotoxic), where cells were treated with a complete culture medium, were defined. The cells were left in contact with the extracts for 48 h. Following that, the medium was replaced by a solution containing 50% resazurin (B21187—Alfa Aesar) solution at 0.04 mg/mL in PBS and 50% complete culture medium, and the plates were incubated for 3 h at 37 °C and 5% CO_2. After that time, using a Biotek (Winooski, VT, USA) ELX800 microplate reader, the absorbance was read at 570 and 600 nm and the corrected absorbance was determined by subtracting the absorbance measured at 600 nm from the one measured at 570 nm and subtracting the medium control. Cell viability is given by the percentage of viable cells in the tested samples relative to the negative control.

2.3.3. Bioactivity Assessment

The bioactivity of the HAp and HAp/BT aerogels was analyzed by evaluating the growth of CaP structures on the surface of the membrane when in contact with simulated body fluid (protocol for SBF solution can be found in Supplementary Materials). The samples evaluated consist of a HAp aerogel without particles, three HAp aerogels with $BaTiO_3$, each with a different particle size, and a control group that was not in contact with SBF. Each sample was submerged for 1, 3, and 7 days and the SBF solution was renewed every 2 days. After removal from SBF, the samples were rinsed in distilled water to remove water-soluble salts, such as NaCl. Finally, the samples were observed in SEM and EDS to identify the growth of CaP structures.

3. Results and Discussion

3.1. *Barium Titanate*

3.1.1. X-ray Diffraction Analysis

The XRD analysis was carried out to identify the tetragonal phase of the $BaTiO_3$ particles and its prevalence over the non-piezoelectric cubic phase. Comparison of the peaks with the ICDD data sheets #00-005-0626 and #01-084-9618, corresponding to tetragonal and cubic BT, respectively, confirmed that the diffractograms in Figure 1 are characteristic of BT in the tetragonal phase (also, no additional peaks due to impurities were detected).

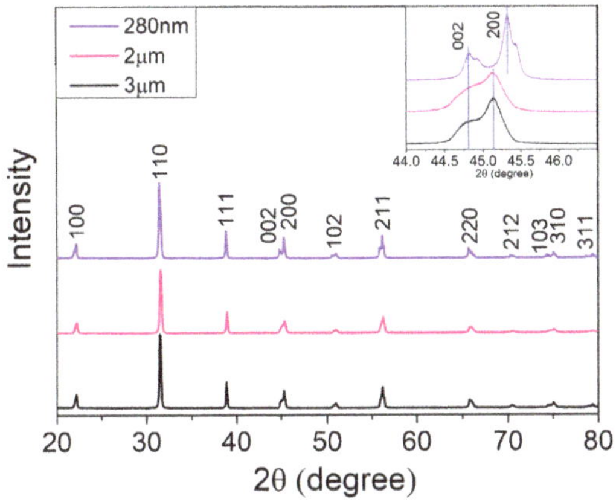

Figure 1. XRD analysis of the $BaTiO_3$ powders.

The transformation from cubic to tetragonal phase exhibits itself in the XRD as the presence of double peaks along the diffraction pattern due to the asymmetry of the unit cell along the *c* axis [23]. For $BaTiO_3$, this is seen in better detail around 2θ = 45°, corresponding to planes (002) and (200), where the existence of a split peak distinguishes tetragonal BT from its cubic phase. The double diffraction peak is more visible on the 280 nm powders, with a peak at 44.8° (002) and 45.4° (200). Regarding the 2 and 3 μm powders, there is a similarity between the intensity of the diffraction split, also corresponding to the same planes. As can be observed, when compared with the nanometric BT, the peaks' separation is much less pronounced for the micrometric particles. These results are in agreement with the DSC data for BT 280 nm powders (the DSC/TG plot is presented in Supplementary Materials in Figure S1), where a small peak is observed at 130.6 °C, the Curie temperature, corresponding to the transition from tetragonal to cubic BT. On the contrary, no peak could be perceived for the 2 and 3 μm particles in the DSC data.

The average sizes of the crystalline grains, τ, present in Table 1, were calculated according to the Scherrer Equation (1), using an X-ray wavelength of $\lambda(CuK\alpha)$ = 1.5418 Å, a shape factor of the crystallite, κ, of 0.9, and a line broadening at half the maximum intensity, β, for each peak [24–26]:

$$\tau = \frac{\kappa\lambda}{\beta\cos\theta} \quad (1)$$

As can be seen, the average crystallite sizes are larger for the micrometric samples, 2 and 3 μm, leading to larger, less defined peaks and higher superposition between each other. Although all the samples seem to exhibit a tetragonal-dominant structure, as stated before, the samples of 2 and 3 μm show a smaller difference between the angles corresponding to the two peaks in Figure 1, when compared to the sample of 280 nm, which is caused by a reduced tetragonal phase and the partial existence of a cubic phase. A higher percentage of tetragonal phase, such as in the particles of 280 nm, would be more suited for a possible polarization of the HAp aerogels [18].

Table 1. Crystallite size of each $BaTiO_3$ sample according to the Scherrer equation.

Sample	Peak, 2θ (°)	FWHM, β (rad)	Crystallite Size, τ (nm)
280 nm	31.43	4.81×10^{-3}	29.9
2 μm	31.54	4.42×10^{-3}	32.5
3 μm	31.49	3.90×10^{-3}	36.9

3.1.2. Fourier-Transform Infrared Spectroscopy Analysis

Complementary to the XRD structural analysis, FTIR spectra of the commercial $BaTiO_3$ particles were obtained, allowing us to identify the presence of functional groups of BT and to identify bonds that are related to synthesis by-products or residual reactants. In this way, possible compounds that might affect the purity and relevant properties of BT were identified. The FTIR spectra in Figure 2, measured in transmittance mode, exhibit sharp peaks at 438 cm^{-1} and around 540 cm^{-1}, indicating the existence of a Ti-O bond, characteristic of BT [27,28]. Moreover, vibrational bands signaled at 860 and 1450 cm^{-1} are compatible with bending vibrations of a C-O bond, which could be linked to the reagents used for the fabrication of the commercial particles, for instance, barium carbonate.

Finally, for nanometric BT, the broad band around 3000 cm^{-1} can be attributed to the presence of an OH group from the H_2O content in the samples according to Singh et al. [28]. In the same region, we can see two peaks for the micrometric BT particles from around 2900 to 3000 cm^{-1}, which are associated with Ba-OH due to a possible incorporation of OH^- in the lattice [27,28], which can affect the tetragonal lattice and, therefore, piezoelectricity [29].

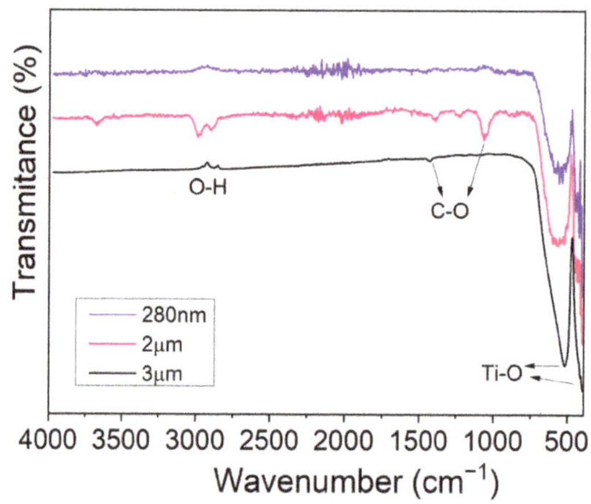

Figure 2. FTIR spectra of 280 nm and 2 and 3 μm $BaTiO_3$ particles, measured in transmittance mode.

3.1.3. Raman Spectroscopy Analysis

To further confirm the existence of a tetragonal phase of the powders, an analysis of the powder's Raman spectroscopy was used. A molecule of BT contains five atoms, leading to 12 optical vibrational modes. Based on the crystallography, Raman-active modes for tetragonal BT are four E(TO + LO), three A_1(TO + LO), and a B_1(TO + LO), while no Raman-active mode is predicted for the cubic phase due to the isotropic distribution of electrostatic forces surrounding Ti^{4+} ions [30]. Accordingly, the Raman spectra presented in Figure 3 are typical of a nonsymmetric structure.

In the Raman spectra, four dominant bands centered near 260, 306, 517, and 715 cm^{-1}, indicating the presence of the tetragonal phase, can be identified. As the literature suggests, the broad Raman band near 260 cm^{-1} corresponds to a Raman mode $A_1(TO_2)$, and a band near 306 cm^{-1} is assigned to the B_1 and E(TO + LO) modes, suggesting an asymmetric vibration of the $[TiO_6]$ octahedra. In addition, the band at 517 cm^{-1} is related to the $A_1(TO_3)$ and E(TO) modes, while a Raman band around 715 cm^{-1} is correlated with the A_1(LO) and E(LO) modes [23,30]. Accordingly, the Raman results clearly show that all $BaTiO_3$ samples possess a distortion of the $[TiO_6]$ octahedra and, consequentially, ferroelectric properties [23] in agreement with XRD and FTIR data presented before.

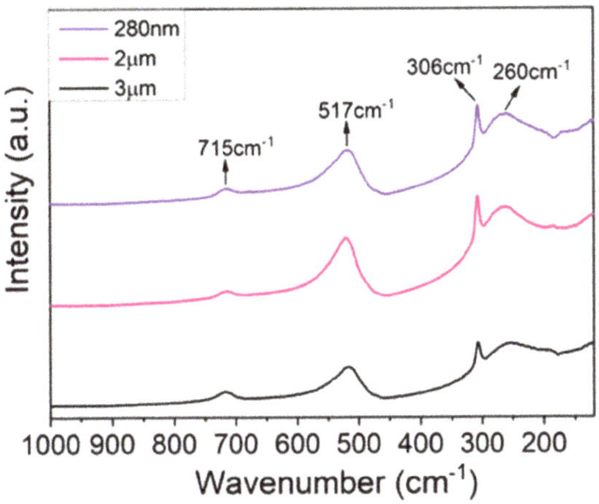

Figure 3. Raman spectroscopy of $BaTiO_3$ particles with sizes 280 nm and 2 and 3 μm.

3.2. *Hydroxyapatite Nanowires*

3.2.1. SEM Analysis

During the synthesis of HAp NWs, the chemical reactions that take place between the reactants are as follows [31]:

$$C_{18}H_{34}O_2 + NaOH \rightarrow C_{18}H_{33}O_2Na + H_2O \quad (2)$$

$$2C_{18}H_{33}O_2Na + CaCl_2 \rightarrow (C_{18}H_{33}O_2)_2Ca + 2NaCl \quad (3)$$

$$NaH_2(PO_4)\cdot 2(H_2O) + 2OH^- \rightarrow PO_4{}^{3-} + 4H_2O + Na^+ \quad (4)$$

$$10(C_{18}H_{33}O_2)_2Ca + 6PO_4{}^{3-} + 2OH^- \rightarrow Ca_{10}(PO_4)_6(OH)_2 + 20C_{18}H_{33}O_2 \quad (5)$$

It can be noticed that the calcium oleate acts as both the calcium source and the precursor for the formation of HAp NWs, while NaH_2PO_4 acts as the phosphorus source. At the beginning of the reaction system, calcium oleate is formed in chemical reactions (2) and (3). In chemical reaction (4), sodium dihydrogen phosphate dihydrate is hydrolyzed to form $PO_4{}^{3-}$ ions. During the crystallization process in reaction (5), hydroxyapatite nuclei are formed under solvothermal conditions. These nuclei later grow into longer structures, in a relatively long period of time, under relatively high temperatures and pressures, as previously mentioned in Section 2.

The existence of nanowires was first confirmed by observing, with SEM, the resulting HAp slurry, after being dried at 40 °C. For synthesis temperatures of 120 °C (data presented in Supplementary Figure S2), no structures were present for either 24 or 30 h of synthesis. Probably, the temperature is too low for the occurrence of the solvothermal reaction. Therefore, this solvothermal synthesis temperature was automatically rejected for the formation of HAp NWs.

Higher temperatures, such as 165 and 180 °C, were evaluated with synthesis times of 5, 7, 18, and 24 h, with resulting samples presented in Figure 4. For shorter synthesis times, such as 5 h, exhibited in Figure 4a,c for 165 and 180 °C, respectively, several structures appear to be starting to form; however, they are almost completely blended with each other. Only in Figure 4d, at 7 h, do some longer structures appear in the vicinity of more differentiated rodlike structures, with lengths around 6 μm. However, these longer structures also appear merged, preventing differentiation from each other. In the case of longer synthesis times, the sample of Figure 4f, of 24 h and 180 °C, shows mostly rod structures with around 10 μm length, whereas for 18 h, both samples, in Figure 4b,e, produced at 165 °C and 180 °C, respectively, display rodlike structures surrounded by wires. However, it should be noted that, with temperatures of 165 °C, the wires appear to be somewhat merged similarly to the sample of 7 h and 180 °C, present in Figure 4d.

Figure 4. SEM imaging of hydroxyapatite synthesized at (**a**) 165 °C for 5 h; (**b**) 165 °C for 18 h; (**c**) 180 °C for 5 h; (**d**) 180 °C for 7 h; (**e**) 180 °C for 18 h; (**f**) 180 °C for 24 h.

3.2.2. X-ray Diffraction Analysis

Following the evaluation of the NWs morphology, some of the solvothermal synthesis temperatures were discarded. Therefore, an evaluation of the crystallinity of HAp and the existence of other compounds was conducted on the samples represented in Figure 4: 165 °C (5 and 18 h) and 180 °C (5, 7, 18, and 24 h). To evaluate the existence of HAp and other impurity compounds, the diffraction peaks were compared to the ICDD sheet of hexagonal $Ca_{10}(PO_4)_6(OH)_2$ and JCPDS cards of reagents used during the synthesis, such as $CaCl_2$, $NaH_2PO_4{\cdot}2H_2O$, and NaOH.

Firstly, for either temperature, a synthesis time of 5 h does not lead to the formation of crystalline hydroxyapatite as there is still the presence of wide peaks and humps, more noticeable in Figure 5a, typical of an amorphous structure. The position of the XRD hump in amorphous HAp is around 30°, while in the obtained pattern, a hump is clearly visible at around 20° and it is also possible to perceive a small hump around 30°. The 20° hump could be related to amorphous dicalcium phosphate anhydrous, DCPA (monetite), and $CaHPO_4$, according to Borkiewicz [32]. However, both XRD diffractograms in Figure 5 exhibit some characteristic crystalline HAp peaks, such as those around 31.7° and 45.4°, that can indicate the beginning of HAp synthesis. It can be noted that, for the same synthesis time, more HAp characteristic peaks were formed at a higher temperature of 180 °C, present in Figure 5b, than at 165 °C, in Figure 5a, essentially showing that higher temperatures accelerate the formation of HAp crystallites, as expected.

The remaining diffraction results, presented in Figure 6, clearly show the presence of crystalline hydroxyapatite in all the samples, with all peaks corresponding to the ICDD datasheet, although some differences between the relative intensities of the identified peaks can be found. For instance, in the case of the sample of 180 °C for 24 h, the two highest intensities, (211) and (300), are switched when compared to tabulated data or other samples in Figure 6. This phenomenon could be explained by a preferential formation of crystalline grains along the longitudinal direction of the nanowires, which appear to be the direction perpendicular to the (300) plane, increasing the intensity of the measured peak at 32.8° and highlighting that change in peak intensities for the larger structures formed for the synthesis time of 24 h. In addition, no characteristic peaks of the reagents or by-products were found in the diffractograms shown in Figure 6.

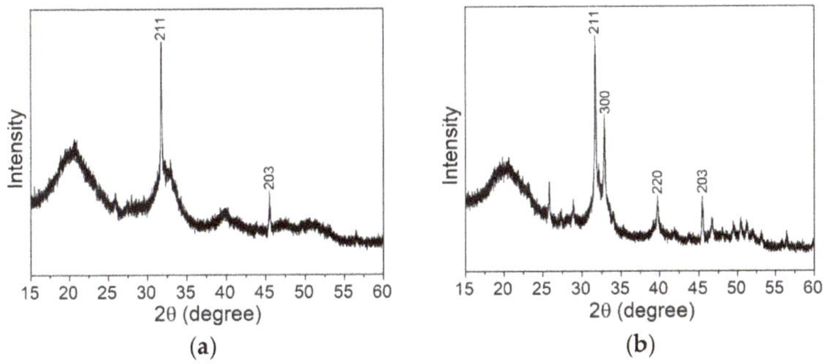

Figure 5. XRD diffractogram for HAp NWs synthesized for 5 h at (**a**) 165 °C and (**b**) 180 °C.

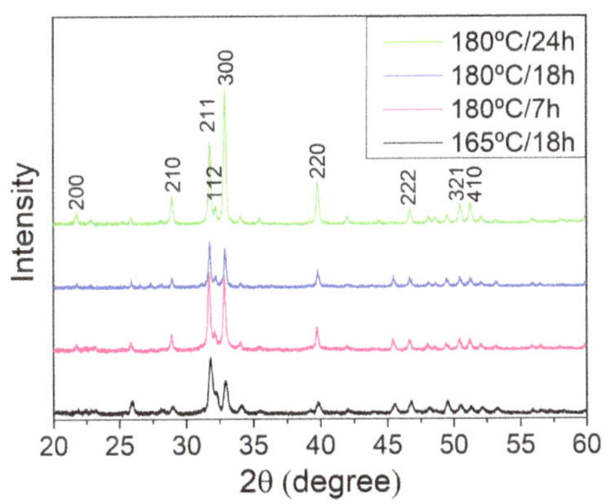

Figure 6. XRD diffractogram for HAp NWs synthesized at different temperatures and times.

The chosen synthesis time and temperature for the HAp NWs, which will be used to obtain the HAp aerogels, were 180 °C for 18 h due to both the sample's XRD diffractograms, typical of crystalline HAp, and the morphology, observed in Section 3.2.1, which seems to present a higher quantity of and longer wirelike structures when compared to the other samples. This more intertwined structure between the HAp wires could be an advantage in the mechanical stability of the aerogels.

The average size of the crystalline grains, τ, present in the HAp 180 °C for 18 h sample was calculated to be 39.05 nm, according to Equation (1), where $2\theta = 31.74°$, with β of 3.694×10^{-3} rad, $\lambda(CuK\alpha) = 1.5418$ Å, and κ of 0.9 [24,26,32].

3.2.3. Fourier-Transform Infrared Spectroscopy Analysis

In Figure 7a, an FTIR analysis of hydroxyapatite synthesized at 180 °C for 18 h is shown. The spectrum shows the typical HAp features, containing sharp O–H and P–O bands. The bands of significant intensity, around 1026 and 1100 cm^{-1}, and the bands near 560, 600, and 960 cm^{-1} correspond to a symmetric stretching vibration of the PO_4^{3-}

tetrahedron. Moreover, the band around 633 cm^{-1} is attributed to the stretching modes of hydroxyl groups in HAp [20,32,33].

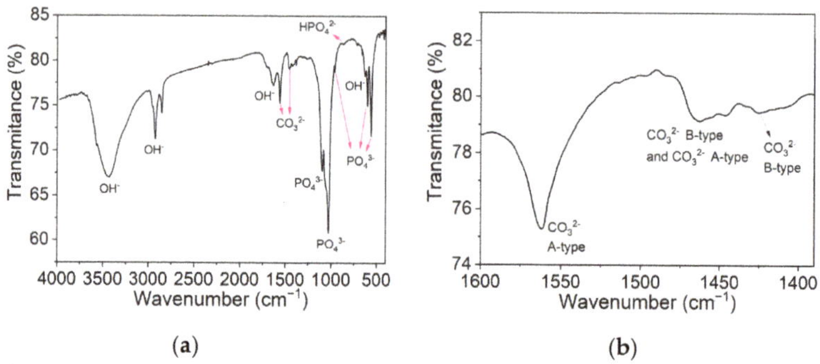

Figure 7. FTIR spectra of HAp synthesized at 180 °C for 18 h for wavenumbers between (**a**) 4000 and 400 cm^{-1}; (**b**) 1600 and 1350 cm^{-1}.

The small band at 870 cm^{-1} indicates the presence of $HPO_4{}^{2-}$ ions. This observation suggests the existence of non-stoichiometric HAp and the presence of carbonated apatite, which can be confirmed by the characteristic bands between wavenumbers 1460 and 1530 cm^{-1}. As seen in Figure 7b, two types of substitutions occurred, A-type carbonate substitutions, where carbonate ions substitute hydroxyl groups, and B-type carbonate substitutions, resulting from carbonate ions that have substituted phosphate ions [31–34]. It can be concluded that the resulting product from the solvothermal synthesis at 180 °C for 18 h is AB-type carbonated apatite. The B-type carbonate substitutions in the apatite lattice are known to increase the extent of solubility in weak acids, a characteristic that can help the substitution of the aerogel with new bone when using this material in bone regeneration [31,32,34]. Adsorbed water can also be seen on the wide bands that appear on the spectra from 2600 to around 3600 cm^{-1} and at 1640 cm^{-1} [33].

3.3. *HAp and HAp/BT Aerogels*

3.3.1. Porosity of the Aerogels

The porosity of the aerogels was calculated using Equation (6), where EW is the expected weight of a HAp solid material, calculated using hydroxyapatite's theoretical density, 3.16 g/cm^3, and the average aerogel's volume of 0.53 cm^3, and RW represents the aerogel's real weight, measured for each sample and which is, on average, 5.35 mg [11].

$$P(\%) = \frac{RW}{EW} \times 100 \tag{6}$$

The samples produced presented an average porosity of (99.68 ± 0.02)% for both HAp and HAp/BT aerogels, with no significant variation between the different aerogel types [11,35].

3.3.2. SEM Analysis

The microscopic homogeneity and porosity of the HAp and HAp/BT aerogels was investigated by imaging the surface of the aerogel using SEM. As can be seen in Figure 8, all samples show similar heterogenous porosity throughout, with pores at a micrometer level, relevant for cell penetration in this scaffold. Furthermore, as can be seen, the aerogels' 3D network created by the nanowires gives rise to a high surface area. This is further verified by the very high value obtained for the porosity value and, consequently, results in a low density for the aerogel (0.1 mg/cm^3).

The length of the wires was not possible to measure since, many times, the interwinding prevented the start and/or end of the wire from being visible in the images. Still, some shorter wires, with a length of approximately 10 µm, were observed. The diameter of the wires varied, though it appeared to be less than 1 µm.

Regarding the aerogels with BT shown in Figure 8, the particles appeared to be dispersed evenly throughout the aerogel, although in certain areas there seemed to be some agglomerates.

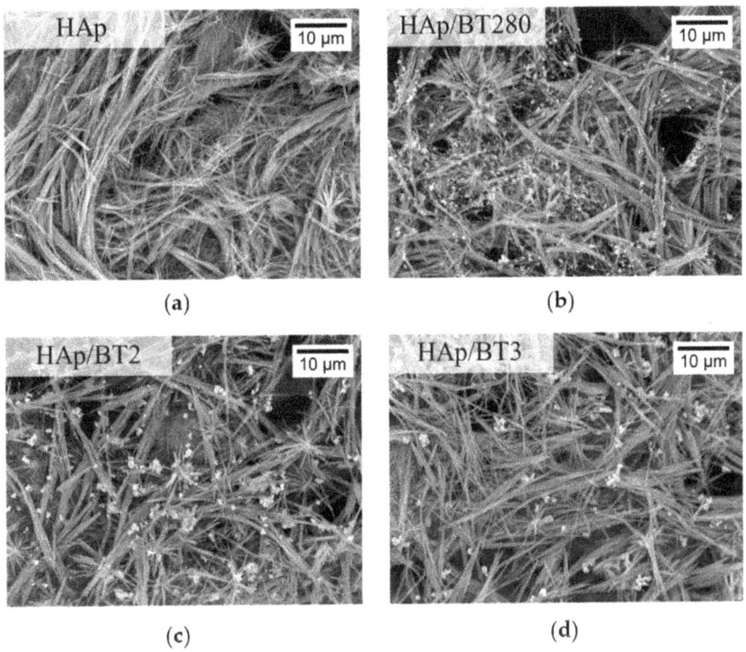

Figure 8. SEM imaging of aerogel samples: (**a**) HAp; (**b**) HAp/BT280; (**c**) HAp/BT2; (**d**) HAp/BT3.

3.3.3. Cytotoxicity Assays

In vitro colorimetric assays using resazurin, in conformity with the standard ISO-10993-5 test protocol [22], allowed a study of the potential cytotoxic effect of BT particles, of sizes 2 µm, 3 µm, and 280 nm, and of the HAp, HAp/BT280, HAp/BT2, and HAp/BT3 aerogels. Given the low mass of the aerogels, the concentrations chosen were significantly lower, 1 mg/mL, than the ones used for the BT particles, 60 mg/mL.

As seen in the graph in Figure 9, all BT powders revealed high cell viability, with a relative cell population above 100%, which is the result of not only cell survival but also cell proliferation. Therefore, it is safe to assume that the BT powders are non-cytotoxic for all the concentrations studied.

In the case of the aerogels, the relative cell population for different concentrations of the extracts is shown in Figure 10. The same pattern was attested in every sample. For the highest concentration used (C0 = 1 mg/mL), the HAp and HAp/BT3 samples were considered severely cytotoxic while the other samples were moderately cytotoxic. For subsequent dilutions, C0/2 to C0/8 (concentration range of 0.5–0.125 mg/mL), all samples showed a non-cytotoxic behavior and can be considered biocompatible. On the other hand, the results for the dilutions, showing a relative cell population of 140%, are unusually high to be justified purely by cell viability and proliferation. It could be speculated that the reduction of resazurin is being affected by other parameters and that the cell populations might have been overestimated.

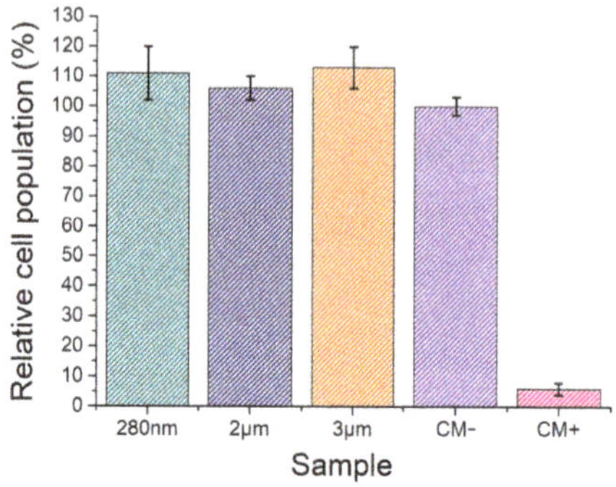

Figure 9. Relative cell populations obtained in the cytotoxicity tests of $BaTiO_3$ particles, 280 nm, 2 μm, 3 μm, negative (CM−), and positive (CM+) control groups. Populations obtained using the extracts prepared at a concentration of 60 mg/mL. Results are average and experimental standard deviation.

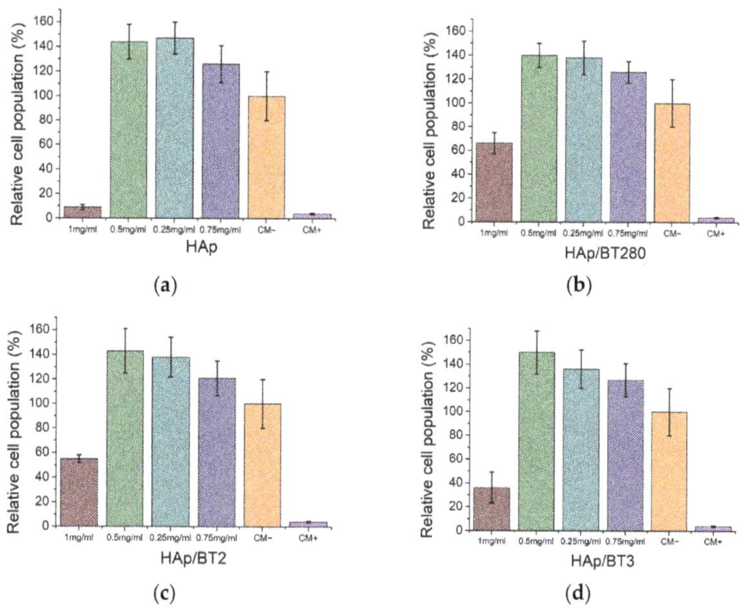

Figure 10. Relative cell populations obtained in the cytotoxicity tests of the aerogels: (**a**) HAp; (**b**) HAp/BT280; (**c**) HAp/BT2; (**d**) HAp/BT3, and negative (CM−) and positive (CM+) control groups. Populations obtained using the extracts prepared at a concentration of 1 mg/mL and three serial dilutions. Results are average and experimental standard deviation.

According to Klimek et al. [36], the apparent cytotoxicity of the HAp might be caused by a massive uptake of Ca^{2+} and HPO_4^{2-} ions from the medium, related to the high surface area of the ceramics, resulting in extracts that are not optimal for cell cultures. This phenomenon will lead to an unexpected cytotoxic effect despite these ion reactions being correlated with increased bioactivity. Moreover, in vivo, tissue liquids continuously circulate around implanted biomaterial and supplement all the adsorbed ions in the implantation area. This could be the case for the ceramics produced, since Gustavsson et al. [37] also reported ion adsorption by calcium-deficient hydroxyapatite with a carbonated surface chemistry, similar to the AB-type carbonated apatite's chemistry observed in FTIR in the Section 3.2.3. When the extracts' concentration range is 0.5–0.125 mg/mL, the uptake of ions might not be enough to lead to cell death. Nevertheless, it can still influence the natural behavior of the viable cells, leading them to overproduce enzymes that reduce resazurin and thus obtain an overestimation of the cell population [38].

3.3.4. Bioactivity Assays in Simulated Body Fluid

The bioactivity was analyzed by submerging the aerogels HAp and HAp/BT in an SBF solution and placing them in an orbital shaker at 37 °C, mimicking average body temperature, for 1, 3, and 7 days, and by preparing a control group which was not submerged. All samples were analyzed through SEM/EDS.

In SEM images, there is no clear evidence of the deposition of apatite crystals for all aerogel samples and for all of the immersion times evaluated, as seen in Figure 11, which corresponds to 7 days in SBF. As expected, the EDS mapping data (presented in Supplementary Figures S3 and S4) reveal the existence of Ca and P in the wire structures, barium, and titanium on the BT particles (visible in the agglomerate of Figure S3). There is also some residue of Na and Cl, with some crystals typical of the cubic lattice of sodium chloride present. This residue might hinder the observation of other structures in some areas of the SEM images. As stated in Section 2, all samples were rinsed in distilled water after removal from SBF with the purpose of removing this type of water-soluble residue. However, there are some materials and morphologies that do not allow the full dissolution of NaCl. The rough morphology and chemistry of the sample, which naturally contains CaP, could also be preventing the observation of deposited apatites in the case of relatively small crystals being formed.

Figure 11. SEM images after 7 days in SBF of the aerogel samples: (**a**) HAp; (**b**) HAp/BT280; (**c**) HAp/BT2; (**d**) HAp/BT3.

Regarding the calcium phosphate deposition on the BT particles, from the SEM imaging in Figure 11, it can also be noted that, after 7 days, no apparent apatite deposition occurred around the barium titanate particles present in the aerogel (EDS mapping presented in Supplementary Materials Figure S3).

Since the SEM imaging and EDS mapping do not confirm if CaP deposition occurred, an EDS analysis was used, and the Ca/P ratio for all samples was estimated to observe if a trend in Ca^{2+} and $PO_4{}^{3-}$ deposition was occurring.

In Figure 12, the evolution of the ratios with the immersion time is similar between samples, with an increase on the first three days and a decrease on the seventh. The atomic ratio Ca/P was estimated by using four different areas in each sample, and the average and standard deviation (uncertainty) were calculated. The uncertainty values were similar for all samples and around 0.5. Although the value of the third day is unusually high (Ca/P » 3), the same trend in behavior is still perceived for all the samples. This trend is a clue that, at first, there is a calcium ion deposition taking place. After a maximum is reached (day 3), there is a significant decrease in the ratio.

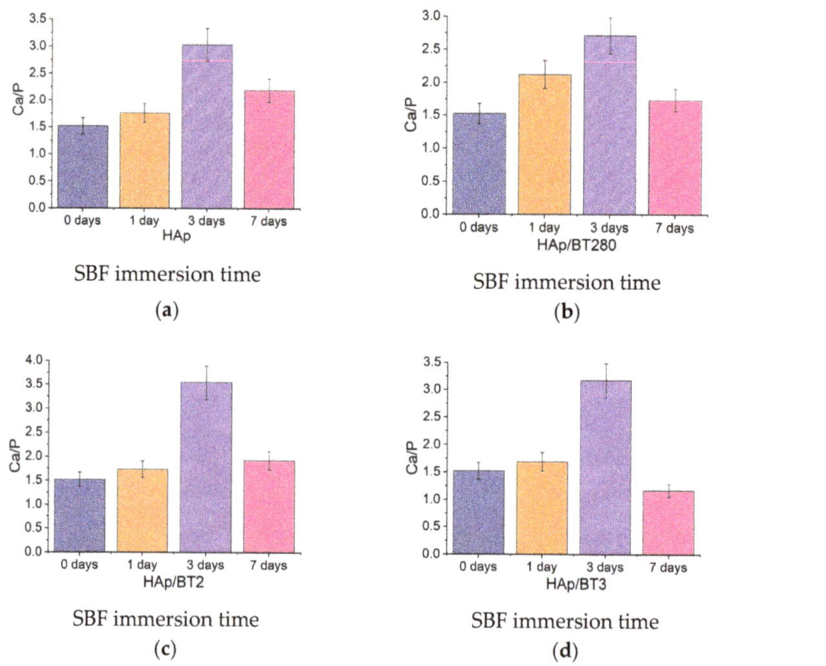

Figure 12. Ca/P, calculated by quantifying the EDS analysis after SBF immersion of the (**a**) HAp; (**b**) HAp/BT280; (**c**) HAp/BT2; and (**d**) HAp/BT3.

Since the hydroxyapatite produced was calcium deficient, due to a Ca/P ratio of 1.5 before SBF immersion, it is possible that an initial deposition of Ca^{2+} ions is occurring, increasing the Ca/P ratio. In SEM images, no changes were observed at the surface of the samples (no flower-like apatitic crystals, aggregates of such crystals, or any continuous layer of CaP were detected). So, it is reasonable to assume that when the deposition of Ca^{2+} ions stops, due to lower concentration of ions in SBF and/or positive surface charges, the Ca^{2+} returns to the solution and $PO_4{}^{3-}$ ions become trapped at the surface. This results in the deposition of calcium ions, followed by their dissolution, and the deposition of phosphate ions, with no formation on the surface of apatitic structures. Also, at the maximum SEM magnification used, the crystals formed could be too small and not distinguishable from the original NW surface to be visible.

The recent literature reviews by Khare et al. [39] and Sood et al. [40] highlight studies involving hydroxyapatite-based materials incorporating particles of BT. Some research focuses on foams with similar porous sizes to the aerogels produced in this work and that showed enhanced bioactivity [41]. If the ceramics are electrically polarized, it is expected to improve further this important property [17,42]. In consequence, a better assessment of the advantages of using a HAp aerogel with $BaTiO_3$ particles must include electrically polarized samples. Subsequently, future work on the aerogel will be made to test the bioactivity and, furthermore, in vitro assays will be performed to quantify cell adhesion and proliferation and the results of the polarized and the non-polarized material will be compared. On the other hand, as Park et al. observed [43], in vivo assays might immediately reveal the advantage of having a dipolar material for the non-polarized piezoelectric material developed.

4. Conclusions

The main goal of this work was to produce hydroxyapatite aerogels with and without $BaTiO_3$ and to study their features (including bioactivity and biocompatibility) with the purpose of creating a new material with osteointegration potential.

The barium titanate powders that were used were tetragonal, and consequently piezoelectric and presenting permanent dipoles. A structural analysis verified that the samples synthesized at 180 °C for 18 h resulted in wirelike structures of crystalline AB-type carbonated hydroxyapatite. Additionally, an EDS analysis showed a calcium deficiency in HAp. The aerogel structures showed an average porosity of 99.68%, with interlinked HAp wires, and a relatively good distribution of BT particles.

Cytotoxicity assays revealed that all the BT particles were non-cytotoxic for a concentration of 60 mg/mL. For the aerogel, a concentration range of 0.5–0.125 mg/mL for all samples showed cell proliferation above 100% and thus, they were biocompatible.

In the SEM images of bioactivity, it was difficult to differentiate between the possibly deposited apatite and the surface of the HAp wires. However, an EDS analysis showed an occurring trend of calcium ions deposition followed by its dissolution and the deposition of phosphate ions.

The HAp-based nanowire aerogels with piezoelectric nano- and microparticles of barium titanate were successfully produced and showed properties, such as bioactivity and biocompatibility, that point to a new biomaterial with osteointegration potential. The aerogel could be suitable for non-load-bearing applications, such as cavity filling. To fully disclose the bone regeneration potential of the composite, further studies are essential and must include electrical polarization, mechanical compressive behavior, and in vivo assays.

Supplementary Materials: The following supporting information can be downloaded at: https://www.mdpi.com/article/10.3390/biomimetics9030143/s1, Protocol for SBF and Figure S1. DSC/TG for sample $BaTiO_3$-280nm from 25 to 220 °C. Figure S2. SEM images of Hydroxyapatite synthesized at 120 °C for (a) and (b) 24 hours; (c) and (d) 30 hours. Figure S3. Quantifying EDS analysis of the HAp/BT280 aerogel after 7 days in SBF. Figure S4. Mapping EDS analysis of the HAp/BT280 aerogel after 7 days in SBF.

Author Contributions: Conceptualization, M.C.L., J.P.M.R.B. and C.T.; methodology, M.C.L., J.P.M.R.B. and C.T.; investigation, C.T. and T.V.; validation, M.C.L. and J.P.M.R.B.; formal analysis, C.T., T.V., J.P.M.R.B., J.C.S. and M.C.L.; writing—original draft preparation, C.T.; writing—review and editing, M.C.L., J.P.M.R.B., T.V. and J.C.S.; supervision, M.C.L. and J.P.M.R.B.; funding acquisition, J.C.S., J.P.M.R.B. and M.C.L. All authors have read and agreed to the published version of the manuscript.

Funding: This work was financed by national funds from FCT—Fundação para a Ciência e a Tecnologia, I.P., in the scope of the projects LA/P/0037/2020, UIDP/50025/2020, and UIDB/50025/2020 of the Associate Laboratory Institute of Nanostructures, Nanomodelling and Nanofabrication—i3N.

Institutional Review Board Statement: Not applicable.

Data Availability Statement: Data are contained within the article and Supplementary Materials.

Conflicts of Interest: The authors declare no conflicts of interest.

References

1. Wei, Y.; Sun, Y. Aging of the Bone. *Adv. Exp. Med. Biol.* **2018**, *1086*, 189–197. [CrossRef]
2. Tresguerres, F.G.F.; Torres, J.; López-Quiles, J.; Hernández, G.; Vega, J.A.; Tresguerres, I.F. The Osteocyte: A Multifunctional Cell within the Bone. *Ann. Anat. Anat. Anzeiger* **2020**, *227*, 151422. [CrossRef]
3. Orthopedic Biomaterials Market Size, Trends and Global Forecast to 2032. Available online: https://www.thebusinessresearchcompany.com/report/orthopedic-biomaterials-global-market-report (accessed on 4 October 2023).
4. Eliaz, N.; Metoki, N. Calcium Phosphate Bioceramics: A Review of Their History, Structure, Properties, Coating Technologies and Biomedical Applications. *Materials* **2017**, *10*, 334. [CrossRef] [PubMed]
5. Gomes, D.S.; Santos, A.M.C.; Neves, G.A.; Menezes, R.R. A Brief Review on Hydroxyapatite Production and Use in Biomedicine. *Ceramica* **2019**, *65*, 282–302. [CrossRef]
6. Samavedi, S.; Whittington, A.R.; Goldstein, A.S. Calcium Phosphate Ceramics in Bone Tissue Engineering: A Review of Properties and Their Influence on Cell Behavior. *Acta Biomater.* **2013**, *9*, 8037–8045. [CrossRef]
7. Sopyan, I.; Mel, M.; Ramesh, S.; Khalid, K.A. Porous Hydroxyapatite for Artificial Bone Applications. *Sci. Technol. Adv. Mater.* **2007**, *8*, 116–123. [CrossRef]
8. Abbasi, N.; Hamlet, S.; Love, R.M.; Nguyen, N.T. Porous Scaffolds for Bone Regeneration. *J. Sci. Adv. Mater. Devices* **2020**, *5*, 1–9. [CrossRef]
9. Xiong, Z.C.; Yang, Z.Y.; Zhu, Y.J.; Chen, F.F.; Zhang, Y.G.; Yang, R.L. Ultralong Hydroxyapatite Nanowires-Based Paper Co-Loaded with Silver Nanoparticles and Antibiotic for Long-Term Antibacterial Benefit. *ACS Appl. Mater. Interfaces* **2017**, *9*, 22212–22222. [CrossRef]
10. Zhu, Y.; Lu, B. Deformable Biomaterials Based on Ultralong Hydroxyapatite Nanowires. *ACS Biomater. Sci. Eng.* **2019**, *5*, 4951–4961. [CrossRef]
11. Huang, G.-J.; Yu, H.-P.; Wang, X.-L.; Ning, B.-B.; Gao, J.; Shi, Y.-Q.; Zhu, Y.-J.; Duan, J.-L. Highly Porous and Elastic Aerogel Based on Ultralong Hydroxyapatite Nanowires for High-Performance Bone Regeneration and Neovascularization. *J. Mater. Chem. B* **2021**, *9*, 1277–1287. [CrossRef]
12. Ball, J.P.; Mound, B.A.; Nino, J.C.; Allen, J.B. Biocompatible Evaluation of Barium Titanate Foamed Ceramic Structures for Orthopedic Applications. *J. Biomed. Mater. Res. Part A* **2014**, *102*, 2089–2095. [CrossRef]
13. Fukada, E.; Yasuda, I. On the Piezoelectric Effect of Bone. *J. Phys. Soc. Jpn.* **1957**, *12*, 1158–1162. [CrossRef]
14. Fukada, E.; Yasuda, I. Piezoelectric Effects in Collagen. *Jpn. J. Appl. Phys.* **1964**, *3*, 117–121. [CrossRef]
15. Jacob, J.; More, N.; Kalia, K.; Kapusetti, G. Piezoelectric Smart Biomaterials for Bone and Cartilage Tissue Engineering. *Inflamm. Regen.* **2018**, *38*, 2. [CrossRef] [PubMed]
16. Pienkowski, D.; Pollack, S.R. The Origin of Stress-generated Potentials in Fluid-saturated Bone. *J. Orthop. Res.* **1983**, *1*, 30–41. [CrossRef] [PubMed]
17. Metwally, S.; Stachewicz, U. Surface Potential and Charges Impact on Cell Responses on Biomaterials Interfaces for Medical Applications. *Mater. Sci. Eng. C* **2019**, *104*, 109883. [CrossRef]
18. Zhang, Y.-Q.; Geng, Q.; Li, C.; Wang, H.-C.; Ren, C.; Zhang, Y.-F.; Bai, J.-S.; Pan, H.-B.; Cui, X.; Yao, M.-X.; et al. Application of Piezoelectric Materials in the Field of Bone: A Bibliometric Analysis. *Front. Bioeng. Biotechnol.* **2023**, *11*, 1210637. [CrossRef] [PubMed]
19. Diez-Escudero, A.; Espanol, M.; Ginebra, M.P. High-Aspect-Ratio Nanostructured Hydroxyapatite: Towards New Functionalities for a Classical Material. *Chem. Sci.* **2023**, *15*, 55–76. [CrossRef]
20. Zheng, Y.; Ma, W.; Yang, Z.; Zhang, H.; Ma, J.; Li, T.; Niu, H.; Zhou, Y.; Yao, Q.; Chang, J.; et al. An Ultralong Hydroxyapatite Nanowire Aerogel for Rapid Hemostasis and Wound Healing. *Chem. Eng. J.* **2021**, *430*, 1385–8947. [CrossRef]
21. Zheng, L.; Zhang, S.; Ying, Z.; Liu, J.; Zhou, Y.; Chen, F. Engineering of Aerogel-Based Biomaterials for Biomedical Applications. *Int. J. Nanomed.* **2020**, *15*, 2363. [CrossRef]
22. Maçon, A.L.B.; Kim, T.B.; Valliant, E.M.; Goetschius, K.; Brow, R.K.; Day, D.E.; Hoppe, A.; Boccaccini, A.R.; Kim, I.Y.; Ohtsuki, C.; et al. A Unified in Vitro Evaluation for Apatite-Forming Ability of Bioactive Glasses and Their Variants. *J. Mater. Sci. Mater. Med.* **2015**, *26*, 115. [CrossRef]
23. Qian, H.; Zhu, G.; Xu, H.; Zhang, X.; Zhao, Y.; Yan, D.; Hong, X.; Han, Y.; Fu, Z.; Ta, S.; et al. Preparation of Tetragonal Barium Titanate Nanopowders by Microwave Solid-State Synthesis. *Appl. Phys. A* **2020**, *126*, 294. [CrossRef]
24. Langford, J.I.; Wilson, A.J.C. Scherrer after Sixty Years: A Survey and Some New Results in the Determination of Crystallite Size. *J. Appl. Crystallogr.* **1978**, *11*, 102–113. [CrossRef]
25. Londoño-Restrepo, S.M.; Jeronimo-Cruz, R.; Millán-Malo, B.M.; Rivera-Muñoz, E.M.; Rodriguez-García, M.E. Effect of the Nano Crystal Size on the X-Ray Diffraction Patterns of Biogenic Hydroxyapatite from Human, Bovine, and Porcine Bones. *Sci. Rep.* **2019**, *9*. [CrossRef] [PubMed]
26. Ingham, B.; Toney, M.F. X-ray Diffraction for Characterizing Metallic Films. In *Metallic Films for Electronic, Optical and Magnetic Applications*; Elsevier: Amsterdam, The Netherlands, 2014; pp. 3–38. ISBN 9780857090577.
27. Bhuiyan, M.R.A.; Alam, M.M.; Momin, M.A.; Uddin, M.J.; Islam, M. Synthesis and Characterization of Barium Amidoborane. *Int. J. Mater. Mech. Eng.* **2012**, *1*, 21–24.
28. Singh, M.; Yadav, B.C.; Ranjan, A.; Kaur, M.; Gupta, S.K. Synthesis and Characterization of Perovskite Barium Titanate Thin Film and Its Application as LPG Sensor. *Sens. Actuators B Chem.* **2017**, *241*, 1170–1178. [CrossRef]

29. Buscaglia, V.; Randall, C.A. Size and Scaling Effects in Barium Titanate. An Overview. *J. Eur. Ceram. Soc.* **2020**, *40*, 3744–3758. [CrossRef]
30. Lazarević, Z.; Romčević, N.; Vijatović, M.; Paunović, N.; Romčević, M.; Stojanović, B.; Dohčević-Mitrović, Z. Characterization of Barium Titanate Ceramic Powders by Raman Spectroscopy, Proceedings of the Tenth Annual Conference of the Materials Research Society of Serbia, September 2008. *ACTA Phys. Pol. A* **2009**, *115*, 808–810. [CrossRef]
31. Jiang, Y.-Y.; Zhu, Y.-J.; Chen, F.; Wu, J. Solvothermal Synthesis of Submillimeter Ultralong Hydroxyapatite Nanowires Using a Calcium Oleate Precursor in a Series of Monohydroxy Alcohols. *Ceram. Int.* **2015**, *41*, 6098–6102. [CrossRef]
32. Borkiewicz, O.; Rakovan, J.; Cahill, C.L. Time-Resolved in Situ Studies of Apatite Formation in Aqueous Solutions. *Am. Mineral.* **2010**, *95*, 1224–1236. [CrossRef]
33. Berzina-Cimdina, L.; Borodajenko, N. Research of Calcium Phosphates Using Fourier Transform Infrared Spectroscopy. In *Infrared Spectroscopy—Materials Science, Engineering and Technology*; Theophanides, T., Ed.; InTech: Rijeka, Croatia, 2012.
34. Madupalli, H.; Pavan, B.; Tecklenburg, M.M.J. Carbonate Substitution in the Mineral Component of Bone: Discriminating the Structural Changes, Simultaneously Imposed by Carbonate in A and B Sites of Apatite. *J. Solid State Chem.* **2017**, *255*, 27–35. [CrossRef] [PubMed]
35. Zhang, Y.-G.; Zhu, Y.-J.; Xiong, Z.-C.; Wu, J.; Chen, F. Bioinspired Ultralight Inorganic Aerogel for Highly Efficient Air Filtration and Oil-Water Separation. *ACS Appl. Mater. Interfaces* **2018**, *10*, 13019–13027. [CrossRef] [PubMed]
36. Klimek, K.; Belcarz, A.; Pazik, R.; Sobierajska, P.; Han, T.; Wiglusz, R.J.; Ginalska, G. "False" Cytotoxicity of Ions-Adsorbing Hydroxyapatite—Corrected Method of Cytotoxicity Evaluation for Ceramics of High Specific Surface Area. *Mater. Sci. Eng. C* **2016**, *65*, 70–79. [CrossRef] [PubMed]
37. Gustavsson, J.; Ginebra, M.P.; Engel, E.; Planell, J. Ion Reactivity of Calcium-Deficient Hydroxyapatite in Standard Cell Culture Media. *Acta Biomater.* **2011**, *7*, 4242–4252. [CrossRef] [PubMed]
38. O'Brien, J.; Wilson, I.; Orton, T.; Pognan, F. Investigation of the Alamar Blue (Resazurin) Fluorescent Dye for the Assessment of Mammalian Cell Cytotoxicity. *Eur. J. Biochem.* **2000**, *267*, 5421–5426. [CrossRef]
39. Khare, D.; Basu, B.; Dubey, A.K. Electrical Stimulation and Piezoelectric Biomaterials for Bone Tissue Engineering Applications. *Biomaterials* **2020**, *258*, 120280. [CrossRef]
40. Sood, A.; Desseigne, M.; Dev, A.; Maurizi, L.; Kumar, A.; Millot, N.; Han, S.S. A Comprehensive Review on Barium Titanate Nanoparticles as a Persuasive Piezoelectric Material for Biomedical Applications: Prospects and Challenges. *Small* **2023**, *19*, 2206401. [CrossRef]
41. Zhang, Y.; Chen, L.; Zeng, J.; Zhou, K.; Zhang, D. Aligned Porous Barium Titanate/Hydroxyapatite Composites with High Piezoelectric Coefficients for Bone Tissue Engineering. *Mater. Sci. Eng. C* **2014**, *39*, 143–149. [CrossRef]
42. Baxter, F.R.; Bowen, C.R.C.R.C.; Turner, I.G.I.G.; Dent, A.C.E.E. Electrically Active Bioceramics: A Review of Interfacial Responses. *Ann. Biomed.* **2010**, *38*, 2079–2092. [CrossRef]
43. Park, J.B.; Kelly, B.J.; Kenner, G.H.; von Recum, A.F.; Grether, M.F.; Coffeen, W.W. Piezoelectric Ceramic Implants: In Vivo Results. *J. Biomed. Mater. Res.* **1981**, *15*, 103–110. [CrossRef]

biomimetics

MDPI

Review

Reconstruction of Critical Sized Maxillofacial Defects Using Composite Allogeneic Tissue Engineering: Systematic Review of Current Literature

Shaqayeq Ramezanzade [1], Mahsa Aeinehvand [2], Heliya Ziaei [2], Zohaib Khurshid [3,4], Seied Omid Keyhan [5], Hamid R. Fallahi [5], James C. Melville [6], Morvarid Saeinasab [7] and Farshid Sefat [7,*]

1 Section for Clinical Oral Microbiology, Department of Odontology Cariology and Endodontics, University of Copenhagen, DK-2200 Copenhagen, Denmark
2 Maxillofacial Surgery & Implantology & Biomaterial Research Foundation, Tehran P.O. Box 14155-6559, Iran
3 Department of Prosthodontics and Dental Implantology, King Faisal University, Al-Ahsa 31982, Saudi Arabia
4 Center of Excellence for Regenerative Dentistry, Department of Anatomy, Faculty of Dentistry, Chulalongkorn University, Bangkok 10330, Thailand
5 Maxillofacial Surgery & Implantology & Biomaterial Research Foundation, Isfahan P.O. Box 61355-45, Iran
6 Oral, Head & Neck Oncology and Microvascular Reconstructive Surgery, Department of Oral and Maxillofacial Surgery, School of Dentistry, University of Texas Health Science Center at Houston, Houston, TX 77030, USA
7 Department of Biomedical and Electronic Engineering, School of Engineering, University of Bradford, Bradford BD7 1DP, UK
* Correspondence: f.sefat1@bradford.ac.uk

Abstract: The current review aimed to assess the reliability and efficacy of tissue-engineered composite grafts in the reconstruction of large maxillofacial defects resulting from trauma or a benign pathologic disease. A systematic review of the literature was conducted using PubMed/Medline, Embase, and Scopus up to March 2022. The eligibility criteria included patients who had been treated with composite allogeneic tissue engineering for immediate/delayed reconstruction of large maxillofacial defects with minimum/no bone harvesting site. In the initial search, 2614 papers were obtained, and finally, 13 papers were eligible to be included in the current study. Most included papers were case reports or case series. A total of 144 cases were enrolled in this systematic review. The mean age of the patients was 43.34 (age range: 9–89). Most studies reported a successful outcome. Bone tissue engineering for the reconstruction and regeneration of crucial-sized maxillofacial defects is an evolving science still in its infancy. In conclusion, this review paper and the current literature demonstrate the potential for using large-scale transplantable, vascularized, and customizable bone with the aim of reconstructing the large maxillofacial bony defects in short-term follow-ups.

Keywords: tissue engineering; maxillofacial defects; composite graft

Citation: Ramezanzade, S.; Aeinehvand, M.; Ziaei, H.; Khurshid, Z.; Keyhan, S.O.; Fallahi, H.R.; Melville, J.C.; Saeinasab, M.; Sefat, F. Reconstruction of Critical Sized Maxillofacial Defects Using Composite Allogeneic Tissue Engineering: Systematic Review of Current Literature. *Biomimetics* **2023**, *8*, 142. https://doi.org/10.3390/biomimetics8020142

Academic Editors: Ryszard Uklejewski and Mariusz Winiecki

Received: 15 February 2023
Revised: 24 March 2023
Accepted: 26 March 2023
Published: 30 March 2023

1. Introduction

The oral and maxillofacial area is a complex region including osseocartilaginous elements, neural and vascular systems, skin, and other lining and covering tissues, teeth, and organs for the senses [1]. There are several causes of significant defects in this region, such as traumatic avulsion, Osteoradionecrosis (ORN), bisphosphonate-related osteonecrosis of the jaws (BRONJ), the resection of benign/malignant tumors and cysts, etc. [2]. The natural repair mechanisms for large maxillofacial defects are insufficient and slow-paced [3]. Therefore, adjunct bone regeneration procedures are crucial to ensure sufficient bone formation within a short time.

Materials of natural origin, derived from a living source without making any modifications consist of four major groups: autografts, allografts, xenografts, and phytogenic materials [4]. The current routine materials of reconstruction include autologous cortico-cancellous bone, vascularized free flap transfer, alloplastic materials with prosthetic appliances, and composite materials [5].

The techniques advocated for each case depend on the associated soft tissue, the pattern of vascularity, defect size, the types of tissue, and patient preference [6,7].

Reconstruction of large maxillofacial defects with conventional materials and techniques of autogenous bone collection presents a set of challenges for the surgeon in the maxillofacial field; the amount of intraoral bone is mostly limited and therefore is not suitable for harvesting and grafting large defects [4]. Likewise, the need for another surgical site results in burdensomely long and complex operations, hospital stay, higher rates of post-surgical complications, and the morbidities of the bone harvesting sites.

A widely used alternative option for bone regeneration is the use of alloplastic materials, which eliminates the need for a donor site and improves surgical efficiency. This approach is also much safer in medically compromised patients, in which the risks of additional graft harvesting surgery outweigh the benefits. The macroporosity of 100 to 400 mm on the surface of such materials acts as trabecular bone and therefore promotes osteoconduction [8]. At the same time, the lack of cellular components required for osteogenesis and weak activity in vascularly compromised environments are counted as major flaws [1].

Mesenchymal stem cells (MSCs) derived from different parts of the human body such as bone marrow, adipose tissue, peripheral blood, etc., have shown an enhancement in bone regeneration when seeded on a scaffold compared to an unseeded scaffold alone [9].

Using an allogenic graft as a biologic scaffold in conjunction with harvested mesenchymal stem cells and recombinant human bone morphogenic protein-2 (rhBMP-2) creates a favorable microenvironment for bone formation. This review aimed to assess the reliability and efficacy of tissue-engineered composite grafts in the reconstruction of large maxillofacial defects of trauma or a benign pathologic disease.

2. Materials and Methods

2.1. Protocol Registration

The search protocol was specified and registered at PROSPERO (prospective international register of systematic reviews) with registration number: CRD42021242399. In addition, the PRISMA 2020 guidelines for conducting this systematic review were followed [10].

2.2. PICO Question

Patient: patients with large maxillofacial defects requiring bone regeneration.

Intervention: surgical bone grafting procedures using composite allogeneic tissue engineering.

Comparison: Conventional autogenous bone grafts/None (non-comparative studies).

2.3. Outcome

1. The complication rates reported.
2. The success rate measured as the amount of new bone volume gained (assessed either directly by percentage bone fill or assessed radiographically).
3. Patient-centered outcomes: satisfaction rate.

2.4. Search Strategy

PubMed/Medline, Embase, and Scopus were searched systematically with no time and language restrictions (up to March 2022) [2]. Also, the reference list of included papers was hand-searched for potential additional papers. Table 1 illustrates the search strategy for each database.

Table 1. Search strategy.

PubMed		
1	("Bioengineering" [Mesh]) OR ("Bioengineering material" [Title/Abstract]) OR ("osteogenic scaffold" [Title/Abstract]) OR ("tissue engineering" [Title/Abstract]) OR ("Tissue Engineering" [Mesh]) OR ("Bone Morphogenetic Proteins" [Mesh]) OR ("Bone Morphogenetic Proteins" [Title/Abstract]) OR ("Mesenchymal Stem Cells" [Mesh]) OR ("Bone Mesenchymal Stem Cells" [Title/Abstract]) OR ("beta-tri calcium phosphate" [Title/Abstract]) OR ("Bone Morphogenetic Protein 2" [Mesh]) OR (rhBPM2) OR (rhBPM-2)	149,201
2	(Mandible[Title/Abstract]) OR (Mandibular[Title/Abstract]) OR (Maxilla[Title/Abstract]) OR (Maxillary[Title/Abstract]) OR (Maxillofacial[Title/Abstract])	175,429
3	("Reconstructive Surgical Procedures" [Mesh]) OR (Reconstruct[Title/Abstract]) OR (Augment[Title/Abstract])	558,576
1 AND 2 AND 3	(TITLE-ABS-KEY (bioengineering) OR TITLE-ABS-KEY ("osteogenic scaffold") OR TITLE-ABS-KEY ("osteogenic scaffolds") OR TITLE-ABS-KEY ("tissue engineering") OR TITLE-ABS-KEY ("Bone Morphogenetic Proteins") OR TITLE-ABS-KEY ("Bone Morphogenetic Protein") OR TITLE-ABS-KEY ("Mesenchymal Stem Cells") OR TITLE-ABS-KEY ("Bone Mesenchymal Stem Cells") OR TITLE-ABS-KEY ("beta-tri calcium phosphate") OR TITLE-ABS-KEY ("Bone Morphogenetic Protein 2") OR TITLE-ABS-KEY (rhbpm2) OR TITLE-ABS-KEY (rhbpm-2)) AND (TITLE-ABS-KEY (mandible) OR TITLE-ABS-KEY (mandibular) OR TITLE-ABS-KEY (maxilla) OR TITLE-ABS-KEY (maxilla) OR TITLE-ABS-KEY (maxillary) OR TITLE-ABS-KEY (maxillofacial)) AND (TITLE-ABS-KEY ("Reconstructive Surgical Procedures") OR TITLE-ABS-KEY (reconstruct) OR TITLE-ABS-KEY (augment) OR TITLE-ABS-KEY (reconstruction) OR TITLE-ABS-KEY (augmentation))	735
Scopus		
1	TITLE-ABS-KEY (bioengineering) OR TITLE-ABS-KEY ("osteogenic scaffold") OR TITLE-ABS-KEY ("osteogenic scaffolds") OR TITLE-ABS-KEY ("tissue engineering") OR TITLE-ABS-KEY ("Bone Morphogenetic Proteins") OR TITLE-ABS-KEY ("Bone Morphogenetic Protein") OR TITLE-ABS-KEY ("Mesenchymal Stem Cells") OR TITLE-ABS-KEY ("Bone Mesenchymal Stem Cells") OR TITLE-ABS-KEY ("beta-tri calcium phosphate") OR TITLE-ABS-KEY ("Bone Morphogenetic Protein 2") OR TITLE-ABS-KEY (rhbpm2) OR TITLE-ABS-KEY (rhbpm-2)	273,893
2	TITLE-ABS-KEY (mandible) OR TITLE-ABS-KEY (mandibular) OR TITLE-ABS-KEY (maxilla) OR TITLE-ABS-KEY (maxillary) OR TITLE-ABS-KEY (maxillofacial)	275,089
3	TITLE-ABS-KEY ("Reconstructive Surgical Procedures") OR TITLE-ABS-KEY (reconstruct) OR TITLE-ABS-KEY (augment) OR TITLE-ABS-KEY (reconstruction) OR TITLE-ABS-KEY (augmentation)	1,114,947
1 AND 2 AND 3	(TITLE-ABS-KEY (bioengineering) OR TITLE-ABS-KEY ("osteogenic scaffold") OR TITLE-ABS-KEY ("osteogenic scaffolds") OR TITLE-ABS-KEY ("tissue engineering") OR TITLE-ABS-KEY ("Bone Morphogenetic Proteins") OR TITLE-ABS-KEY ("Bone Morphogenetic Protein") OR TITLE-ABS-KEY ("Mesenchymal Stem Cells") OR TITLE-ABS-KEY ("Bone Mesenchymal Stem Cells") OR TITLE-ABS-KEY ("beta-tri calcium phosphate") OR TITLE-ABS-KEY ("Bone Morphogenetic Protein 2") OR TITLE-ABS-KEY (rhbpm2) OR TITLE-ABS-KEY (rhbpm-2)) AND (TITLE-ABS-KEY (mandible) OR TITLE-ABS-KEY (mandibular) OR TITLE-ABS-KEY (maxilla) OR TITLE-ABS-KEY (maxilla) OR TITLE-ABS-KEY (maxillary) OR TITLE-ABS-KEY (maxillofacial)) AND (TITLE-ABS-KEY ("Reconstructive Surgical Procedures") OR TITLE-ABS-KEY (reconstruct) OR TITLE-ABS-KEY (augment) OR TITLE-ABS-KEY (reconstruction) OR TITLE-ABS-KEY (augmentation))	1227
Embase		
1	bioengineering:ti,ab,kw OR 'osteogenic scaffold':ti,ab,kw OR 'tissue engineering':ti,ab,kw OR 'bone morphogenetic protein':ti,ab,kw OR 'mesenchymal stem cell':ti,ab,kw OR 'beta-tri calcium phosphate':ti,ab,kw OR 'bone morphogenetic protein 2':ti,ab,kw OR rhbpm2:ti,ab,kw	115,437
2	mandible:ti,ab,kw OR 'jaw disease':ti,ab,kw OR mandibular:ti,ab,kw OR maxilla:ti,ab,kw OR maxillary:ti,ab,kw OR 'maxillofacial disorder':ti,ab,kw OR maxillofacial:ti,ab,kw	202,168
3	'reconstructive surgery':ti,ab,kw OR reconstruct:ti,ab,kw OR reconstruction:ti,ab,kw OR augment:ti,ab,kw OR augmentation:ti,ab,kw	473,086
1 AND 2 AND 3		656

2.5. Inclusion Criteria

The inclusion criteria of the current review were as follows:

1. Original studies, written in English, including randomized controlled trials (RCTs), Clinical trials, observational studies (cohorts and case series) as well as case reports on human patients who had been treated with composite allogeneic tissue engineering for immediate/delayed reconstruction of large maxillofacial defects with minimum/no bone harvesting site.

2. The composite allogeneic tissue engineering was defined as a combination of allogenic bone (scaffolding), bone morphogenic aspirate (source of stem cells), rhBMP-2, and platelet-rich plasma/platelet-rich fibrin (cell signaling for the promotion of stem cell migration and differentiation into osteoblasts).
3. No minimum follow-up was established.
4. Studies must report on at least one of the outcomes of interest:
    - The complication rates were reported. Either early post-surgical complications or long-term post-surgical complications.
    - The success rate is measured as the amount of new bone volume gained (assessed either directly by gross observation or assessed radiographically).
    - Patient-centered outcomes: satisfaction rate and esthetic and functional results.

The Exclusion criteria were as follows (the reasons for excluding articles are also recorded in Table 2):

1. Nonhuman study and cadaver studies.
2. Studies involving significant autogenous bone grafts from sites like the ilium, rib, fibula, or calvarium.

Table 2. Excluded articles with reasons.

Articles (First Author, Year, Title)	Reason for Exclusion
N.M.A. Lopes, 2012, Use of rhBMP-2 to reconstruct a severely atrophic mandible: A modified approach	tricalcium phosphate instead of allogenic bone
Schuckert KH, 2009, Mandibular Defect Reconstruction Using Three-Dimensional Polycaprolactone Scaffold in Combination with Platelet-Rich Plasma and Recombinant Human Bone Morphogenetic Protein-2: De Novo Synthesis of Bone in a Single Case	de novo not allogenic
Jörg Wiltfang, 2016, man as a Living Bioreactor: Prefabrication of a Custom Vascularized Bone Graft in the Gastrocolic Omentum	bovine bone not allograft
Rômulo Maciel Lustosa, 2014, Mandible reconstruction using rhBMP-2: case report and literature review	bovine bone xenograft not allograft
G. K. Sándor,2013, Adipose stem cell tissue-engineered construct used to treat large anterior mandibular defect: a case report and review of the clinical application of good manufacturing practice-level adipose stem cells for bone regeneration (β-tricalcium phosphate)	(β-TCP) granules not allogenic
K. Mesimäki, 2009, Novel maxillary reconstruction with ectopic bone formation by GMP adipose stem cells	beta-tricalcium phosphate not allogenic
B. Zamiri, 2013, Reconstruction of human mandibular continuity defects with allogenic scaffold and autologous marrow mesenchymal stem cells	ex-vivo MSC
L. M. S. Zanettini, 2018, use of Recombinant Human Bone Morphogenetic Protein-2 Associated With Lyophilized Bovine Bone in Reconstruction of Atrophic Maxilla	bovine not allograft
M. Albanese, 2012, Fresh-frozen human bone graft to repair defect after mandibular giant follicular cyst removal: A case report. Cell and Tissue Banking.	no rhBMP
R. Bertolai, 2015, Bone graft and mesenchimal stem cells: Clinical observations and histological analysis. Clinical Cases in Mineral and Bone Metabolism.	mesenchymal stem cells engineered freeze-dried bone allografts/no rhBMP
C. M. Clokie, 2008, Reconstruction of 10 major mandibular defects using bioimplants containing BMP-7	no rhBMP

Table 2. *Cont.*

Articles (First Author, Year, Title)	Reason for Exclusion
M. Cicciù, 2012,Protein-Signaled Guided Bone Regeneration Using Titanium Mesh and Rh-BMP2 in Oral Surgery: A Case Report Involving Left Mandibular Reconstruction after Tumor Resection	no allograft
S. C. Desai, 2013, Use of Recombinant Human Bone Morphogenetic Protein 2 for Mandible Reconstruction	no BMA
C. M. Misch, 2015, Vertical Bone Augmentation Using Recombinant Bone Morphogenetic Protein, Mineralized Bone Allograft, and Titanium Mesh: A Retrospective Cone Beam Computed Tomography Study	not large defect
B. B. Kim, 2014, Hybrid mandibular reconstruction technique: Preliminary case series of prosthetically-driven vascularized fibula free flap combined with tissue engineering and virtual surgical planning	no rhBMP2
Mark C. Fagan, 2008, Simultaneous hard and soft tissue augmentation for implants in the esthetic zone: Report of 37 consecutive cases.	no rhBMP2
B. Haj Yahya, 2020, Non-Autogenous Innovative Reconstruction Method Following Mandibulectomy	defect size is small
H. I. Canter, 2007, Reconstruction of mandibular defects using autografts combined with demineralized bone matrix and cancellous allograft	bone harvest included
C. Loperfido, 2014, Severe mandibular atrophy treated with a subperiosteal implant and simultaneous graft with rhBMP-2 and mineralized allograft: a case report	no stem cell
A. Deshmukh, 2015, Bilateral maxillary sinus floor augmentation with tissue-engineered autologous osteoblasts and demineralized freeze-dried bone [2]	defect size is small
M. S. Block, 2010, Use of Living Cell Construct to Enhance Bone Reconstruction: Preliminary Results	no rhBMP

2.6. Study Selection Process

In order to determine proper materials, two reviewers conducted a duplicate searching process using the inclusion and exclusion criteria independently. Instances of divergence of opinion were resolved by consulting a third investigator (Sh.R.). The full-text version of papers was obtained for all titles that appeared to meet the inclusion criteria or in case of any hesitancy. Then, each paper was studied at least twice by two reviewers (M.A. and H.Z.).

2.7. Data Extraction

Whenever applicable, two authors (M.A. and H.Z.) retrieved the following data from the finally included studies based on a predefined paper checklist, and three other authors (Sh.R., Z.Kh., and J.M.) supervised the extraction process for accuracy. Since poorly reported outcomes of included materials could thread the validity of our work, we contacted the corresponding author of the study via email, sending up to two emails, in case of missing data or any hesitancy. The following data were extracted:

First author, year of publication, country of origin, study type, mean age, sex, number of cases, mean follow-up (range), gained bone volume, rates of complications, donor-site morbidities, and success rates and main outcomes.

2.8. Risk of Bias Assessment

Two examiners (M.A. and H.Z.) conducted the quality assessment according to the following quality assessment tools and supervised by a third author (Sh.R.) for accuracy. Any disagreement was resolved by consensus (Table 3).

The methodological quality and synthesis of case series and case reports by Murad et al. were used for bias assessment [11–14]. Summing the scores and presenting an aggregate score was not appropriate, and making an overall judgment about the quality should be based on the most critical questions.

Table 3. Risk of bias assessment for case series and case reports.

First Author (Year of Publication)	1. Does the Patient(s) Represent(s) the Whole Experience of the Investigator (Centre) or Is the Selection Method UNCLEAR to the Extent That Other Patients with Similar Presentation May Not Have Been Reported?	2. Was The Exposure Adequately Ascertained?	3. Was the Outcome Adequately Ascertained?	4. Were Other Alternative Causes that May Explain the Observation Ruled Out?	5. Was There a Challenge/Rechallenge Phenomenon?	6. Was There a Dose–Response Effect?	7. Was Follow-up Long Enough for Outcomes to Occur?	8. Is the Case(s) Described with Sufficient Details to Allow Other Investigators to Replicate the Research or to Allow Practitioners to Make Inferences Related to Their Own Practice?
James C. Melville, 2016, Houston, English [15]	N	Y	Y	N	Y	NA	Y	Y
James C. Melville, 2017, Houston, English [16]	N	Y	Y	N	Y	NA	Y	Y
J. C. Melville, 2014, Houston, English [17]	N	Y	Y	N	Y	NA	Y	Y
James C. Melville, 2019, Houston, English [18]	N	Y	Y	N	Y	NA	Y	Y
Matthias Schlund, 2019, Oman, English [19]	N	Y	Y	N	Y	NA	Y	N
Kamel Alraei, 2020, Saudi Arabia, English [20]	N	Y	Y	N	Y	NA	Y	Y
Jeanette Johnson, 2016, Texas, English [21]	N	Y	Y	N	Y	NA	Y	Y
N. Ali, 2018, Texas, English [22]	N	Y	Y	N	Y	NA	Y	Y
Todd G. Carter, 2008, Seattle, English [23]	N	Y	Y	Y	Y	NA	Y	Y
Weiss R., 2022, USA, English [24]	N	Y	Y	N	Y	NA	Y	Y
Melville et al., 2019, USA, English [25]	N	Y	Y	Y	Y	NA	Y	Y

N: no, Y: yes, NA: not applicable, The timing for the outcomes to occur is considered for short-term outcomes.

2.9. Data Analysis

Individual patient data were aggregated, and descriptive statistics were performed (MS Excel 2016).

3. Results

Figure 1 illustrates the PRISMA flow diagram for the study selection process at each level [10]. In the initial search, 2618 papers were obtained through PubMed, Scopus, and Embase. After duplication removal, 1401 papers remained the titles and abstracts of which were assessed for eligibility. A total of 1359 papers were removed by reading the title and abstract. Full texts were retrieved for the remaining 34 papers. Of those, 21 papers were excluded with reason. Finally, 11 papers were found to be eligible to be included in the current study.

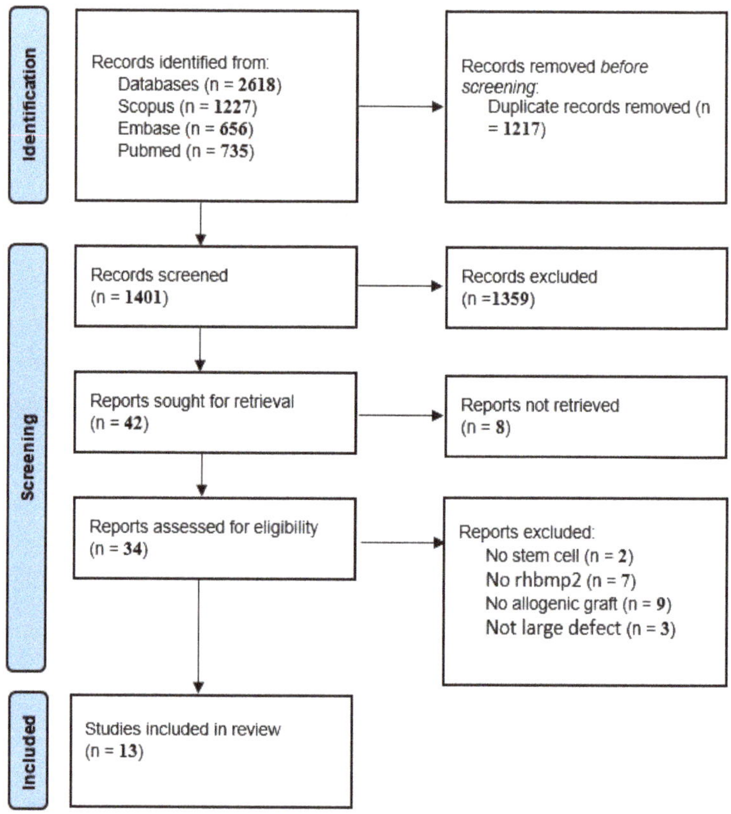

Figure 1. The PRISMA flowchart of included studies.

Study Characteristics

The characteristics of the included materials are shown in Table 4. Regarding the study type, eight case reports, three case series, and two clinical trials were included. A total of 144 cases were enrolled in this systematic review. The mean age of the patients was 43.34 (age range: 9–89). All incoming articles reported age. Seventy-one cases were male, and seventy-three were female. The mean follow-up time was 24.2 months with a range of 6 to 60 months. The included materials were published between the years 2008 and 2022 in the following countries: USA, India, Oman, Turkey, Israel, and Saudi Arabia.

Table 4. Characteristics of included materials.

First Author/Year/ Country of Origin/ Language	Type of Study	Number of Cases/Duration of Follow Up	Mean Age/Sex	Summary of Method	cause of Defect/Size of Defect/Filling Rate of Defect (Bone Gain)	Mesenchymal Stem Cells Harvesting Site	Success Rate
James C. Melville, 2016, Houston, English [15]	Case report	5 cases/ 12–14 months	18–66 years old/ 3 men and 2 women	the freeze-dried cortical-cancellous bone in combination with large rhBMP-2 (12 mg)/ absorbable collagen sponge (ACS) and 120 mL of BMAC obtained from anterior or posterior hip were mixed homogeneously. 10 mL of crushed cortical cancellous bone for each 1 cm length of the defect.	benign mandibular tumors with no history of chemotherapy or radiation to the mandible (Ossifying fibroma, Desmoplastic, Juvenile ossifying fibroma)/3.5 to 8.0 cm/10–14.5 mm and regenerated bone height was in the range of 22–26 mm	BMAC was harvested from either the bilateral anterior iliac crest or unilateral posterior iliac crest	100% success
James C. Melville, 2017, Houston, English [16]	Case report	1 case/6 months	45 years old/woman	radial forearm fasciocutaneous flap combined with a tissue-engineered bone graft consisting of allogeneic bone, rhBMP-2, and BMAC.	trauma (deficient projection of the left malar region, loss of left maxillary ridge alveolar bone, loss of dentition and upper eyelid ptosis, and lower eyelid ectropion)	BMAC form the iliac crest	100% success
Robert E Marx, 2014, USA, English [26,27]	Case report	40 case/6 months	mean age 57 years (19–78 years)/ 22 men and 12 women	in situ tissue-engineered graft containing 54 ± 38 CD34+ cells/mL along with 54 ± 38 CD44+, CD90+, and CD105+ cells/mL together with rhBMP-2 in an absorbable collagen sponge (1 mg/cm of defect) and crushed cancellous allogeneic bone.	/6- to 8-cm continuity defects/trabecular bone area of 36 ± 10%, versus 67 ± 13% for group B	four puncture sites in the bilateral anterior iliac crest	Group A: 40% success rate Group B: 100% success rate

Table 4. *Cont.*

First Author/Year/ Country of Origin/ Language	Type of Study	Number of Cases/Duration of Follow Up	Mean Age/Sex	Summary of Method	cause of Defect/Size of Defect/Filling Rate of Defect (Bone Gain)	Mesenchymal Stem Cells Harvesting Site	Success Rate
J. C. Melville, 2014, Houston, English [17]	retrospective study	9 cases/4 years	mean age 23.7-year-old (3 patients under the age of 17)/5 men, 4 women	freeze-dried cortical cancellous bone was obtained used in combination with 12 mg of rhBMP-2/ACS and 120 cc of Bone Marrow Aspirate Concentrate	ameloblastoma, OKC, Myxoma, Ossifying Fibroma and Central Giant Cell Tumor/4 cm to 12 cm	anterior hip	100% success rate
Robert E Marx, 2013, USA, English [28]	clinical trial	20 cases/6 months	Mean age 58 and 62/5 men, 15 women	two types of grafts in large vertical maxillary defects: a composite graft of recombinant human bone morphogenetic protein-2/acellular collagen sponge (rhBMP-2/ACS), crushed cancellous freeze-dried allogeneic bone (CCFDAB), and platelet-rich plasma (PRP); and size-matched 100% autogenous grafts	horizontal defects/1 cm vertical deficiency and 1 cm horizontal deficiency spanning at least a four-tooth length/2-mm-diameter bone core (bone area 59 ± 12% and 54 ± 10%)	tibia plateau or anterior ilium	composite graft: 97.4% success rate autogenous grafts: 100% success rate

Table 4. *Cont.*

First Author/Year/ Country of Origin/ Language	Type of Study	Number of Cases/Duration of Follow Up	Mean Age/Sex	Summary of Method	cause of Defect/Size of Defect/Filling Rate of Defect (Bone Gain)	Mesenchymal Stem Cells Harvesting Site	Success Rate
James C. Melville, 2019, Houston, English [18]	retrospective case	34 cases/5 years	mean age 37.79 ± 20.4 (9–89 years old)/ 19 men, 15 women	first, BMAC was obtained from the patient's anterior or posterior hip using the Harvest Bone Marrow Aspirate Concentrate System. Second, a medium to large rhBMP-2 kit was used, according to size of the defect (a large kit was used for defects greater than 6 cm). Third, corticocancellous bone (MTF Biologics. Edison, NJ, USA) was milled down to a 1.0- to 2.0-mm particulate graft. Finally, a non-resorbable titanium mesh or resorbable poly (L-lactide) (PLLA) or poly (D, L-lactide) (PDLLA) membrane was used as the containment system for the graft.	ablative tumor surgery or traumatic accidents (ameloblastoma, ossifying fibroma, odontogenic keratocyst (OKC), and sclerosing osteomyelitis, odontogenic myxoma, giant cell tumor associated with hyperparathyroidism, and central giant cell granuloma)/continuity defect 5.61 ± 2.92 and Noncontinuity defect 4.77 ± 3.33/mean height 2.12 ± 0.44 Mean width 1.53 ± 0.55	anterior or posterior hip	Continuity defect 90% Noncontinuity defect 100%
Matthias Schlund, 2019, Oman, English [19]	clinical report	1 cases/1 years	33-year-old/men	using a fresh-frozen humeral allograft as scaffold seeded with progenitor cells collected through iliac bone marrow aspirate and vascularized with a radial forearm free flap	severe craniofacial trauma resulting in several fractures of the facial skeleton including a comminuted mandibular fracture from left parasymphysis to left angle	iliac bone marrow aspirate	100% success rate

Table 4. *Cont.*

First Author/Year/ Country of Origin/ Language	Type of Study	Number of Cases/Duration of Follow Up	Mean Age/Sex	Summary of Method	cause of Defect/Size of Defect/Filling Rate of Defect (Bone Gain)	Mesenchymal Stem Cells Harvesting Site	Success Rate
Kamel Alraei, 2020, Saudi Arabia, English [20]	case report	1 case/4 years	27 years old/female	reconstruction of the mandibular defect using rhBMP-2 combined with bone marrow aspirate concentrate (BMAC), and an allograft with a titanium mesh. placed a total of 12 mg of rhBMP-2 in four absorbable collagen sponges	a calcifying, cystic, odontogenic tumor (a Pindborg tumor)/2.7 mm	60 cc of bone marrow from left iliac crest	Success rate 100%
Jeanette Johnson, 2016, Texas, English [21]	Case report	1 case/1 years	11 years old/female	reconstruction using a costochondral rib graft, allogeneic bone, bone marrow aspirate concentrate, and recombinant human morphogenetic protein-2.	unilocular radiolucent lesion of the posterior left mandible/ 3.4 × 4.2 × 3.1 cm	bone marrow aspirate concentrate	Success rate 100%
N. Ali, 2018, Texas, English [22]	Retrospective study	24 cases/3 years	Mean age 32.1 years old (11–65 years old)/ 14 men and 10 women	freeze-dried cancellous bone was obtained from MTF and used in combination with 12 mg of rhBMP-2/ACS, and 120 cc of BMAC was obtained from the patients	ablative tumor and trauma/ 5–17 cm/ 3.0 cm bone height and >1.0 cm bone width	anterior or posterior hip	88% success rate

Table 4. *Cont.*

First Author/Year/ Country of Origin/ Language	Type of Study	Number of Cases/Duration of Follow Up	Mean Age/Sex	Summary of Method	cause of Defect/Size of Defect/Filling Rate of Defect (Bone Gain)	Mesenchymal Stem Cells Harvesting Site	Success Rate
Todd G. Carter, 2008, Seattle, English [23]	Case report	5 cases/22 months	42,43, 41, 81/2 men, 3 women	reconstruction with rhBMP-2. Because, in case 1, rhBMP-2 absorbed on a collagen sponge alone failed to regenerate bone, autogenous bone marrow, and allogenic bone were combined with rhBMP-2-impregnated collagen sponges to increase the osteogenic.	open and displaced left mandibular angle fracture, with an infection that required incision and drainage, multiple facial lacerations and a comminuted mandible fracture, osteomyelitis of the right mandible, radiolucent lesions in the mandible/4 cm, 4.5 cm	bone marrow from left iliac crest, bone marrow cells from the patient's right anterior iliac crest.	-
Weiss R., 2022, USA, English [29]	Case report	2 cases/4 months	62-year-old woman/ 24-year-old woman	reconstruction with (1) corticocancellous bone chips (2) bone marrow aspirate, (3) rhBMP-2 A combined intraoral and extraoral approach was used to allow access. Cadaveric rib allograft was secured to the inferior aspect of the reconstruction plate. Nerve allografts were secured to the inferior alveolar nerve stumps and a water-tight closure	excisional biopsy of ameloblastoma 8 years earlier, segmental defect $6.0 \times 5.0 \times 3.7$ cm/no past medical history, $4.4 \times 2.1 \times 2.1$ cm	bone marrow was harvested via trochar and cannula from the anterior iliac crest	100%

Table 4. *Cont.*

First Author/Year/ Country of Origin/ Language	Type of Study	Number of Cases/Duration of Follow Up	Mean Age/Sex	Summary of Method	cause of Defect/Size of Defect/Filling Rate of Defect (Bone Gain)	Mesenchymal Stem Cells Harvesting Site	Success Rate
Melville et al., 2019, USA, English [30]	Case report	1 case/6 months	45-year-old female	exposing and preparing the alveolar defect as well as harvesting the BMAC from the iliac crest and harvesting the radial forearm. crushed corticocancellous bone and rhBMP-2 were mixed with BMAC. The tissue-engineered graft was placed and packed onto the defect. The radial forearm was then sutured around and over the graft.	post-traumatic maxillary alveolar ridge defect/size not mentioned	bone marrow can be aspirated from the anterior ilium	100% success

Among the articles reviewed, two methods of immediate surgery and two-stage surgery were performed.

In 2016, a retrospective case study by Melville et al. [14] treated five patients with large mandibular defects caused by tumor ablation. The average amount of mandible defects was between 3.5 and 8 cm, which was treated with a combined method of freeze-dried cortical-cancellous bone and rhBMP-2 and BMAC in one session at the same time as removing the tumor with an intra-oral approach. With the same method of treatment, Melville et al. reported the treatment of a large maxillary defect that had been damaged due to trauma. This was a novel technique for large maxillary defects combining conventional techniques and tissue engineering techniques to create a custom-made graft utilizing in situ tissue engineering [16]. Schlund and colleagues reported a similar technique in a 33-year-old patient; they vascularized the allogenic graft with a radial forearm free flap to overcome poor vascularization in tissue-engineered allogeneic bone [19].

N. Ali et al. [22] reported a success rate of 88% by treating 24 surgical cases with a combination of allogeneic transplantation and Melville-like proteins and stem cells.

J. Johnson et al. [21] and her colleagues combined costochondral rib graft, allogeneic bone, BMAC, and recombinant human morphogenetic protein-2 in an 11-year-old patient with a 3.4 × 4.2 × 3.1 cm defect. A 100% success has been reported in a 1-year study.

Using the abovementioned combination, Kamal et al. [20] used titanium mesh for better bond results. Melville et al. [18] in another retrospective study treated 34 cases of tumor-like ameloblastoma, ossifying fibroma, odontogenic keratocyst (OKC), etc., using a non-resorbable titanium mesh or resorbable poly(L-lactide) (PLLA) or poly(D, L-lactide) (PDLLA) membrane addition of the mentioned combination bond, it was found that the graft is exceptionally vulnerable to bacterial contamination and also any patients with a history of uncontrolled health disease, chemotherapy, or radiation therapy negatively affect the graft's viability.

RE Marx [27,28] and colleagues in two separate studies with two techniques and changing the amount of stem cells and their type compared the results to autogenous transplantation. In these two studies, the success rate was 97.4% to 40% compared to autogenous transplantation. The noticeable complication with this technique was edema, which was graded as nearly twice that of the autogenous graft and lasted nearly twice as long. They also stated that, in a series of cases, there is still a need to synchronize this technique with the autogenic technique and significantly more swelling.

4. Discussion

Alveolar defects caused by oncologic resection or trauma often involve extensive volumetric bone loss in the vertical and horizontal dimensions. If remaining untreated, they can lead to noticeable quality-of-life, nutritional, and speech issues [31,32]. The reconstruction of these hard tissue defects for shaping the appropriate facial form and functional rehabilitation poses a significant challenge for oral and maxillofacial surgeons. Successful reconstruction with a reasonably high long-term success rate (up to 70%) has been achieved with autogenous bone grafts. For decades, autografts as a natural biomaterial have been considered the gold standard due to superior osteoinductivity, osteoconductivity, and osteogenesis, compared with other types of materials [20,33]. Osteo-cutaneous free flaps, especially fibula free flaps, are the most common autografts used for crucial-sized grafts [19].

They also have shown histocompatibility and avoidance of immune rejection [34]. Morbidities in graft harvesting sites and bone transplantation sites, increased surgical time, and prolonged hospitalization are the main drawbacks of the conventional technique. Extensive graft harvesting from extra-oral sites has an increased risk of hematoma, pain and sensory disturbances, herniation of abdominal content, pelvic instability, and infections. Likewise, in huge defects, longer than 6 cm, the increased failure rate is not out of the question [35,36].

A promising alternative method to address a wide range of maxillofacial scenarios is tissue engineering. The strategies used in tissue engineering based on the use of cells, scaffolds, and bioactive molecules encompass tissue and organ regeneration [37]. Tissue engineering for maxillofacial bone defects is most successful in osteogenesis when mimicking both the macro- and micro-environment. Current literature supports the use of a biomimetic, bioactive osseointegrative customized scaffold according to the defect accompanied by growth factors and stem cells [38]. In recent years, a new tissue engineering technique using a combination of allogenic bone, BMAC, and rhBMP-2 has been introduced and advocated by Melville et al. for immediate reconstruction of large maxillofacial defects with less invasiveness, less intraoperative time, lower cost, and minimum/no donor site morbidities than conventional autografting methods [15].

The three basics of successful regeneration in this technique are allogeneic bone (scaffolding), BMAC (stem cells), and rhBMP-2 and platelet-rich plasma/platelet-rich fibrin (cell signaling for the promotion of stem cell migration and differentiation into osteoblasts).

The scaffold is a three-dimensional framework on which cells can adhere and proliferate. A good scaffold to reconstruct bone is bone; allografts such as humeral bone have the desired strength to bear the mastication loads [19,39].

BMAC is a rich source of MSCs and osteoprogenitor cells, cytokines, and growth factors that can be derived from the tibia or iliac crests and delivered to bony defects [27]. BMAC is an affordable, easy-to-harvest, and safe technique to collect a considerable number of mesenchymal stem cells, and the results with this technique are comparable with an autograft alone [40,41].

rhBMP-2 added to allografts was used for alveolar reconstruction defects and sinus floor augmentation successfully, but recent experiments also suggest several clinical benefits of the off-label use of rhBMP-2 in the reconstruction of critical-sizedmaxillofacial defects.

The included materials reported both immediate and delayed reconstruction with a transoral or extraoral approach. Although the preferred technique was mostly an immediate reconstruction, in cases with potential extensive soft tissue loss after surgery, delayed reconstruction was conducted. Sufficient soft tissue is a crucial factor for success rate; the required soft tissue volume would allow for a primary watertight tension-free closure to prevent bacterial contamination. If the amount of soft tissue seemed insufficient, a vascularized free flap was performed and later followed by delayed tissue-engineered reconstruction. Special care must be taken with an intraoral approach as tissue-engineered bone grafts are highly vulnerable to salivary leakage and bacterial contamination of the graft [18].

Literature on the use of rhBMP-2 in large maxillofacial defects in children is scarce. Only three papers included children (9–18 years) in their studies [17,18,21]. Although the predictability and safety of the combination of allogenic bone, BMAC, and BMP for reconstruction after resection of benign tumors in adult patients have been demonstrated, their use in children is still in dispute. The US Food and Drug Administration has warned about the use of BMP in patients with developing skeletons [42]. Use should be judicious as complications and long-term outcomes cannot be validated until larger studies on pediatrics be conducted.

Recently, a similar study has been conducted with an aim of reconstruction of critical-size tibia defects in a sheep model. The bone substitute in combination with endothelial progenitor cells (EPC), mesenchymal stem cells (MSC), and with (or without) growth factors BMP-2 was prevascularized and transplanted into a critical-size bone defect in 17 sheep models. They used an AV loop as an even less invasive approach, compared to a forearm free flap, for axial vascularization. During the first and third months after transplantation at the defect, good success was achieved [43].

There has been some evidence for the potential of immediate tissue engineering techniques to be an alternative treatment for the current gold standards, transplantation of vascularized autologous bone harvested from unharmed areas, in routine practice.

Limitations of the Technique and Future Implications

Despite the growing body of literature and advances in the reconstruction of critical-sized maxillofacial defects, the technique has a number of shortcomings and several challenges ahead.

One negative point of added rhBMP-2 is inevitable post-surgical swelling and edema attributed to the inflammatory cytokine-like nature of rhBMP-2, on which steroids have little impact. Therefore, pre-surgical precautions are warranted. rhBMP-2, when combined with absorbable collagen sponges as a carrier, provides a continuous release of the protein into the bone formation environment for three weeks after the surgery. The complication is blocked vascular growth and soft tissue compression at the bone regeneration site [44].

Nevertheless, some contraindications are listed for BMAC harvestings, such as cases with congenital disorders, metabolic diseases, malignancy, or a history of trauma in the harvesting site. Caution should be exercised in young patients (<18 years) [24]. Non-vascularized allografts have a high vulnerability to bacterial contamination [19].

Inadequate defect fit of scaffolds and personalized, customized substitute devices is another challenge to be faced, as the process may require multiple steps/device parts [15]. In addition, current techniques only allow for homogeneous bony structures regeneration, while clinical scenarios in craniofacial defects caused by tumor ablation or trauma mostly require the engineering of multiple tissues, which include soft, hard, and nerve tissue.

The temporomandibular joint as an osteochondral unit containing bone, cartilage, and transitional layers is one example. In complex cases of mandibular or zygomatic arch defects, reconstruction of the temporomandibular joint (TMJ) may be required. The TMJ has poor regenerative capacity due to the avascular nature of cartilage.

The emergence of three-dimensional printing (3DP) technologies made notable progress in the regeneration of complex heterogeneous defects. This technology enables individualized substitute device construction [25]. The multilayer scaffold design creates a vascular network for better oxygen diffusion and waste exchange in heterogeneous defects [31].

The current literature is encouraging but as yet is too scarce to allow a firm conclusion to be drawn. With the knowledge of the possibility afforded for the future reconstruction of large maxillofacial defects, further studies with large sample sizes and long-term follow-ups are warranted to validate the routine use of this technology in the maxillofacial field.

5. Conclusions

The reviewed technique combines the allogenic graft as a biologic scaffold with bone marrow aspirate and rhBMP-2 to create a custom-made graft.

The current literature demonstrates the potential for using large-scale transplantable, vascularized, and customizable bone with the aim of reconstructing large maxillofacial bony defects in short-term follow-ups. This approach might be an alternative to the current therapeutic clinical options that include vast autogenous bone harvest and many patient morbidities, although, further clinical trials with larger sample sizes in long-term follow-ups are needed to draw a firm conclusion.

Author Contributions: Conceptualization, S.R., M.A., H.Z. and H.R.F.; methodology, S.R., M.A. and H.Z.; literature search, S.R., M.A. and H.Z., validation, M.A., H.Z., Z.K. and S.O.K.; data extraction, S.R., M.A. and H.Z.; investigation, S.R., H.Z., H.R.F. and J.C.M.; data curation, S.R., M.A., H.Z. and M.S.; writing—original draft preparation, S.R., M.A., H.Z. and J.C.M.; writing—review and editing, S.R., M.A., Z.K., H.R.F. and M.S.; visualization, Z.K., S.O.K., J.C.M. and F.S.; supervision, S.R., Z.K., J.C.M. and F.S.; project administration, S.O.K. and F.S. All authors have read and agreed to the published version of the manuscript.

Funding: This research received no external funding.

Institutional Review Board Statement: Not applicable.

Data Availability Statement: Not applicable.

Conflicts of Interest: The authors declare no conflict of interest.

References

1. Schimming, R.; Schmelzeisen, R. Tissue-engineered bone for maxillary sinus augmentation. *J. Oral Maxillofac. Surg.* **2004**, *62*, 724–729. [CrossRef]
2. Akinbami, B.O. Reconstruction of Continuity Defects of the Mandible with Non-vascularized Bone Grafts. Systematic Literature Review. *Craniomaxillofac. Trauma Reconstr.* **2016**, *9*, 195–205. [CrossRef]
3. Langer, R.; Vacanti, J.P. Tissue engineering. *Science* **1993**, *260*, 920–926. [CrossRef]
4. Zhao, R.; Yang, R.; Cooper, P.R.; Khurshid, Z.; Shavandi, A.; Ratnayake, J. Bone Grafts and Substitutes in Dentistry: A Review of Current Trends and Developments. *Molecules* **2021**, *26*, 3007. [CrossRef] [PubMed]
5. Spicer, P.; Young, S.; Kasper, F.K.; Athanasiou, K.A.; Mikos, A.G.; Wong, M.E.-K. Tissue engineering in oral and maxillofacial surgery. In *Principles of Tissue Engineering*; Elsevier: Tenta, Egypt, 2014; pp. 1487–1506.
6. Jensen, O.T.; Sennerby, L. Histologic analysis of clinically retrieved titanium microimplants placed in conjunction with maxillary sinus floor augmentation. *Int. J. Oral Maxillofac. Implant.* **1998**, *13*, 513–521.
7. Lorenzetti, M.; Mozzati, M.; Campanino, P.P.; Valente, G. Bone augmentation of the inferior floor of the maxillary sinus with autogenous bone or composite bone grafts: A histologic-histomorphometric preliminary report. *Int. J. Oral Maxillofac. Implant.* **1998**, *13*, 69–76.
8. Vignesh, U.; Mehrotra, D.; Howlader, D.; Kumar, S.; Anand, V. Bone Marrow Aspirate in Cystic Maxillofacial Bony Defects. *J. Craniofacial Surg.* **2019**, *30*, e247–e251. [CrossRef]
9. Viña, J.A.; El-Alami, M.; Gambini, J.; Borras, C.; Viña, J.; Peñarrocha, M.A. Application of mesenchymal stem cells in bone regenerative procedures in oral implantology. A literature review. *J. Clin. Exp. Dent.* **2014**, *6*, e60–e65. [CrossRef]
10. Page, M.J.; McKenzie, J.E.; Bossuyt, P.M.; Boutron, I.; Hoffmann, T.C.; Mulrow, C.D.; Shamseer, L.; Tetzlaff, J.M.; Moher, D. Updating guidance for reporting systematic reviews: Development of the PRISMA 2020 statement. *J. Clin. Epidemiol.* **2021**, *134*, 103–112. [CrossRef] [PubMed]
11. Khurshid, Z.; Tariq, R.; Asiri, F.Y.; Abid, K.; Zafar, M.S. Literature search strategies in dental education. *J. Taibah Univ. Med. Sci.* **2021**, *16*, 799–806. [CrossRef]
12. Wiltfang, J.; Rohnen, M.; Egberts, J.-H.; Lützen, U.; Wieker, H.; Açil, Y.; Naujokat, H. Man as a living bioreactor: Prefabrication of a custom vascularized bone graft in the gastrocolic omentum. *Tissue Eng. Part C Methods* **2016**, *22*, 740–746. [CrossRef]
13. Deshmukh, A.; Kalra, R.; Chhadva, S.; Shetye, A. Bilateral maxillary sinus floor augmentation with tissue-engineered autologous osteoblasts and demineralized freeze-dried bone. *Contemp. Clin. Dent.* **2015**, *6*, 243. [CrossRef]
14. Murad, M.H.; Sultan, S.; Haffar, S.; Bazerbachi, F. Methodological quality and synthesis of case series and case reports. *BMJ Evid.-Based Med.* **2018**, *23*, 60–63. [CrossRef] [PubMed]
15. Melville, J.C.; Nassari, N.N.; Hanna, I.A.; Shum, J.W.; Wong, M.E.; Young, S. Immediate Transoral Allogeneic Bone Grafting for Large Mandibular Defects. Less Morbidity, More Bone. A Paradigm in Benign Tumor Mandibular Reconstruction? *J. Oral Maxillofac. Surg.* **2017**, *75*, 828–838. [CrossRef]
16. Melville, J.C.; Tursun, R.; Green, J.M., III; Marx, R.E. Reconstruction of a post-traumatic maxillary ridge using a radial forearm free flap and immediate tissue engineering (bone morphogenetic protein, bone marrow aspirate concentrate, and cortical-cancellous bone): Case report. *J. Oral Maxillofac. Surg.* **2017**, *75*, 438.e1–438.e6. [CrossRef] [PubMed]
17. Melville, J.; Marx, R.; Tursun, R.; Moody, M.; Hew, D.; Schacht, S.; Starley, E.; Broumand, V.; Peleg, M.; Sawatari, Y. The utilization of allogeneic bone, bone morphogenetic protein and bone marrow aspirate concentrate for immediate reconstruction of benign tumor continuity defects. *J. Oral Maxillofac. Surg.* **2014**, *72*, e204–e205. [CrossRef]
18. Melville, J.C.; Tran, H.Q.; Bhatti, A.K.; Manon, V.; Young, S.; Wong, M.E. Is Reconstruction of Large Mandibular Defects Using Bioengineering Materials Effective? *J. Oral Maxillofac. Surg.* **2020**, *78*, 661.e1–661.e29. [CrossRef]
19. Schlund, M.; Nicot, R.; Depeyre, A.; Alkasbi, J.; Ferri, J. Reconstruction of a large posttraumatic mandibular defect using bone tissue engineering with fresh-frozen humeral allograft seeded with autologous bone marrow aspirate and vascularized with a radial forearm flap. *J. Craniofacial Surg.* **2019**, *30*, 2085–2087. [CrossRef] [PubMed]
20. Alraei, K.; Sharqawi, J.; Harcher, S.; Ghita, I. Efficacy of the Combination of rhBMP-2 with Bone Marrow Aspirate Concentrate in Mandibular Defect Reconstruction after a Pindborg Tumor Resection. *Case Rep. Dent.* **2020**, *2020*, 8281741. [CrossRef]
21. Johnson, J.; Jundt, J.; Hanna, I.; Shum, J.W.; Badger, G.; Melville, J.C. Resection of an ameloblastoma in a pediatric patient and immediate reconstruction using a combination of tissue engineering and costochondral rib graft: A case report. *J. Am. Dent. Assoc.* **2017**, *148*, 40–43. [CrossRef]
22. Ali, N.; Young, S.; Shum, J.W.; Hanna, I.; Wong, M.E.; Melville, J.C. The Efficacy of Bioengineering (Stem Cells, Allogeneic Bone, and rhBMP-2) for Reconstruction of Large Mandibular Continuity Defects: A Retrospective Study of 24 Patients over a 3-Year Period. *J. Oral Maxillofac. Surg.* **2018**, *76*, e75. [CrossRef]
23. Carter, T.G.; Brar, P.S.; Tolas, A.; Beirne, O.R. Off-label use of recombinant human bone morphogenetic protein-2 (rhBMP-2) for reconstruction of mandibular bone defects in humans. *J. Oral Maxillofac. Surg.* **2008**, *66*, 1417–1425. [CrossRef] [PubMed]
24. Melville, J.C.; Mañón, V.A.; Blackburn, C.; Young, S. Current Methods of Maxillofacial Tissue Engineering. *Oral Maxillofac. Surg. Clin. N. Am.* **2019**, *31*, 579–591. [CrossRef]
25. Bauermeister, A.J.; Zuriarrain, A.; Newman, M.I. Three-Dimensional Printing in Plastic and Reconstructive Surgery: A Systematic Review. *Ann. Plast. Surg.* **2016**, *77*, 569–576. [CrossRef] [PubMed]

26. Mayfield, C.K.; Ayad, M.; Lechtholz-Zey, E.; Chen, Y.; Lieberman, J.R. 3D-Printing for Critical Sized Bone Defects: Current Concepts and Future Directions. *Bioengineering* **2022**, *9*, 680. [CrossRef]
27. Marx, R.E.; Harrell, D.B. Translational research: The CD34+ cell is crucial for large-volume bone regeneration from the milieu of bone marrow progenitor cells in craniomandibular reconstruction. *Int. J. Oral Maxillofac. Implant.* **2014**, *29*, e201–e209. [CrossRef] [PubMed]
28. Marx, R.E.; Armentano, L.; Olavarria, A.; Samaniego, J. rhBMP-2/ACS grafts versus autogenous cancellous marrow grafts in large vertical defects of the maxilla: An unsponsored randomized open-label clinical trial. *Int. J. Oral Maxillofac. Implant.* **2013**, *28*, e243-51. [CrossRef]
29. Weiss, R.O.; Wong, P.E.; Reddy, L.V. (Eds.) *Immediate Reconstruction of Segmental Mandibular Defects via Tissue Engineering*; Baylor University Medical Center Proceedings; Taylor & Francis: Montgomery, TX, USA, 2022.
30. Melville, J.C.; Tran, H.Q.; Shum, J.W.; Tursun, R.; Marx, R.E. Reconstruction of Post-Traumatic Maxillary Ridges Using a Radial Forearm Free Flap and Allogeneic Tissue-Engineered Bone Grafts. In *Regenerative Medicine and Plastic Surgery: Elements, Research Concepts and Emerging Technologies*; Springer: Cham, Switzerland, 2019; pp. 349–355.
31. Nerem, R.M.; Seliktar, D. Vascular tissue engineering. *Annu. Rev. Biomed. Eng.* **2001**, *3*, 225–243. [CrossRef]
32. Hertrampf, K.; Wenz, H.J.; Lehmann, K.M.; Lorenz, W.; Koller, M. Quality of life of patients with maxillofacial defects after treatment for malignancy. *Int. J. Prosthodont.* **2004**, *17*, 657–665.
33. Elsalanty, M.E.; Genecov, D.G. Bone grafts in craniofacial surgery. *Craniomaxillofac. Trauma Reconstr.* **2009**, *2*, 125–134. [CrossRef]
34. Misch, C.M.; Jensen, O.T.; Pikos, M.A.; Malmquist, J.P. Vertical bone augmentation using recombinant bone morphogenetic protein, mineralized bone allograft, and titanium mesh: A retrospective cone beam computed tomography study. *Int. J. Oral Maxillofac. Implant.* **2015**, *30*, 202–207. [CrossRef] [PubMed]
35. Foster, R.D.; Anthony, J.P.; Sharma, A.; Pogrel, M.A. Vascularized bone flaps versus nonvascularized bone grafts for mandibular reconstruction: An outcome analysis of primary bony union and endosseous implant success. *Head Neck* **1999**, *21*, 66–71. [CrossRef]
36. Pogrel, M.A.; Podlesh, S.; Anthony, J.P.; Alexander, J. A comparison of vascularized and nonvascularized bone grafts for reconstruction of mandibular continuity defects. *J. Oral Maxillofac. Surg.* **1997**, *55*, 1200–1206. [CrossRef] [PubMed]
37. Fisher, J.P.; Mikos, A.G.; Bronzino, J.D.; Peterson, D.R. *Tissue Engineering: Principles and Practices*; CRC Press: Boca Raton, FL, USA, 2012.
38. Nelms, L.; Palmer, W.J. Tissue engineering in mandibular reconstruction: Osteogenesis-inducing scaffolds. *Plast. Aesthetic Res.* **2019**, *6*, 21. [CrossRef]
39. Stevens, B.; Yang, Y.; Mohandas, A.; Stucker, B.; Nguyen, K.T. A review of materials, fabrication methods, and strategies used to enhance bone regeneration in engineered bone tissues. *J. Biomed. Mater. Res. Part B Appl. Biomater.* **2008**, *85*, 573–582. [CrossRef] [PubMed]
40. Forriol, F.; Denaro, L.; Longo, U.G.; Taira, H.; Maffulli, N.; Denaro, V. Bone lengthening osteogenesis, a combination of intramembranous and endochondral ossification: An experimental study in sheep. *Strateg. Trauma Limb Reconstr.* **2010**, *5*, 71–78. [CrossRef]
41. Gianakos, A.L.; Sun, L.; Patel, J.N.; Adams, D.M.; Liporace, F.A. Clinical application of concentrated bone marrow aspirate in orthopaedics: A systematic review. *World J. Orthop.* **2017**, *8*, 491–506. [CrossRef]
42. Molinari, R.W.; Molinari, C. The Use of Bone Morphogenetic Protein in Pediatric Cervical Spine Fusion Surgery: Case Reports and Review of the Literature. *Glob. Spine J.* **2016**, *6*, e41–e46. [CrossRef]
43. Kengelbach-Weigand, A.; Thielen, C.; Bäuerle, T.; Götzl, R.; Gerber, T.; Körner, C.; Beier, J.P.; Horch, R.E.; Boos, A.M. Personalized medicine for reconstruction of critical-size bone defects—A translational approach with customizable vascularized bone tissue. *NPJ Regen. Med.* **2021**, *6*, 49. [CrossRef] [PubMed]
44. Sheikh, Z.; Javaid, M.A.; Hamdan, N.; Hashmi, R. Bone Regeneration Using Bone Morphogenetic Proteins and Various Biomaterial Carriers. *Materials* **2015**, *8*, 1778–1816. [CrossRef]

biomimetics

Article

3D Electrospun Polycaprolactone Scaffolds to Assess Human Periodontal Ligament Cells Mechanobiological Behaviour

Rémy Gauthier [1,*], Nina Attik [2,3], Charlène Chevalier [2], Vincent Salles [2,4,5], Brigitte Grosgogeat [2,3,6], Kerstin Gritsch [2,3,6] and Ana-Maria Trunfio-Sfarghiu [7]

1 UCBL, MATEIS UMR CNRS 5510, Bât. Saint Exupéry, Univ Lyon, CNRS, INSA de Lyon, 23 Av. Jean Capelle, 69621 Villeurbanne, France
2 UMR CNRS 5615, Laboratoire des Multimatériaux et Interfaces, Univ Lyon, Université Claude Bernard Lyon 1, 69622 Villeurbanne, France
3 Faculté d'Odontologie, Univ Lyon, Université Claude Bernard Lyon 1, 69008 Lyon, France
4 Institute of Industrial Science, The University of Tokyo, Tokyo 153-8505, Japan
5 LIMMS, CNRS-IIS UMI 2820, The University of Tokyo, Tokyo 153-8505, Japan
6 Hospices Civils de Lyon, Service d'Odontologie, 69008 Lyon, France
7 INSA-Lyon, CNRS UMR5259, LaMCoS, Univ Lyon, 69621 Villeurbanne, France
* Correspondence: remy.gauthier@cnrs.fr

Abstract: While periodontal ligament cells are sensitive to their 3D biomechanical environment, only a few 3D in vitro models have been used to investigate the periodontal cells mechanobiological behavior. The objective of the current study was to assess the capability of a 3D fibrous scaffold to transmit a mechanical loading to the periodontal ligament cells. Three-dimensional fibrous polycaprolactone (PCL) scaffolds were synthetized through electrospinning. Scaffolds seeded with human periodontal cells (10^3 mL^{-1}) were subjected to static ($n = 9$) or to a sinusoidal axial compressive loading in an in-house bioreactor ($n = 9$). At the end of the culture, the dynamic loading seemed to have an influence on the cells' morphology, with a lower number of visible cells on the scaffolds surface and a lower expression of actin filament. Furthermore, the dynamic loading presented a tendency to decrease the Alkaline Phosphatase activity and the production of Interleukin-6 while these two biomolecular markers were increased after 21 days of static culture. Together, these results showed that load transmission is occurring in the 3D electrospun PCL fibrous scaffolds, suggesting that it can be used to better understand the periodontal ligament cells mechanobiology. The current study shows a relevant way to investigate periodontal mechanobiology using 3D fibrous scaffolds.

Keywords: periodontal regeneration; human periodontal ligament cells; mechanobiology; dynamic mechanical loading; polycaprolactone fibrous scaffold

Citation: Gauthier, R.; Attik, N.; Chevalier, C.; Salles, V.; Grosgogeat, B.; Gritsch, K.; Trunfio-Sfarghiu, A.-M. 3D Electrospun Polycaprolactone Scaffolds to Assess Human Periodontal Ligament Cells Mechanobiological Behaviour. *Biomimetics* **2023**, *8*, 108. https://doi.org/10.3390/biomimetics8010108

Academic Editors: Ryszard Uklejewski and Mariusz Winiecki

Received: 23 February 2023
Revised: 2 March 2023
Accepted: 4 March 2023
Published: 7 March 2023

1. Introduction

Periodontitis is a chronic inflammatory disease affecting the tissues surrounding and supporting the teeth and can lead to teeth loss if left untreated. It is a highly prevalent disease and more than 10% of the world's population suffer from severe cases of periodontitis [1]. The periodontium is composed of both hard (alveolar bone and cementum) and soft tissues (periodontal ligament and gingiva). In order to obtain subsequent structural and functional regeneration, a tissue-engineering approach is required to allow for an accurately compartmentalized healing response. In this context, periodontal regeneration strategies remain currently a real clinical challenge to avoid mobility and maintain a good quality of teeth life [2].

The periodontium complex structure, with alternating hard and soft tissues, makes its tissue regeneration a challenge. In addition to a complex microbiota, that is the main cause for periodontal disease initiation [3], the biomechanical environment offered by the oral cavity is known to influence the cells behavior and the tissue organization [4]. Amongst others, a healing strategy used in a clinic consists in the placement of an impermeable and resorbable or non-resorbable [5] membrane preventing further migration of epithelial cells, hence allowing for the periodontal cells to synthetize new periodontal tissues [6,7]. Fibrous scaffolds are also under investigation for such guided tissue regeneration strategies [8,9]. However, such guided tissue regeneration methods do not offer a suitable structural guidance for periodontal cells to synthetize the adapted extracellular matrix (ECM) [10].

Cells behavior depends on their substrate's properties (structural, mechanical, etc.) [11–13]. This is why it is necessary to understand the behavior of periodontal cells seeded on scaffolds with different properties to develop new tissue engineering strategies [14]. Such strategies associate specific scaffolds and stimuli to promote a targeted tissue regeneration [10]. While alveolar bone and cementum are part of periodontal tissues, the current study focuses on a periodontal ligament (PDL) regeneration. The PDL is known to play a major role on the periodontal integrity. Through its biomechanical properties, it transmits to the underlying alveolar bone a suitable mechanical stress allowing for bone structural adaptation through a well-known mechanobiological pathway [15–17]. It is therefore of major importance to understand PDL tissue engineering-based regeneration.

In that context, it is clear that the properties of the scaffold used are determinant in the regenerative potential of the tissue engineering strategy. While hydrogel-based scaffolds have been largely used, due to their capacity to offer structural guidance and allow for drugs or bioactive molecules delivery, they might not be suitable in terms of biomechanical performance [10]. In vivo, periodontal cells are immerged in a ground substance, which can justify the major use of a hydrogel with similar properties for periodontal regeneration investigations [18–20]. However, if this ground substance is assumed to play a role in the viscous properties of the PDL [21], its elastic properties are mainly associated with its fibrous matrix made of collagen bundles [22,23]. Regarding that, in vivo, the PDLCs are closely bonded to the collagen bundles [24]. This suggests that the deformation of these bundles will directly influence the deformation of the PDLCs.

These mechanical parameters are of importance when developing new tissue engineering strategies. PDLCs are mechanosensitive cells: a mechanical stimulation has a significant influence on their behavior [4]. The influence of a mechanical loading on scaffold-free PDLCs has already largely been studied. Such interesting investigations have shown that PDLCs are sensitive to a static pressure [25,26] or static tensile loading [27,28], a cyclic pressure [29], cyclic compressive [30], or cyclic tensile [31,32] loading, as well as to fluid flow [33,34]. These studies are of great relevance as these natures of loadings are observed in vivo. A pressure is applied through the compression of the ground substance [35], compressive or tensile loadings through the deformation of the collagen bundles [23], and a fluid flow is associated with the flow of the ground substance through the collagen porosities [21]. Still, most of these studies have considered a 2D approach whereas it is known that cellular behavior differs in 3D [36,37]. Furthermore, these previous studies used scaffold-free cell cultures, whereas in the PDL, the loading is not directly applied, but transmitted through the different ECM components.

To the authors' knowledge, only one study focused on the behavior of PDLCs subjected to a mechanical loading in a 3D scaffold [20]. While this study presents an interesting 3D loading device, the scaffold used is collagen gel based ($\approx$1 kPa) that is much softer than the PDL collagen bundles ($\approx$1 MPa [23]), and could thus not represent the accurate load to the seeded cells. It is thus necessary to develop 3D scaffolds with biomechanical properties representative of those of the PDL. Interestingly, polycaprolactone (PCL) appears as a good candidate to build complex architectures that have been shown to be interesting in terms of the PDLCs 3D static culture [38]. Furthermore, PCL-based fibrous matrices are known to have local and apparent elastic properties close to that of the PDL collagen bundles [4]. Nevertheless, it is not known whether or not a PCL fiber is able to transmit the mechanical loading to the cells as it is conducted in vivo by the collagen bundles. In this context, the aim of the present study was to investigate the influence of a mechanical loading on the behavior of PDLCs seeded in a 3D PCL fibrous scaffold in order to suggest it as relevant scaffold for a PDL mechanobiology investigation to develop durable and predictable periodontal regeneration strategies.

2. Materials and Methods

- Scaffold preparation

A fibrous matrix of PCL (Purasorb PC 12, inherent viscosity 1.2 dL/g, T_f = 60 °C, Sigma-Aldrich, St. Louis, MO, USA) were fabricated using electrospinning in ambient conditions of humidity and temperature. A 0.15 g·mL^{-1} solution of PCL was dissolved in 1,1,1,3,3,3-Hexafluoro-2-propanol (HFIP, Sigma-Aldrich, France) and was prepared at least 24 h before the spinning process. The homogeneous PCL solution was then poured into a syringe connected to a 21-gauge blunt needle. The PCL solution was extruded at 2 mL·h^{-1} and the needle tip was placed at 6 cm from a grounded target consisting of a metal cylinder 6 mm in diameter. A motor was used to rotate this cylindric target at 100 rounds per minute. A potential difference of 4.4 kV was applied using a high-voltage generator (Iseg) connected to the needle (Figure 1). The spinning process was performed at 20 ± 2 °C and 30 ± 2% of relative humidity (lab conditions). The solvated PCL was extruded at 2 mL·h^{-1} using a syringe pump (KDS-100, KD Scientific, Holliston, MA, USA) for 20 min.

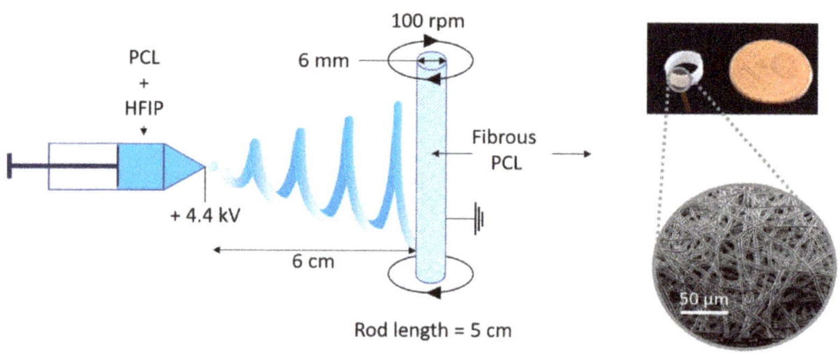

Figure 1. Schematic of the electrospinning set up and illustration of the obtained scaffold.

The obtained tube of PCL, with an inner diameter of 6 mm and a random fibers orientation, was then cut in ring samples, 3 mm in height, using a low speed saw (ISOMET 4000, Buehler, Lake Bluff, IL, USA) equipped with a micrometer screw to allow for an accurate parallelism between the two faces of the rings. Such a shape was chosen in order to mimic the 3D PDL anatomical geometry around the dental root and thus reproduce the apparent deformation and stress during the mechanical loading (Figure 1). The scaffolds were then dried at 40 °C in an oven for 6 days to ensure a complete HFIP evaporation.

- Scaffold characterization

The morphology of the fibrous matrix was characterized through an X-ray micro-computed tomography µCT using an EasyTom Nano device (RX Solutions, Overland Park, KS, USA) with an accelerating voltage of 40 kV with a 0.7 µm isotropic voxel size. An analysis was performed to investigate the inner fibrous structure of the scaffolds. A total of 3000 projections were recorded in a continuous rotation mode, with an exposure time set at 0.4 s. The fibers diameter and porosity were analyzed using the BoneJ ImageJ plugin [39].

- Cell culture

Human PDLCs were derived from human primary cell cultures (#2630, ScienCell, Carlsbad, CA, USA). Cells were suspended in a culture medium using Fibroblast Medium (#2301, ScienCell, USA) containing 10% fetal bovine serum (#0500, ScienCell, USA), 1% penicillin/streptomycin (#0503, ScienCell, USA), 1 % fibroblast growth supplement (#2352, ScienCell, USA), and 0.1% amphotericin B (15290026, ThermoFisher, Waltham, MA, USA) at

37 °C, 5% CO_2, and 97% of relative humidity for 4 days before the experiment. The culture medium was changed every 3 or 4 days. The media in contact with cells was withdrawn and stored at −20 °C for further analyses.

The PCL scaffolds samples were placed in the culture plate perpendicularly in a 24 well plate (as in Figure 1). Before seeding, the samples were sterilized under UV light for 30 min and incubated in fetal bovine serum (FBS) at 37 °C for 2 h to promote cell adhesion [40]. After the FBS removal, 500 μL of the cell suspension (10^3 mL^{-1}) was seeded on the scaffolds drop by drop on the external and internal faces of the scaffold and cultured as such for 4h. Then, an additional 500 μL of the culture media was added and the cell-scaffolds were cultured as such for 24 h prior to the start of the experiment. A total of 9 scaffolds were used for each experiment: 3 samples seeded with PDLCs and subjected to a dynamic mechanical stimulation (Dyn), 3 samples seeded with PDLCs and maintained under static conditions (Stat), and 3 non-seeded samples were maintained under static conditions (NS_Stat) in a culture media without cells. The experiment was conducted independently three times resulting in a total of 9 samples per condition.

At the end of the culture time (21 days) cellularized and non-cellularized scaffold samples were fixed in a 3.7 vol% paraformaldehyde (ACROS Organics) for 30 min for further analyses.

- Dynamic bioreactor and loading sequence.

The loading device used in the current study is a custom-made bioreactor allowing for the application of a cyclic compressive load that can be fitted within a cell culture incubator [41]. This bioreactor allows to test several samples immerged in the same culture medium. The loading chamber consists of a:

A lower part, a stainless-steel plate of 50 mm in diameter on which the samples are placed. This plate is moved up and down thanks to a piezoelectric actuator that performs a sinusoidal movement of a controlled frequency.

An upper part, made of Plexiglas®, is fixed to the device in order to ensure the compression and also to ensure the visualization of the samples in the chamber. For our tests we used 3 samples for each dynamic experiment.

The loading sequence consisted in a first compressive static pre-loading of 10 days followed by a dynamic compression for 11 days, resulting in a total of 21 days of culture.

At day 0, the scaffolds were subjected to a static loading. The contact between the upper plate and the PCL matrices was considered as the first variation in the recorded force. From this starting point, the steel plate raised to 500 μm, inducing a 17% static compressive deformation of the scaffolds ($\varepsilon_{\%} = (L{-}L_0)/L_0$). At day 3, the plate was moved upward by 250 μm more, inducing a total axial displacement of 750 μm associated with a final scaffold deformation of 25%. This static pre-load was controlled in terms of displacement and not force due to the different initial stiffnesses of the different PCL samples. This was conducted in order to ensure the contact between the upper plate and the scaffold during the whole experiment. Preliminary results did show that the contact was lost when the initial deformation was too low.

At day 10, the dynamic compressive loading was applied to the tested scaffolds. This consisted in an axial displacement of 100 μm of the steel lower plate, ranging from −50 μm to +50 μm with respect to the position at the end of the static pre-loading step. A total displacement of 100 μm was defined because it has been shown that a mini-pig molar (considered as a relevant model for mastication) displacement, under a chewing representative force, can be up to 140 μm [42,43]. The loading frequency was set to 0.2 Hz. Finally, the loading was chosen to be continuous, with no resting period, even if it is known that resting times have an influence on cell responses [44]. This extreme loading sequence was defined this way in order to focus the experiment on the influence of the mechanical loading, and to analyze if the PCL is able to transmit the mechanical loading to the cells. The loading was only stopped for 10 min at days 13, 17, and 21 for the culture medium renewal.

- Mechanical characterization

The stiffness of all the different scaffolds were measured at day 0 (d_0) and day 21 (d_{21}). The matrices were subjected to a sinusoidal displacement of 100 μm and 0.2 Hz for at least 10 cycles. First we checked that our sinusoidal force signal is a mainly elastic signal (Pi/2 offset from the displacement signal). Then we were able to calculate the stiffness using the following equation:

$$K = \frac{F}{\Delta d} \tag{1}$$

where K (expressed in $N \cdot \mu m^{-1}$ in the current study) is the stiffness, F (N) is the force amplitude for one cycle, and Δd = 100 μm is the total displacement. The stiffness was measured for each cycle during the compressive phase, and the final value was the averaged value between all the cycles. To compare the different conditions between them, the relative differences between d_{21} and d_0 were calculated as follows:

$$\Delta_{Rel} = \frac{K_{Final} - K_{Initial}}{K_{Initial}} \times 100\ (\%) \tag{2}$$

The relative difference was measured instead of each absolute value because a matrix stiffness largely depends on its geometrical properties. By studying the relative difference, such geometrical parameters vanish and the different matrices can be compared.

- Cell morphology by Confocal Laser Scanning Microscopy (CLSM)

After the fixation process, all the cell-seeded scaffolds were permeabilized using Triton X100 (Fisher Bioreagents, 1 vol% in PBS), then blocked and kept in PBS containing 1 vol% BSA (Bovine Serum Albumin, Corning,) at 4 °C. Actin filaments were then stained with Alexa Fluor™ 488 Phalloidin (A12379, Thermo Fisher Scientific, Waltham, MA, USA) at a 1:100 volume ratio in PBS (green fluorescence) at room temperature for 40 min. The cells' nuclei were stained with DAPI (4′,6-diamidino-2-phenylindole, blue fluorescence, P3566, Thermo Fisher Scientific, France) at a 1:3000 volume ratio in PBS at room temperature for 10 min. The samples were then cut in order to hold them flat on a glass slide and observed their external and internal sides using an LSM 800 confocal laser scanning microscope (Carl Zeiss Microscopy, GmbH, Oberkochen, Germany).

- Extracellular quantitative assay of ALP activity and inflammatory biomarkers

At d_{10} and d_{21}, the culture media of both conditions were defrosted and used for Alkaline Phosphatase (ALP) activity assessment for static and dynamic conditions in function of time. This corresponds to the time step before the beginning of the dynamic mechanical stimulation (day 10) and the final time step after the mechanical stimulation (day 21). The ALP activity was quantified using the kit K412-500 according to the manufacturer's instructions (BioVision Incorporated, Waltham, MA, USA). Briefly, 50 μL of 5 mM ALP reaction solution (2 4-nitropheyl-phosphate disodium salt hexahydrate tablets dissolved in 5.4 mL ALP Assay Buffer) was added to 80 μL of each culture media and incubated at room temperature. After 1 h, the reaction was stopped by adding 20 μL of Stop solution. The absorbance of the colored reaction product (pNPP) within the solution was then measured at 405 nm using a micro-plate reader (Infinite® M 200 PRO, NanoQuant plate, Tecan, Männedorf, Switzerland). ALP activity was then calculated using the following equation:

$$ALP\ activity = \frac{n_{pNPP}}{\Delta t \times V} \tag{3}$$

where n_{pNPP} (μmol) is the quantity of *pNPP*, Δt = 60 min is the reaction time, and V = 80 μL is the volume of media added to the solution. Finally, this value was multiplied by the volume of media collected from the wells for the static conditions and from the bioreactor for the dynamic conditions. This was done because the volume from the bioreactor was different from the wells volume due to the experimental set up.

At day 10 and day 21, the culture media were also used to quantify the concentration of IL-6 inflammatory mediator. ELISA kits were used to quantify the concentration of IL-6 (Elabscience, Houston, TX, USA) following the manufacturers' instructions. Briefly, the samples were added to the ELISA wells and combined with IL-6 specific antibody. After adding the stop solution, the samples' absorbance was measured at 405 nm using a micro-plate reader. The concentration (pg/mL) of IL-6 was then quantified by comparing the samples absorbance to the reference curves. As for the ALP, the total IL-6 amounts for the final concentration were normalized by the volume collected from the wells and from the bioreactor for the static and dynamic conditions, respectively.

As the seeded scaffolds subjected to the dynamic condition were gathered within the same chamber, due to the experimental set up, it was not possible to distinguish the three different scaffolds regarding these biochemical measurements. In order to compare the static and the dynamic conditions, the removed medium from the three separated static wells were mixed together. The measurements were then performed in triplicate for the ALP and the IL-6. The measured values were then normalized by the number of days between the medium changes, for example day 10 the medium was changed 3 days before, and for day 21, the medium was changed 4 days before.

- Statistical analyses

Due to the low number of samples per condition, the results were analyzed by applying the non-parametric Mann-Whitney test using R© (The R foundation for Statistical Computing, Austria). Regarding the mechanical characterization, the relative difference in stiffness was measured for each of the 9 scaffolds (3 samples × 3 experimental series). For the ALP and IL-6 measurements, only the three averaged values of the three experimental series were considered for the statistical analyses.

3. Results

- Scaffold characterization

The μCT scans allowed to observe and analyze the matrices at different length scales (Figure 2). The electrospinning protocol allowed to obtain 600 μm thick scaffolds (Figure 2 left) with randomly oriented fibers (Figure 2 middle and right). The fibers had an average diameter of 2.3 ± 0.5 μm with maximum diameter of 5.8 μm. The pores had an average diameter of 6.4 ± 2.9 μm with a maximum diameter of 22.2 μm. The pore volume fraction was 91%.

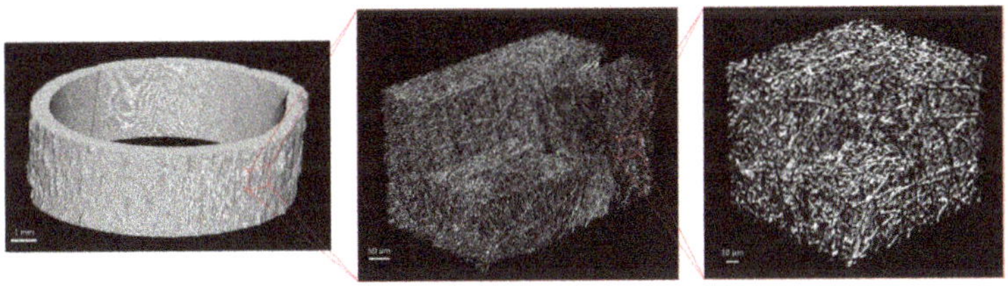

Figure 2. Three μCT 3D renderings at three magnifications: **left**: the whole matrix, **middle**: a large volume of interest (500 × 300 × 300 μm^3), and **right**: a small volume of interest (100 × 100 × 100 μm^3).

The diameter of the fibers was homogeneous throughout the volume, with only some isolated areas of large fibers of 6 μm (Figure 3 left). Conversely, the porosity was not homogeneously distributed over the thickness of the sample. Larger pores, from 15 to 20 μm in diameter, were observed in the outer half of the thickness of the sample whereas small pores, from 2 to 10 μm, were found through the inner part of the matrices (Figure 3 right).

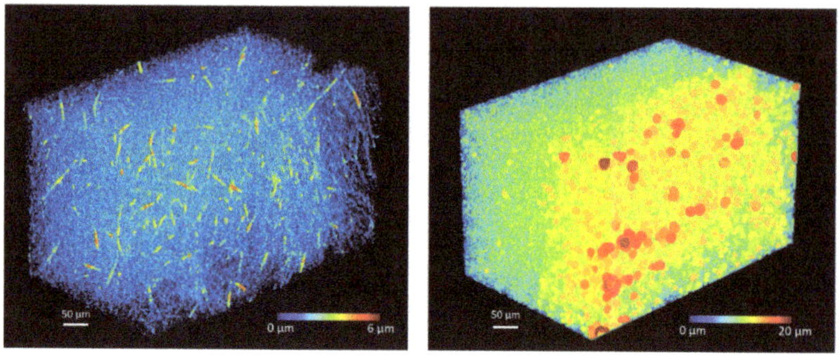

Figure 3. Distribution of **left**: fibers diameter and **right**: pores diameter on a 500 × 300 × 300 µm^3 volume of interest.

- Mechanical characterization

After 21 days of culture, the stiffness of the static cell-seeded scaffolds decreased by 37% on average for the three series, and the stiffness of the matrices subjected to a dynamic mechanical loading decreased by 38% on average. The matrices without cells but incubated in static conditions showed an averaged decrease of 11% (Figure 4). A significant difference in the decrease in stiffness was observed between the cellularized Stat and the non-cellularized scaffolds ($p = 0.01$) and between the Dyn and the NS-Stat scaffolds (0.003) when considering each sample. No difference was measured between the Stat and Dyn cell-seeded scaffolds.

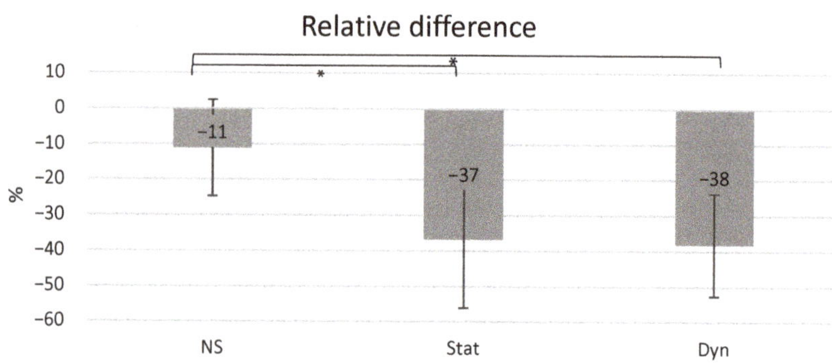

Figure 4. Averaged value of the stiffness relative differences for the NS (Non-seeded), Stat (static condition), and Dyn (dynamic conditions) obtained by analyzing nine samples (three matrices and three experimental campaigns) between d_{10} and d_{21}. *: significantly different (p-value < 0.05).

The individual values for each scaffold and each condition revealed an inhomogeneous distribution, particularly the stiffness difference value for NS_2 during the third experimental series seems aberrant (Table 1). It has to be noticed that this particular value has been removed for the averaged value and for the statistical analysis presented in Figure 4. The absolute stiffness values for all the samples are provided in the Supplementary Materials (Table S1).

Table 1. Values of stiffness relative difference (in %) between d_0 and d_{21} for all the samples of the three experimental campaigns. The red value has been considered as aberrant and removed from the statistical analysis. The subscripts 1, 2 and, 3 correspond to the three samples in each series. Samples from the same column (with the same name and subscript) are not associated.

	NS-Stat$_1$	NS-Stat$_2$	NS-Stat$_3$	Stat$_1$	Stat$_2$	Stat$_3$	Dyn$_1$	Dyn$_2$	Dyn$_3$
1st series	−20	−1	−5	−60	−34	−37	−38	−35	−49
2nd series	−25	−35	11	−73	−36	−19	−22	−58	−7
3rd series	−8	63	−7	−33	−2	−8	−41	−51	−43

- Cell morphology by CLSM

For the static conditions, the CLSM images showed cells that were spreading over the scaffold exterior surface, often in clusters. Their actin filaments within their cytoskeleton (green) were well organized (Figure 5 left). Conversely, few nuclei with only a little actin expression were observed on the dynamic scaffold surfaces (Figure 5 right).

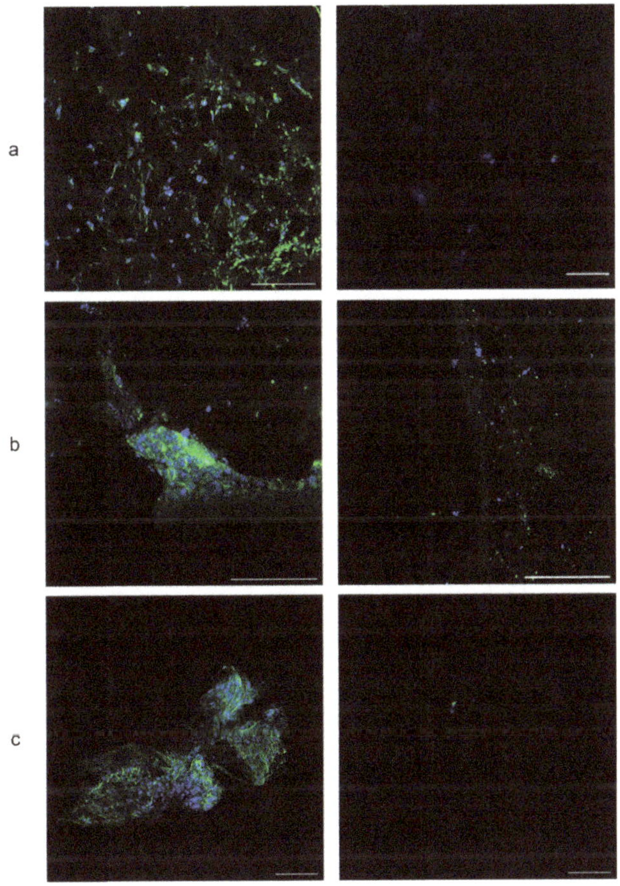

Figure 5. Human PDLCs morphology and spreading by CLSM. (**a**/**b**/**c**): 1st/2nd/3rd series under static (**left**) and dynamic (**right**) conditions. Actin cytoskeleton in green fluorescence and nuclei in blue fluorescence. Note that for series 2 (**b**), the gain for the Alexa 488 Phalloidin fluorescence was put on high in order to detect a signal for the dynamic condition. This resulted in the autofluorescence of the scaffolds fibers. Scale bars: 100 μm.

After this result, and to ensure that cells were present within the scaffolds subjected to the used dynamic loading, one Stat and Dyn sample from each series was treated with Trypsin for 15 min at 37 °C in order to enable cell detachment and cell harvesting from the scaffold samples. The cell media was then added to stop the trypsin action, the cell suspension was centrifuged (5 min at 1200 rpm) and resuspended to allow for cell counting using a scepter™ Handheld Automated Cell Counter (Merck Millipore, Germany). Results showed that cell proliferation occurred in both the static and dynamic conditions, with a higher number of cells in the scaffolds subjected to the static condition (Table 2).

Table 2. Cell number within one matrix from each condition and each campaign, and mean and standard deviation over the three series.

10^4 mL^{-1}	Stat	Dyn
1st series	17.9	2.6
2nd series	7.5	7.2
3rd series	19.7	7.6
Mean	15.0	5.8
SD	5.4	2.3

- Extracellular quantitative assay of ALP activity and inflammatory biomarkers

The results for the biochemical parameters (ALP activity and IL-6) presented correspond to a tendency, as with only three samples (three series) no statistical signifcance was obtained.

Regarding the ALP activity, the static group presented an increased tendency between days 10 and 21 while the opposite was found for the dynamic group. In addition, it was observed that the ALP activity for the static group appeared lower than the dynamic group at day 10, while at day 21, both groups presented a similar activity (Figure 6). The numerical values and the bar charts for each experimental series are shown in the Supplementary Materials (Table S2 and Figure S1, respectively).

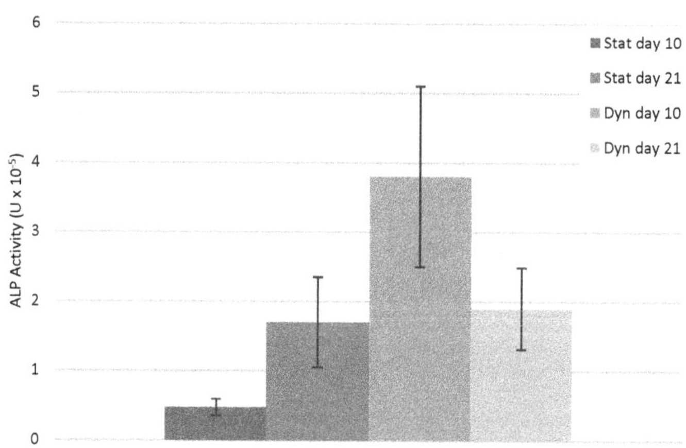

Figure 6. ALP activity (U) averaged over the three series for the static and dynamic groups at days 10 and 21.

The variation in IL-6 amounts between day 10 and 21 appeared similar as the ALP activity variation for both groups, with an increase in the production of IL-6 for the static group and a decrease for the dynamic group. Inversely, the amount of IL-6 for the static group appeared similar to the dynamic group at day 10, but higher at day 21 (Figure 7). The numerical values and the bar charts for each experimental series are shown in the Supplementary Materials (Table S3 and Figure S2, respectively).

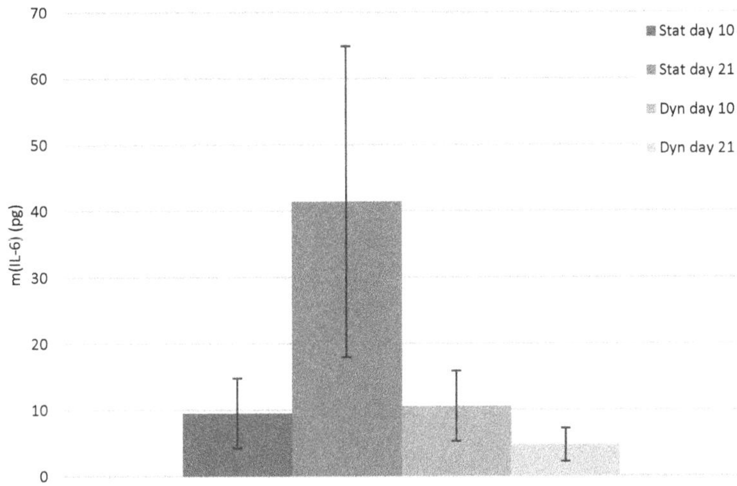

Figure 7. IL-6 amount (pg) averaged over the three series for the static and dynamic groups at days 10 and 21.

4. Discussion

The objective of the present study was to investigate the behavior of human periodontal ligament cells (PDLCs) subjected to a mechanical loading within a 3D fibrous electrospun scaffold as a biomechanical model of the fibrous periodontal ligament [4]. While the periodontal ligament architecture consists of collagen bundles and that periodontal biomechanics is known to influence its regeneration, there is currently only a few mechanobiological studies using fibrous scaffolds. The obtained results demonstrated that the tested biomechanical loading significantly influenced the PDLCs behavior. Understanding how PDLCs answer to a mechanical loading is of interest as they are considered to have a high capacity of regeneration [45].

First the current study does not allow for a conclusion about the influence of the mechanical loading on the cellularized scaffolds elastic properties. The stiffness of the scaffolds with cells significantly decreased for both the static and dynamic conditions with the same magnitude. On the other hand, the scaffold without cells presented a significantly lower decrease in stiffness. This result suggests a biological degradation of the scaffolds PCL fibers. In order to isolate the influence of the mechanical loading on the PCL fibers elasticity, further investigations on the ageing of scaffolds without cells and subjected to a cyclic loading have to be performed. Some results from the literature showed a mechanical fatigue for non-seeded PCL samples subjected to tensile [46] or compressive [47] cyclic loading. More particularly, Panadero et al. measured a decrease of nearly 17% of the maximum stress for the PCL sample after only 100 mechanical cycles [47]. While their samples are not fibrous, these results suggest that the samples of the current study may undergo a mechanical fatigue during the cyclic compression, and that both mechanical fatigue and biological degradation may contribute to the stiffness decrease mechanism.

Conversely, the influence of the scaffold cyclic deformation was observed on the cells' morphology and organization. While a connected and spread network was observed on the scaffold's surface in static conditions, isolated cells were hardly found for the dynamic condition. Unfortunately, the PCL is opaque and does not allow to observe the cells network deep within the architecture using a standard confocal microscope. In order to ensure the presence of cells within the scaffolds, cell counting was performed after cell harvesting using Trypsin to unleash the cells from the PCL fibers. The results demonstrated that while cells were not observed on the surface of the dynamic group, cell proliferation was still measured. To explain this result, the hypotheses are whether that the cyclic loading promotes the cells to penetrate within the 3D architecture scaffolds or that the cells lying on the scaffold external surface do not withstand the mechanical loading due to less focal adhesion. Interestingly, it is known that cells are able to migrate within the 3D architecture of a PCL fibrous scaffold under static conditions [48] and that the cell-scaffold interactions are different according to the 2D or 3D structures [49]. In the current study, cell sheathes observed on the external surface of the scaffolds can be considered as lying on a 2D surface when compared to the cells that have migrated deeply within the 3D architecture. This supports the second hypothesis stating that only cells that have migrated within the 3D scaffold architecture have withstood the cyclic loading. As in vivo, periodontal cells are organized three-dimensionally on the 3D collagen bundle network [50], it appears as relevant to investigate the cellular behavior of cells within the scaffold 3D architecture.

Furthermore, it can be observed that the actin production is inhibited during the cyclic mechanical loading. It has already been shown that a multiaxial dynamic mechanical loading had a detrimental influence on the actin fibers organization [51]. Moreover, while short term loading may promote actin production, it decreases with loading time, as it has been shown on myocyte 3D cultures [52]. Interestingly, it has been shown that actin is not produced, and even degraded, during loading, but that if resting periods are allowed, actin production is promoted sometime after the loading [53].

Regarding cell signaling, the current study shows that the fibrous 3D PCL scaffold is able to transmit the mechanical load to the seeded human PDLCs during a cyclic mechanical stimulation. The mechanical loading tended to decrease the cells ALP activity while this biomolecular signaling was increased under static conditions. It is in accordance with the previous results showing that a cyclic loading downregulates the expression of periodontal cell osteogenic markers, such as ALP [54,55]. This result is in line with the low strain needed for bone tissue formation [56]. Conversely, under static conditions, where cells are not subjected to any strain, the ALP activity was increased with time, which is in accordance with a previous study performed on osteoblasts [57]. It is important to state that after the static pre-load, the ALP activity was higher than for the static condition without pre-load after the same duration of culture (10 days). This strongly supports that it is the continuous and cyclic nature of the applied mechanical loading that may be involved in the decrease in the cell mineralization potential, which is in line with the low motion requirement for bone formation. Similarly, the cyclic 3D compressive loading showed a detrimental influence on the production of IL-6 by human PDLCs. Previous studies showed a decrease in IL-6 expression with the increasing duration of a cyclic 2D compressive loading [30]. Together, these results support that the PCL fibers are able to transmit the mechanical loading toward the cells. Still, it has to be noticed that due to a low number of samples, no statistical significance was obtained. The reason for such low number of samples relies in the setup of the used in-house bioreactor. In this mechanobiological device, the three investigated matrices are immerged in the same culture media, which prevents the analyzing of the biomolecular signaling for each of them. Still, a clear tendency is observed and supports for further detailed investigations.

To the authors knowledge, this is the first time that such periodontal mechanobiological investigations have been performed on a PCL 3D fibrous scaffold. In contrast with the collagen or hydrogel scaffolds that are generally used to study periodontal mechanosensitivity, PCL presents properties that could predict the periodontal collagen bundles that are able to transmit the mechanical deformation to the cells [4]. Still, the current study provides preliminary data and cannot be directly translated to clinics. While the present study shows that PDLCs seeded in 3D fibrous scaffolds are sensitive to the loading applied to the cellularized scaffolds, different parameters can be discussed. Particularly, the continuous cyclic mechanical loading applied in the current study is not representative of an in vivo loading, where resting periods are recorded. As the aim was to assess the potential of a 3D scaffold to transmit the mechanical load to cells, an intensive loading was used. Furthermore, the current PCL scaffolds have a random architecture, but periodontal collagen bundles present a specific organization, depending on the biomechanical loading [4]. Similarly, the structural organization of the fibers within the scaffolds was not controlled while, in vivo, the PDL collagen bundles present specific orientations depending on the anatomical location [24]. Moreover, it can be noticed that the 3D scaffolds used in the present study have quite an inhomogeneous structural organization from the external to the internal scaffold surfaces. This may be due to a variation in the electric field during the electrospinning process as the electrospun scaffold was deposited on the steel rotating target. However, despite these limitations, the influence of the mechanical loading on the cell's morphology and biomolecular signaling was observed in the current study, suggesting that PCL 3D fibrous scaffolds can be used for further investigations on a PDLCs mechanobiological behavior.

A similar methodology can thus be applied, with efforts on the scaffolds architecture and applied loading, to further investigate human PDLCs mechanobiology in a biomechanical environment close to an in vivo periodontal environment. More particularly, it is known that the configuration of the PDL, trapped between the dense and impermeable bone and cementum, has an influence on how cells are mechanically stimulated during loading [4]. The lower permeability prevents water from flowing out of the ligament, with a direct influence on its internal pressure and fluid-flow relaxation [58]. Tremendous efforts are currently made to develop PCL-based bi- or multiphasic scaffolds in order to mimic the cement-PDL-bone configuration in terms of mechanical and structural properties of the periodontal tissue [38,59]. By tuning the scaffolds architecture, it is possible to rigorously control how cells are mechanically stimulated during loading, for example through numerical methods as it is used in bone regeneration applications [60]. In that context, and in regard of the results provided in the current study, further investigations on complex 3D PCL-based scaffolds seeded with PDLCs have to be performed to better understand the periodontal ligament mechanobiology. The current study shows a new relevant way to treat periodontal mechanobiology using 3D fibrous scaffolds.

5. Conclusions

The current study showed the potentiality to use polycaprolactone-based electrospun 3D fibrous scaffolds instead of collagen to undergo human periodontal cell mechanobiological investigations. Under mechanical loading the cells seeded within the 3D polycaprolactone fibrous scaffold presented a different behavior in terms of their cellular organization and signaling when compared to the control static condition. This suggests that a polycaprolactone 3D fibrous scaffold is able to transmit the load to the cells and might represent an interesting experimental model to analyze the periodontal ligament cells mechanobiological behavior. Considering that polycaprolactone is a suitable material to mimic the periodontal ligament collagen bundles that transmits the mechanical deformation in vivo, these results highlight that electrospun polycaprolactone 3D scaffolds with a tuned architecture are an interesting model to investigate the periodontal ligament mechanobiology. Further works with an enhanced control of the mechanical loading and the scaffold architecture are on-going to provide relevant data for a potential future clinical translation.

Supplementary Materials: The following supporting information can be downloaded at: https://www.mdpi.com/article/10.3390/biomimetics8010108/s1, Figure S1: Bar charts of ALP activity averaged values and SD for the two conditions, the two time-steps, and the three experimental series. Figure S2: Bar charts of mass of IL-6 averaged values and SD for the two conditions, the two time-steps, and the three experimental series. Table S1: Absolute stiffness (in 10^{-3} N·μm^{-1}) values for all the samples at day 0 and day 21 for the three different series. The subscripts 1, 2 and 3 correspond to the 3 samples in each series. Samples from the same column (with the same name and subscript) are not associated. NS-Stat: Non-seeded static scaffolds. Table S2: Mean and standard deviation (SD) of the ALP activity measured on the triplicate for each condition and timestep. Table S3: Mean and standard deviation (SD) of the mass of IL-6 measured on the triplicate for each condition and timestep.

Author Contributions: Conceptualization, R.G., N.A., V.S., B.G., K.G. and A.-M.T.-S.; methodology, R.G., N.A., C.C., V.S. and A.-M.T.-S.; validation, R.G., N.A., B.G., K.G. and A.-M.T.-S.; formal analysis, R.G.; investigation, R.G., N.A. and A.-M.T.-S.; resources B.G. and A.-M.T.-S.; data curation, R.G., B.G. and A.-M.T.-S.; writing—original draft preparation, R.G.; writing—review and editing, R.G., N.A., C.C., V.S., B.G., K.G. and A.-M.T.-S.; supervision, B.G., K.G. and A.-M.T.-S.; project administration, B.G. and A.-M.T.-S.; funding acquisition, R.G., N.A., B.G., K.G. and A.-M.T.-S. All authors have read and agreed to the published version of the manuscript.

Funding: This work was partly founded by the ANR project Toothbox (ANR 16-CE08-0024) and by the Fédération IngéLySE. This work was supported by the JSPS Core-to-Core Program (grant number: JPJSCCA20190006). This work was carried out as part of the MIMIMed@Lyon project supported by the Institut Carnot Ingénierie@Lyon, labelled by the French National Research Agency. Works in the ISO-5 clean room was supported by a projects from INSA Lyon and LaMCOS, supervised by Christelle Der- Loughian and Aline Bel-Brunon, and from Carnot Institute Ingénierie@Lyon (TROPISME project).

Informed Consent Statement: Informed consent was obtained from all subjects involved in the study.

Data Availability Statement: All the data are provided in the current study, in the principal manuscript and the Supplementary Materials. The raw data can be provided by the authors if needed.

Acknowledgments: The authors would like to thank Corbion for providing the PCL used in this study.

Conflicts of Interest: The authors declare no conflict of interest.

References

1. Frencken, J.E.; Sharma, P.; Stenhouse, L.; Green, D.; Laverty, D.; Dietrich, T. Global Epidemiology of Dental Caries and Severe Periodontitis—A Comprehensive Review. *J. Clin. Periodontol.* **2017**, *44*, S94–S105. [CrossRef]
2. Peres, M.A.; Macpherson, L.M.D.; Weyant, R.J.; Daly, B.; Venturelli, R.; Mathur, M.R.; Listl, S.; Celeste, R.K.; Guarnizo-Herreño, C.C.; Kearns, C.; et al. Oral Diseases: A Global Public Health Challenge. *Lancet* **2019**, *394*, 249–260. [CrossRef] [PubMed]
3. Kinane, D.F.; Stathopoulou, P.G.; Papapanou, P.N. Periodontal Diseases. *Nat. Rev. Dis. Prim.* **2017**, *3*, 17038. [CrossRef] [PubMed]
4. Gauthier, R.; Jeannin, C.; Attik, N.; Trunfio-Sfarghiu, A.-M.; Gritsch, K.; Grosgogeat, B. Tissue Engineering for Periodontal Ligament Regeneration: Biomechanical Specifications. *J. Biomech. Eng.* **2021**, *143*, 030801. [CrossRef]
5. Caffesse, R.G.; Nasjleti, C.E.; Morrison, E.C.; Sanchez, R. Guided Tissue Regeneration: Comparison of Bioabsorbable and Non-Bioabsorbable Membranes. Histologic and Histometric Study in Dogs. *J. Periodontol.* **1994**, *65*, 583–591. [CrossRef] [PubMed]
6. Villar, C.C.; Cochran, D.L. Regeneration of Periodontal Tissues: Guided Tissue Regeneration. *Dent. Clin. N. A.* **2010**, *54*, 73–92. [CrossRef]
7. Cho, Y.D.; Kim, K.H.; Lee, Y.M.; Ku, Y.; Seol, Y.J. Periodontal Wound Healing and Tissue Regeneration: A Narrative Review. *Pharmaceuticals* **2021**, *14*, 456. [CrossRef]
8. Zhang, H.; Wang, K.; Gao, T.; Zhang, R.; Cai, Z.; Liu, J.; Ma, H.; Zhang, W. Controlled Release of BFGF Loaded into Electrospun Core–Shell Fibrous Membranes for Use in Guided Tissue Regeneration. *Biomed. Mater.* **2020**, *15*, 035021. [CrossRef]
9. Zhang, E.; Zhu, C.; Yang, J.; Sun, H.; Zhang, X.; Li, S.; Wang, Y.; Sun, L.; Yao, F. Electrospun PDLLA/PLGA Composite Membranes for Potential Application in Guided Tissue Regeneration. *Mater. Sci. Eng. C* **2016**, *58*, 278–285. [CrossRef]
10. Liang, Y.; Luan, X.; Liu, X. Recent Advances in Periodontal Regeneration: A Biomaterial Perspective. *Bioact. Mater.* **2020**, *5*, 297–308. [CrossRef]
11. Discher, D.E.; Janmey, P.; Wang, Y.L. Tissue Cells Feel and Respond to the Stiffness of Their Substrate. *Science* **2005**, *310*, 1139–1143. [CrossRef]
12. Ippolito, A.; Deshpande, V.S. Contact Guidance via Heterogeneity of Substrate Elasticity. *Acta Biomater.* **2021**. [CrossRef]
13. Hadden, W.J.; Young, J.L.; Holle, A.W.; McFetridge, M.L.; Kim, D.Y.; Wijesinghe, P.; Taylor-Weiner, H.; Wen, J.H.; Lee, A.R.; Bieback, K.; et al. Stem Cell Migration and Mechanotransduction on Linear Stiffness Gradient Hydrogels. *Proc. Natl. Acad. Sci. USA* **2017**, *114*, 5647–5652. [CrossRef]
14. Yamada, S.; Shanbhag, S.; Mustafa, K. Scaffolds in Periodontal Regenerative Treatment. *Dent. Clin. N. A.* **2022**, *66*, 111–130. [CrossRef] [PubMed]
15. McCulloch, C.; Lekic, P.; McKee, M.D. Role of Physical Forces in Regulating the Form and Function of the Periodontal Ligament. *Periodontology 2000* **2000**, *24*, 56–72. [CrossRef]
16. Lin, J.D.; Ryder, M.; Kang, M.; Ho, S.P. Biomechanical Pathways of Dentoalveolar Fibrous Joints in Health and Disease. *Periodontology 2000* **2020**, *82*, 238–256. [CrossRef]
17. Dutra, E.H.; Nanda, R.; Yadav, S. Bone Response of Loaded Periodontal Ligament. *Curr. Osteoporos. Rep.* **2016**, *14*, 280–283. [CrossRef] [PubMed]
18. Berkowitz, K. The Structure of the Periodontal Ligament: An Update. *Eur. J. Orthod.* **1990**, *12*, 51–76. [CrossRef] [PubMed]

19. Attik, N.; Garric, X.; Bethry, A.; Subra, G.; Chevalier, C.; Bouzouma, B.; Verdié, P.; Grosgogeat, B.; Gritsch, K. Amelogenin-Derived Peptide (ADP-5) Hydrogel for Periodontal Regeneration: An In Vitro Study on Periodontal Cells Cytocompatibility, Remineralization and Inflammatory Profile. *J. Funct. Biomater.* **2023**, *14*, 53. [CrossRef]
20. Berendsen, A.D.; Smit, T.H.; Walboomers, X.F.; Everts, V.; Jansen, J.A.; Bronckers, A.L.J.J. Three-Dimensional Loading Model for Periodontal Ligament Regeneration in Vitro. *Tissue Eng. Part C Methods* **2009**, *15*, 561–570. [CrossRef]
21. Bergomi, M.; Cugnoni, J.; Botsis, J.; Belser, U.C.; Anselm Wiskott, H.W. The Role of the Fluid Phase in the Viscous Response of Bovine Periodontal Ligament. *J. Biomech.* **2010**, *43*, 1146–1152. [CrossRef]
22. Komatsu, K.; Chiba, M. Synchronous Recording of Load-Deformation Behaviour and Polarized Light-Microscopic Images of the Rabbit Incisor Periodontal Ligament during Tensile Loading. *Arch. Oral Biol.* **2001**, *46*, 929–937. [CrossRef] [PubMed]
23. Pini, M.; Zysset, P.; Botsis, J.; Contro, R. Tensile and Compressive Behaviour of the Bovine Periodontal Ligament. *J. Biomech.* **2004**, *37*, 111–119. [CrossRef] [PubMed]
24. Hirashima, S.; Ohta, K.; Kanazawa, T.; Okayama, S.; Togo, A.; Miyazono, Y.; Kusukawa, J.; Nakamura, K. ichiro Three-Dimensional Ultrastructural Analysis and Histomorphometry of Collagen Bundles in the Periodontal Ligament Using Focused Ion Beam/Scanning Electron Microscope Tomography. *J. Periodontal Res.* **2020**, *55*, 23–31. [CrossRef]
25. Ullrich, N.; Schröder, A.; Jantsch, J.; Spanier, G.; Proff, P.; Kirschneck, C. The Role of Mechanotransduction versus Hypoxia during Simulated Orthodontic Compressive Strain—An in Vitro Study of Human Periodontal Ligament Fibroblasts. *Int. J. Oral Sci.* **2019**, *11*, 33. [CrossRef] [PubMed]
26. Li, S.; Li, Q.; Zhu, Y.; Hu, W. GDF15 Induced by Compressive Force Contributes to Osteoclast Differentiation in Human Periodontal Ligament Cells. *Exp. Cell Res.* **2020**, *387*, 111745. [CrossRef]
27. Long, P.; Liu, F.; Piesco, N.P.; Kapur, R.; Agarwal, S. Signaling by Mechanical Strain Involves Transcriptional Regulation of Proinflammatory Genes in Human Periodontal Ligament Cells in Vitro. *Bone* **2002**, *30*, 547–552. [CrossRef]
28. Sun, C.; Janjic Rankovic, M.; Folwaczny, M.; Otto, S.; Wichelhaus, A.; Baumert, U. Effect of Tension on Human Periodontal Ligament Cells: Systematic Review and Network Analysis. *Front. Bioeng. Biotechnol.* **2021**, *9*, 695053. [CrossRef]
29. Wenger, K.H.; El-Awady, A.R.; Messer, R.L.W.W.; Sharawy, M.M.; White, G.; Lapp, C.A. Pneumatic Pressure Bioreactor for Cyclic Hydrostatic Stress Application: Mechanobiology Effects on Periodontal Ligament Cells. *J. Appl. Physiol.* **2011**, *111*, 1072–1079. [CrossRef]
30. Wu, J.; Li, Y.; Fan, X.; Zhang, C.; Wang, Y.; Zhao, Z. Analysis of Gene Expression Profile of Periodontal Ligament Cells Subjected to Cyclic Compressive Force. *DNA Cell Biol.* **2011**, *30*, 865–873. [CrossRef] [PubMed]
31. Liu, M.; Dai, J.; Lin, Y.; Yang, L.; Dong, H.; Li, Y.; Ding, Y.; Duan, Y. Effect of the Cyclic Stretch on the Expression of Osteogenesis Genes in Human Periodontal Ligament Cells. *Gene* **2012**, *491*, 187–193. [CrossRef] [PubMed]
32. Ma, J.; Zhao, D.; Wu, Y.; Xu, C.; Zhang, F. Cyclic Stretch Induced Gene Expression of Extracellular Matrix and Adhesion Molecules in Human Periodontal Ligament Cells. *Arch. Oral Biol.* **2015**, *60*, 447–455. [CrossRef] [PubMed]
33. Van Der Pauw, M.T.M.; Klein-Nulend, J.; Van Den Bos, T.; Burger, E.H.; Everts, V.; Beertsen, W. Response of Periodontal Ligament Fibroblasts and Gingival Fibroblasts to Pulsating Fluid Flow: Nitric Oxide and Prostaglandin E2 Release and Expression of Tissue Non-Specific Alkaline Phosphatase Activity. *J. Periodontal Res.* **2000**, *35*, 335–343. [CrossRef] [PubMed]
34. Tang, M.; Peng, Z.; Mai, Z.; Chen, L.; Mao, Q.; Chen, Z.; Chen, Q.; Liu, L.; Wang, Y.; Ai, H. Fluid Shear Stress Stimulates Osteogenic Differentiation of Human Periodontal Ligament Cells via the Extracellular Signal-Regulated Kinase 1/2 and P38 Mitogen-Activated Protein Kinase Signaling Pathways. *J. Periodontol.* **2014**, *85*, 1806–1813. [CrossRef]
35. Walker, T.W.; Ng, G.C.; Burke, P.S. Fluid Pressures in the Periodontal Ligament of the Mandibular Canine Tooth in Dogs. *Arch. Oral Biol.* **1978**, *23*, 753–765. [CrossRef]
36. Zonderland, J.; Moroni, L. Steering Cell Behavior through Mechanobiology in 3D: A Regenerative Medicine Perspective. *Biomaterials* **2021**, *268*, 120572. [CrossRef]
37. Yang, L.; Yang, Y.; Wang, S.; Li, Y.; Zhao, Z. In Vitro Mechanical Loading Models for Periodontal Ligament Cells: From Two-Dimensional to Three-Dimensional Models. *Arch. Oral Biol.* **2015**, *60*, 416–424. [CrossRef]
38. Ivanovski, S.; Vaquette, C.; Gronthos, S.; Hutmacher, D.W.; Bartold, P.M. Multiphasic Scaffolds for Periodontal Tissue Engineering. *J. Dent. Res.* **2014**, *93*, 1212–1221. [CrossRef]
39. Doube, M.; Klosowski, M.M.; Arganda-Carreras, I.; Cordelières, F.P.; Dougherty, R.P.; Jackson, J.S.; Schmid, B.; Hutchinson, J.R.; Shefelbine, S.J. BoneJ: Free and Extensible Bone Image Analysis in ImageJ. *Bone* **2010**, *47*, 1076–1079. [CrossRef] [PubMed]
40. Blaudez, F.; Ivanovski, S.; Ipe, D.; Vaquette, C. A Comprehensive Comparison of Cell Seeding Methods Using Highly Porous Melt Electrowriting Scaffolds. *Mater. Sci. Eng. C* **2020**, *117*, 111282. [CrossRef]
41. Hannoun, A.; Perrier-Groult, E.; Cureu, L.; Pasdeloup, M.; Berthier, Y.; Mallein-Gerin, F.; Trunfio-Sfarghiu, A.M. New "Tribo-Bioreactor" for In-Situ Monitoring of the Mechanical Stress Transmission at the Cellular Level: Application to Cartilage Tissue Engineering. *Biotribology* **2021**, *25*, 100158. [CrossRef]
42. Ben-Zvi, Y.; Maria, R.; Pierantoni, M.; Brumfeld, V.; Shahar, R.; Weiner, S. Response of the Tooth-Periodontal Ligament-Bone Complex to Load: A MicroCT Study of the Minipig Molar. *J. Struct. Biol.* **2019**, *205*, 155–162. [CrossRef] [PubMed]
43. Ortún-Terrazas, J.; Cegoñino, J.; Pérez del Palomar, A. In Silico Study of Cuspid' Periodontal Ligament Damage under Parafunctional and Traumatic Conditions of Whole-Mouth Occlusions. A Patient-Specific Evaluation. *Comput. Methods Programs Biomed.* **2020**, *184*, 105107. [CrossRef]

44. Kaneko, D.; Sasazaki, Y.; Kikuchi, T.; Ono, T.; Nemoto, K.; Matsumoto, H.; Toyama, Y. Temporal Effects of Cyclic Stretching on Distribution and Gene Expression of Integrin and Cytoskeleton by Ligament Fibroblasts In Vitro. *Connect. Tissue Res.* **2009**, *50*, 263–269. [CrossRef]
45. Zhu, W.; Liang, M. Periodontal Ligament Stem Cells: Current Status, Concerns, and Future Prospects. *Stem Cells Int.* **2015**, *2015*, 318269. [CrossRef]
46. Klouda, L.; Vaz, C.M.; Mol, A.; Baaijens, F.P.T.; Bouten, C.V.C. Effect of Biomimetic Conditions on Mechanical and Structural Integrity of PGA/P4HB and Electrospun PCL Scaffolds. *J. Mater. Sci. Mater. Med.* **2008**, *19*, 1137–1144. [CrossRef]
47. Panadero, J.A.; Vikingsson, L.; Gomez Ribelles, J.L.; Lanceros-Mendez, S.; Sencadas, V. In Vitro Mechanical Fatigue Behavior of Poly-ε-Caprolactone Macroporous Scaffolds for Cartilage Tissue Engineering: Influence of Pore Filling by a Poly(Vinyl Alcohol) Gel. *J. Biomed. Mater. Res. Part B Appl. Biomater.* **2015**, *103*, 1037–1043. [CrossRef] [PubMed]
48. Rampichová, M.; Chvojka, J.; Buzgo, M.; Prosecká, E.; Mikeš, P.; Vysloužilová, L.; Tvrdík, D.; Kochová, P.; Gregor, T.; Lukáš, D.; et al. Elastic Three-Dimensional Poly (ε-Caprolactone) Nanofibre Scaffold Enhances Migration, Proliferation and Osteogenic Differentiation of Mesenchymal Stem Cells. *Cell Prolif.* **2013**, *46*, 23–37. [CrossRef]
49. Cukierman, E.; Pankov, R.; Stevens, D.R.; Yamada, K.M. Taking Cell-Matrix Adhesions to the Third Dimension. *Science* **2001**, *294*, 1708–1712. [CrossRef] [PubMed]
50. Hirashima, S.; Kanazawa, T.; Ohta, K.; Nakamura, K. Three-Dimensional Ultrastructural Imaging and Quantitative Analysis of the Periodontal Ligament. *Anat. Sci. Int.* **2020**, *95*, 1–11. [CrossRef] [PubMed]
51. Vashaghian, M.; Diedrich, C.M.; Zandieh-Doulabi, B.; Werner, A.; Smit, T.H.; Roovers, J.P. Gentle Cyclic Straining of Human Fibroblasts on Electrospun Scaffolds Enhances Their Regenerative Potential. *Acta Biomater.* **2019**, *84*, 159–168. [CrossRef] [PubMed]
52. Wang, Y.; Zhao, Z.; Li, Y.; Li, Y.; Wu, J.; Fan, X.; Yang, P. Up-Regulated α-Actin Expression Is Associated with Cell Adhesion Ability in 3-D Cultured Myocytes Subjected to Mechanical Stimulation. *Mol. Cell. Biochem.* **2010**, *338*, 175–181. [CrossRef] [PubMed]
53. Ho, F.C.; Zhang, W.; Li, Y.Y.; Chan, B.P. Mechanoresponsive, Omni-Directional and Local Matrix-Degrading Actin Protrusions in Human Mesenchymal Stem Cells Microencapsulated in a 3D Collagen Matrix. *Biomaterials* **2015**, *53*, 392–405. [CrossRef]
54. Cai, X.; Zhang, Y.; Yang, X.; Grottkau, B.E.; Lin, Y. Uniaxial Cyclic Tensile Stretch Inhibits Osteogenic and Odontogenic Differentiation of Human Dental Pulp Stem Cells. *J. Tissue Eng. Regen. Med.* **2011**, *5*, 347–353. [CrossRef] [PubMed]
55. Yamaguchi, M.; Shimizu, N.; Shibata, Y.; Abiko, Y. Effects of Different Magnitudes of Tension-Force on Alkaline Phosphatase Activity in Periodontal Ligament Cells. *J. Dent. Res.* **1996**, *75*, 889–894. [CrossRef] [PubMed]
56. Elliott, D.S.; Newman, K.J.H.; Forward, D.P.; Hahn, D.M.; Ollivere, B.; Kojima, K.; Handley, R.; Rossiter, N.D.; Wixted, J.J.; Smith, R.M.; et al. A Unified Theory of Bone Healing and Nonunion. *Bone Jt. J.* **2016**, *98B*, 884–891. [CrossRef]
57. Wutticharoenmongkol, P.; Sanchavanakit, N.; Pavasant, P.; Supaphol, P. Novel Bone Scaffolds of Electrospun Polycaprolactone Fibers Filled with Nanoparticles. *J. Nanosci. Nanotechnol.* **2006**, *6*, 514–522. [CrossRef] [PubMed]
58. Bergomi, M.; Wiskott, H.W.A.; Botsis, J.; Mellal, A.; Belser, U.C. Load Response of Periodontal Ligament: Assessment of Fluid Flow, Compressibility, and Effect of Pore Pressure. *J. Biomech. Eng.* **2010**, *132*, 014504. [CrossRef]
59. Vaquette, C.; Fan, W.; Xiao, Y.; Hamlet, S.; Hutmacher, D.W.; Ivanovski, S. A Biphasic Scaffold Design Combined with Cell Sheet Technology for Simultaneous Regeneration of Alveolar Bone/Periodontal Ligament Complex. *Biomaterials* **2012**, *33*, 5560–5573. [CrossRef]
60. Bouet, G.; Cruel, M.; Laurent, C.; Vico, L.; Malaval, L.; Marchat, D. Validation of an in Vitro 3D Bone Culture Model with Perfused and Mechanically Stressed Ceramic Scaffold. *Eur. Cells Mater.* **2015**, *29*, 250–267. [CrossRef]

biomimetics

Review

Recent Methods for Modifying Mechanical Properties of Tissue-Engineered Scaffolds for Clinical Applications

Andrew Johnston and Anthony Callanan *

Institute for Bioengineering, School of Engineering, University of Edinburgh, Edinburgh EH9 3DW, UK; s1658384@ed.ac.uk
* Correspondence: anthony.callanan@ed.ac.uk

Abstract: The limited regenerative capacity of the human body, in conjunction with a shortage of healthy autologous tissue, has created an urgent need for alternative grafting materials. A potential solution is a tissue-engineered graft, a construct which supports and integrates with host tissue. One of the key challenges in fabricating a tissue-engineered graft is achieving mechanical compatibility with the graft site; a disparity in these properties can shape the behaviour of the surrounding native tissue, contributing to the likelihood of graft failure. The purpose of this review is to examine the means by which researchers have altered the mechanical properties of tissue-engineered constructs via hybrid material usage, multi-layer scaffold designs, and surface modifications. A subset of these studies which has investigated the function of their constructs in vivo is also presented, followed by an examination of various tissue-engineered designs which have been clinically translated.

Keywords: scaffold; biomimetic; tissue engineering; 3D printing; electrospinning

Citation: Johnston, A.; Callanan, A. Recent Methods for Modifying Mechanical Properties of Tissue-Engineered Scaffolds for Clinical Applications. *Biomimetics* **2023**, *8*, 205. https://doi.org/10.3390/biomimetics8020205

Academic Editors: Ryszard Uklejewski and Mariusz Winiecki

Received: 17 March 2023
Revised: 3 May 2023
Accepted: 12 May 2023
Published: 16 May 2023

1. Introduction

The means of restoring mechanical functionality to damaged biological tissues has proven to be a longstanding challenge to medical practitioners. The complexity of the task is largely due to the intricacy of the native tissue, where biomechanical properties are largely dictated by the extracellular matrix (ECM) [1,2]. This structure consists of a multifaceted network of various constituents such as water, polysaccharides, and proteins such as collagen and elastin [3]. When under load, the ECM relies primarily on these proteins working in conjunction with one another for support [4], with the collagen providing mechanical resistance to deformation under increased load conditions, and elastin being responsible for maintaining the elasticity of the tissue and resistance under relatively low loading [5]. Mechanical integrity and elasticity are not only critical for explicit functions such as joint articulation and other musculoskeletal interactions; they are just as significant for various parasympathetic processes such as peristalsis, respiration, and vasodilation [6–8]. The degradation of tissue biomechanics, whether via natural processes such as ageing, or via exposure to circumstances leading to internal or external injury of soft and hard tissues, can precede issues ranging from loss of mobility and discomfort, to critical conditions such as pulmonary fibrosis and supravalvular aortic stenosis [9–11].

The complexity of repairing extensively damaged tissues is such that contemporary regenerative medicinal practices are limited in treatment options [12]. The transplantation of fresh tissue is seen as the optimal solution, as the new material is theoretically able to fulfil the physiological and mechanical requirements of the original tissue [13]. However, both allografts and autologous grafts bear several limitations. In the case of allografting, graft-versus-host disease, transplant rejection, bleeding, and infection are constant risk factors [14], whereas autologous grafting is not always an option, due to either previous harvesting of the donor site or systemic disease rendering the tissue unsuitable [15]. An alternative therapeutic in promoting the repair of damaged tissues, stem cell treatment, is largely in its infancy as a research area, and faces several significant hurdles, namely, a high risk of immune response, scalability, and overcoming the negative public perception of such a treatment [16,17]. An alternative means of restoring the mechanical properties of biological tissue is therefore necessary in order to address the ever-increasing demand for biological tissue transplantation.

Tissue engineering is a rapidly expanding field which aims to design constructs which supplement or replace damaged biological tissue. The fabrication of a tissue-engineered graft can be undertaken in a variety of ways, including phase separation, bioprinting, gas foaming and electrospinning techniques [18–23]. From these methods, a dense interconnected network of material may be formed, which is termed a scaffold [24]. The purpose of a scaffold is to provide a chemically and mechanically appropriate environment to facilitate cellular growth, which may then be implanted into a given patient [25–27]. The ideal tissue-engineered scaffold would allow the development and restoration of fully functional tissue at the graft site, maintaining structural integrity under physiological stress, and ultimately be removed by the body's natural processes after fulfilling its function [28–30].

A key challenge within this area is ensuring the greatest possible resemblance between the mechanical response of the tissue-engineered graft and the host's native tissue. This is a crucial aspect of a successful design, as biological tissues are load-bearing by nature, regardless of function, from the strongest regions of cortical bone with mechanical strength in the gigapascal range [31] to the most delicate of neural fibres throughout the nervous system whose strength generally lies within the low kilopascal scale [32]. The requirement to maintain the physical integrity of the scaffold while also retaining biocompatibility has given rise to so-called 'biomaterials', which the majority of tissue-engineered scaffolds are composed of and which fall into several broad camps: natural polymers, synthetic polymers, ceramics, decellularized matrices, hydrogels, and metals [33–38]. Natural polymers can include agarose, alginates, and chitosan while synthetic polymers include polycaprolactone, poly(l-lactic acid), and Poly(ethylene glycol) diacrylate. Ceramics can comprise aluminium oxide, bioglass, and hydroxyapatite, while decellurised matrices can be derived from almost any bodily feature, such as bones and organs. Hydrogels are typically composed of naturally derived constituents such as collagen, gelatine and hyaluronic acid, and metals used for scaffolds can include magnesium, tantalum, and titanium [26,36,39–53]. The broad range of characteristics of these materials allows for a vast range of tissue types to be mechanically accounted for; an overview of the typical strength of these material groups in comparison to the stiffness of organic tissues is given in Figure 1 [32,54–67].

One can observe from Figure 1 that particular material types are more suited for certain regions than others; it is clear, for example, that hydrogel alone is an unsuitable replacement for bone tissue. While the consequences of a graft which features an insufficient load-bearing capacity may appear self-evident, the response of surrounding native tissue is often less explicit. For example, a disparity in elastic properties between a tissue-engineered vascular graft and the surrounding native vessel tissue may cause the graft to ineffectively constrict and dilate in tandem with the surrounding vasculature, leading to an increase in smooth muscle cell proliferation, subsequent thrombus formation, and occlusion of the vessel [68,69]. Additionally, when engineered tissue is employed elsewhere in the body, such as for bone or the anterior cruciate ligament, the surrounding tissue will undergo remodelling in accordance with Wolff's Law and its corollary, Davis' Law, which state that both osseous and soft tissue, respectively, will structurally adapt themselves in accordance with the mechanical stresses which they encounter [70,71]. This translates to an overcompensation of the native tissue in the case of an insufficiently robust implant, while a construct which exceeds the mechanical requirements of the native tissue can lead to the degradation of the surrounding architecture, with both scenarios compromising the functionality of the original biological tissue [72,73]. It is therefore crucial that the mechanical properties of engineered tissue are characterised in such a way as to best compliment the surrounding biomechanical environment of the native tissue. This does not necessarily require achieving perfect parity between the stiffness of the existing tissue and a given implant; indeed, several of the studies considered in this review suggest that sufficiently mechanically resilient biocompatible materials are perhaps more suitable for the promotion of tissue regeneration than those who endeavour to match the native tissue's properties in a mechanical context alone. Nevertheless, ensuring a high degree of equivalence between an engineered and a given native tissue's mechanical properties is a functional requirement of a successful engineered graft.

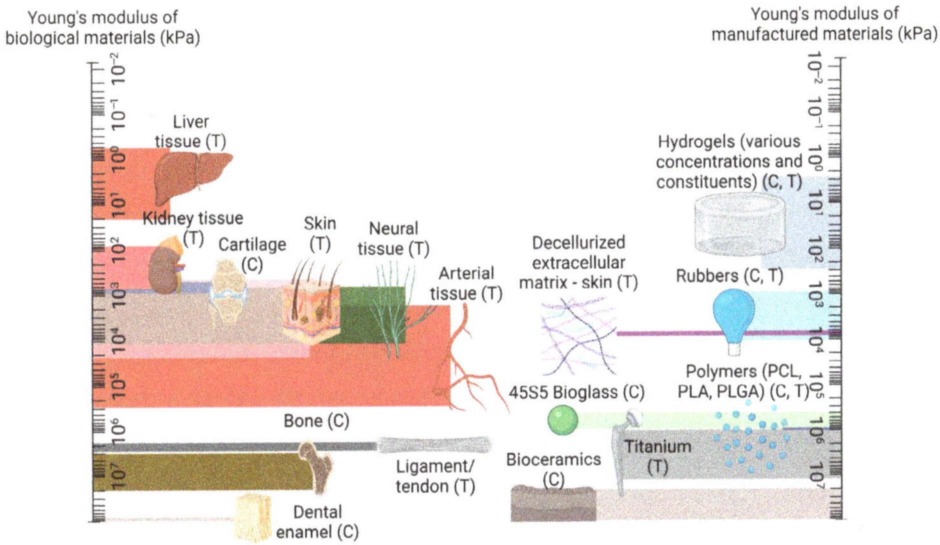

Figure 1. A comparison of mechanical strength between native tissue types, in comparison to manufactured materials, presented on logarithmic scales, with compression and tension testing denoted with (C) and (T) as appropriate. Created using Biorender.com, 15 May 2023.

To address this challenge, tissue engineering researchers are investigating a broad range of solutions to optimise the mechanical response of tissue-engineered grafts, while also maintaining the biocompatibility of the design. The current studies in this field include the use of multiple materials in conjunction with one another, such as polycaprolactone and collagen, bi- and multi-layered scaffolds which often delegate the functional requirements of the construct to each layer, and modifying existing scaffold parameters such that their performance may be tuned to better suit the conditions of the native tissue [74–76]. Success in these areas has led to multiple studies in vivo with such designs, and, in some cases, to full clinical translation, whereby the cases of which are examined in Section 7 of this work.

To support the development of these designs, a broad range of literature is available which describes the mechanical properties of specific native tissue types and engineered tissue scaffolds. These include bone, cartilage, liver, kidney, osteochondral tissue, muscle, and tendon, and utilise a broad range of mechanical property assessment methods, including compressive, dynamic, nanoindentation, shear, and tensile testing, with additional methods including finite element analysis (linear and non-linear) and ultrasound [58,77–82]. Using these techniques, native tissue properties such as Young's modulus, yield stress, failure strength and strain, fracture toughness, and viscoelastic behaviour can be assessed. Review articles of this nature are invaluable as resources for the engineering of specific tissue types. However, a comparative examination of a broad variety of tissues, materials, fabrication, and testing methods would offer a unique insight into the effects of parameters and their effect on the mechanical performance of a particular design, which could potentially be applied to other tissue types. A review of this nature, to the best of the authors' knowledge, has not been presented in the literature before.

This work provides a review of the current techniques by which researchers have successfully modified the mechanical properties of tissue-engineered grafts. The scope of this review will encapsulate all applicable construction methods which are currently employed in this regard, and all tissue types will be considered in accordance with the focus of each study. A Supplementary Table, Table S1, provides a greater deal of insight into the studies considered in this work, with a brief summary of each publication provided therein. The extent of this review is all relevant work published within the years of 2021 to 2023, and it encompasses primary articles reporting on the use of hybrid-, multi-layer-, and surface-modified-type scaffolds, as well as clinically translated designs whose use and outcomes were published within the above chronological range. Articles published prior to these dates were utilised as citations for statements and empirical data where appropriate.

2. Mechanical Properties of Biological Tissues and Their Influence

The mechanical properties of biological materials are a product of a complex array of proteins, cells, and interstitial fluid flow, whose interactions characterise the response of their respective tissue under load [83,84]. As a product of their intricacy, biological tissues exhibit complex mechanical properties such as heterogeneity, viscoelasticity, and anisotropy [83–85]. Accurately assessing the mechanical properties of these tissues is therefore intrinsically challenging, with research groups often attaining a variety of results for the same tissue. Table 1 illustrates this, featuring a short collection of various biological tissues, and a generally broad range of respective mechanical properties, encountered by several research groups for the same mode of analysis.

Table 1. A table of selected tissue types and their mechanical properties.

Tissue Type	Failure Strain	Yield Stress (MPa)	Young's Modulus (MPa)	Shear Modulus (MPa)	Reference
Coronary Artery	0.45	0.39–1.8	1.55 ± 0.26	0.3	[86–90]
Cartilage	0.183	4.58 ± 2.04	0.5–0.9	0.26–0.32	[91–94]
Bone	0.25–0.67	71.56	17,900–19,080	3600	[95–97]
Skin	1.5	17–21	60–70	0.002–0.008	[65,98,99]
Spine/ Sciatic/Ulnar Nerve	0.293–0.73	11.7	0.7–10	0.02–0.054	[100–103]

The diversity in these results can not only be attributed to the structural composition of the tissues alone; factors such as hydration and age are also known to affect the mechanical response of biological tissues [104,105]. This process of tissue-specific cells responding to mechanical stimuli is termed mechanotransduction, which is a means by which organic tissue may convert external loading, such as tension or vibrational waves, to biochemical, biophysical, or molecular signals via a series of events culminating in a cellular response [106]. Figure 2 illustrates this process, beginning with a given external loading event encountered by the body's mechanosensors. This is termed mechanocoupling, and, depending on the load type, is translated by one or more mechanosensors to adjacent cells in the ECM [107]. Via a complex series of intra-cellular interactions, these mechanical signals are converted to various signal types which instruct the relevant cell groups to behave in a particular way; for example, an increase in mechanical load due to extensive physical activity can lead to compensatory growth in relevant skeletal and muscular members [106,108].

It is the cellular response to various conditions which allows the human body to maintain homeostasis, while also quickly adapting to environmental circumstances; vasoconstriction and vasodilation are common examples of this [109]. However, a crucial caveat is that biological tissues have evolved to expect a certain degree of physiological loading. A reduction in the stress and strain forces imparted on, for example, the head region of the femur as a result of a hip implant, can in turn lead to the 'stress-shielding' effect [110].

This is where a non-native element acts as the main load-bearing facilitator during typical movement (e.g., walking or running). As a result, the incentive for cells to remodel and strengthen their respective tissues during mechanotransduction is lowered and local bone density is reduced, potentially leading to implant loosening, stress fracturing, and abnormal bone development [111,112]. It follows that other tissue-engineered designs are bound by the same factors; a mechanically under- or over-strengthened graft is not only impractical, but can cause deleterious effects on neighbouring local tissue groups as outlined above. This highlights the importance of mechanical parameters when designing a tissue-engineered graft.

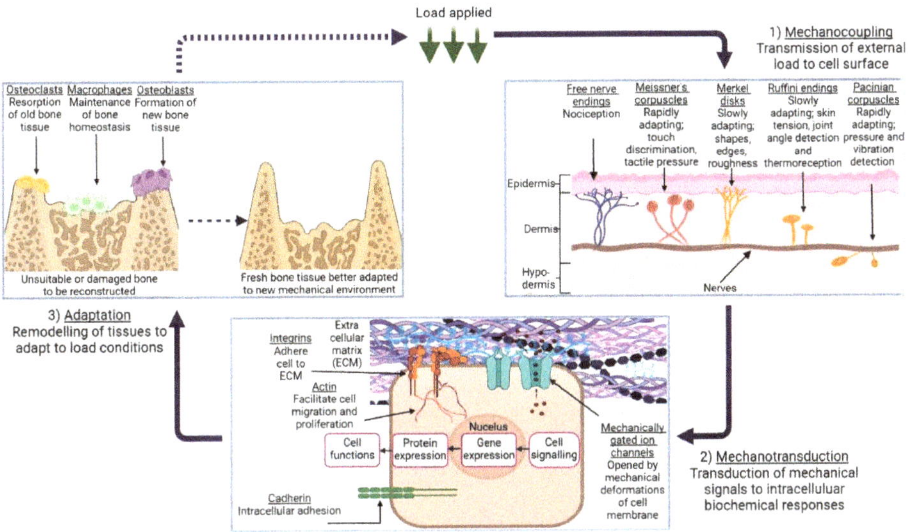

Figure 2. An illustration of the mechanotransduction cycle, in the context of osseous tissue remodelling in response to external loading applied to a given area of skin, via the medium of a generic cuboidal cell. (**1**) demonstrates the various types of mechanoreceptors present in human dermis and their function, while (**2**) depicts the mechanotransduction pathway, followed when the ECM is subject to deformation. (**3**) shows the remodelling process of bone tissue in response to this loading, and the roles which the various cells play in this process. Created using BioRender.com, 24 April 2023.

3. Hybrid Materials

The non-linear biomechanical response of a tissue under load is difficult to reproduce in synthetic materials alone, as this response is characterised by complex biological structures and interactions, as described in Section 2. Current research indicates that a hybrid polymer blend, consisting of both synthetic and natural materials, allows for constructs which better represent this intricate behaviour [113]. This is due to the naturally derived constituent typically retaining its microstructural details, which can promote cell attachment and integration; this is beneficial when employing a scaffold in a biological environment [114,115]. A broad range of hybrid polymer combinations are currently undergoing extensive research, ranging from binary PCL-hydroxyapatite designs [29] to complex three- or four-part hybrid amalgamations [116,117]. Table 2 provides an overview of a selection of such work, including the materials and fabrication methods employed during each study, the intended tissue type, and which mechanical properties were assessed throughout the work. As an extension of this, Table 3 explores the Young's moduli value ranges attained in each of the studies in question, quantifying the effects of these material combinations.

Table 2. Summarisation of materials, fabrication methods, intended tissue type, and which mechanical properties were altered and assessed during studies which employed a hybrid polymer blend.

Tissue Type	Material	Fabrication Method	Mechanical Properties Assessed				Author, Year, Reference
			Tensile/Compressive Modulus	Failure Strain	Ultimate Tensile Strength	Storage/Loss Modulus	
Bone	PCL + HA	3D printing	✓	✓			Rezania et al. (2022) [29]
Bone	Gelatine + HA + hPE	3D printing	✓	✓			Lee et al. (2022) [117]
Bone	Col and Col + PGA and Col + PGA/HA + PGA and Col + PGA/HA + PGA (2-layer membrane)	3D printing	✓	✓		✓	Nguyen et al. (2022) [115]
Bone	HA-NFs + GelMA	Hydrogel solution	✓	✓	✓		Wang et al. (2022) [118]
Bone	Gelatine + GO + PHEMA	Freeze-drying	✓				Tabatabaee et al. (2022) [119]
Cartilage	Alginate (Core) + Chitosan (Shell and gel) + SF(gel)	Core-shell microspheres	✓	✓	✓		Min et al. (2022) [120]
Cartilage	GelMA + HyAMA + chondrospheroids	Hydrogel solution	✓				Wang et al. (2022) [118]
Cartilage	PEC + SF + SA + PrP	Phase separation	✓				Singha et al. (2022) [18]
Cartilage	rSF and rSF + rGO	Electrospinning	✓			✓	Dorishettya et al. (2022) [121]
Cartilage/Bone	Gellan gum + Alginate sodium and Gellan gum + Alginate sodium + TMP-BG	3D Printing	✓	✓	✓	✓	Chen et al. (2022) [122]

Table 2. *Cont.*

Tissue Type	Material	Fabrication Method	Mechanical Properties Assessed				Author, Year, Reference
			Tensile/Compressive Modulus	Failure Strain	Ultimate Tensile Strength	Storage/Loss Modulus	
Kidney	PCL + Laminin + Span80™ emulsion	Electrospinning	✓	✓	✓		Baskapan et al. (2022) [123]
Nerve	PLA + Col	Electrospinning	✓	✓	✓		Xu et al. (2022) [124]
Skin	PCL + SF + SESM + Gelatine and PCL + SF + SESM + MC	Electrospinning	✓	✓	✓		Salehi et al. (2022) [125]
Skin	PCL+ Col and PCL + Col + ZnO and PCL + Col + ZnO + VEGF	Electrospinning	✓	✓	✓		Li et al. (2021) [126]

PCL: Polycaprolactone, HA: Hydroxyapatite, hPE: human placental extracts, PGA: Poly-glutamic acid, Col: collagen, HA-NFs: Hydroxyapatite nanofibers: GelMA: Methacrylated gelatine, GO: graphene oxide, PHEMA: Poly(2-hydroxyethyl methacrylate), TMP-BG: Thixotropic magnesium-phosphate-based gel, SF: silk fibroin, HyAMA: Methacrylated hyaluronic acid, PEC: Polyelectrolyte complexation, SA: sodium alginate, PrP: platelet-rich plasma; rSF: regenerated silk fibroin, SESM: soluble eggshell membrane, MC: Methylcellulose, ZnO: zinc oxide, VEGF: vascular endothelial growth factor, dECM: decellurised extra cellular matrix.

Table 3. An overview of Young's modulus value ranges attained during studies presented in Table 2.

Tissue Type	Material	Fabrication Method	Elastic Modulus Range Attained in Tension (T) or Compression (C)						Author, Year, Reference
			<1 kPa	1–100 kPa	100–1000 kPa	1–100 MPa	100–1000 MPa	>1 GPa	
Bone	PCL + HA	3D printing				✓ (C)	✓ (T)		Rezania et al. (2022) [29]
Bone	Gelatine + HA + hPE	3D printing			✓ (C)				Lee et al. (2022) [117]
Bone	Col and Col + PGA and Col + PGA/HA + PGA and Col + PGA/HA + PGA (2-layer membrane)	3D printing		✓ (C)	✓ (C)				Nguyen et al. (2022) [115]

Table 3. *Cont.*

Tissue Type	Material	Fabrication Method	Elastic Modulus Range Attained in Tension (T) or Compression (C)						Author, Year, Reference
			<1 kPa	1–100 kPa	100–1000 kPa	1–100 MPa	100–1000 MPa	>1 GPa	
Bone	HA-NFs + GelMA	Hydrogel solution		✓ (C)					Wang et al. (2022) [118]
Bone	Gelatine + GO + PHEMA	Freeze-drying				✓ (C)			Tabatabaee et al. (2022) [119]
Cartilage	Alginate (Core) + Chitosan (Shell and gel) + SF(gel)	Core-shell microspheres				✓ (C)			Min et al. (2022) [120]
Cartilage	GelMA + HyAMA + chondrospheroids	Hydrogel		✓ (C)					Wang et al. (2022) [118]
Cartilage	PEC + SF + SA + PrP	Phase separation			✓ (C)				Singha et al. (2022) [18]
Cartilage	rSF and rSF + rGO	Electrospinning			✓ (C)	✓ (C)			Dorishettya et al. (2022) [121]
Cartilage/Bone	Gellan gum + Alginate sodium and Gellan gum + Alginate sodium + TMP-BG	3D Printing		✓ (C)	✓ (C)	✓ (C)			Chen et al. (2022) [122]
Kidney	PCL + Laminin + Span80™ emulsion	Electrospinning		✓ (T)					Baskapan et al. (2022) [123]
Nerve	PLA + Col	Electrospinning			✓ (T)	✓ (T)			Xu et al. (2022) [124]
Skin	PCL + SF + SESM + Gelatine and PCL + SF + SESM + MC	Electrospinning				✓ (T)			Salehi et al. (2022) [125]
Skin	PCL+ Col and PCL + Col + ZnO and PCL + Col + ZnO + VEGF	Electrospinning			✓ (T)	✓ (T)			Li et al. (2021) [126]

The scope for hybrid biomaterial design is vast, which allows researchers to consider a broad range of materials. For example, several research groups have recently investigated the inclusion of multi-walled carbon nanotubes (MWCNTs), which are nanometre-diameter carbon-based tubular constructs, with the aim of utilising the nanotubes' high tensile strength and electrical/thermal conductivity to enhance the overall biomechanical properties of their scaffold. Sang et al. dispersed MWCNTs with concentrations ranging between 1, 3, and 5% through a chitosan/polyethylene glycol scaffold, which was prepared by freeze-drying, for the purpose of neural tissue engineering [127]. The electrical conductivity of neurons is key to their function, and one of the key findings of this analysis was the rise in electrical conductivity of the scaffold in proportion to increasing MWCNT concentrations. Interestingly, the elastic modulus of the hybrid scaffold was also enhanced with an increasing concentration of MWCNTs, from 1.17 kPa to 2.09 kPa, demonstrating the load-bearing abilities of carbon nanotubes. As another example of this, Ramasamy et al. employed MWCNTs for use in a poly(l-lactic acid)–boron nitride piezoelectric nanosheet hybrid, where it was found that the combination of these materials led to a significantly stronger graft when compared to unmodified poly(l-lactic acid) nanofibers quantified by a 666% increase in tensile strength in contrast [128].

Hybrid plant-based designs are also promising in this area; a recent study by Jiawei et al. investigated the inclusion of hydroxyapatite into sugarcane stem, whose lignin had been removed (i.e., delignified) via soaking in a $NaClO_2$/acetate buffer solution. The degree of delignification induced a proportionate increase in the sugarcane stem porosity, albeit at the cost mechanical performance under compressive load, with the greatest extent—8 h of immersion—reducing the compressive modulus of 14.6 MPa for pure sugarcane stem to 4.82 MPa. An optimal soaking time of 4 h was established, which promoted an elongated elastic response from the scaffold under load whilst maintaining structural integrity [129].

4. Multi-Layer Scaffolds

The challenge of tuning the elastic response of a scaffold in tandem with balancing its biocompatibility characteristics has led several research groups to fabricate scaffolds consisting of multiple layers. This approach allows for the functionalisation of each stratum independently and more closely represents the layered nature of most biological tissue, such as arterial walls, osseous tissue, and oesophageal tissue [130–132]. The potential impact of various material and fabrication combinations has allowed researchers to explore a broad range of designs consisting of materials such as silk, graphene, and titanium [133–135]. Table 4 provides a brief overview of the groups which have investigated the potential of bi- and multi-layered scaffold designs, while Table 5 illustrates the Young's modulus ranges attained during these studies.

Table 4. Summarisation of materials per layer, fabrication methods, intended tissue type, and which mechanical properties were altered and assessed during studies which employed a multi-layer scaffold.

Tissue Type	Material per Layer	Fabrication Method	Mechanical Properties Modified and Assessed				Author, Year, Reference
			Tensile/ Compressive Modulus	Failure Strain	Ultimate Tensile Strength	Storage/Loss Modulus	
Bone	PLLA/Cell sheet	Electrospinning	✓	✓			Tevlek et al. (2021) [136]
Cartilage	PCL + Gelatine/Gelatine + Alginate	Electrospinning	✓				Semitela et al. (2021) [137]
Cartilage	Alg + HA 50:50 (~70 layers) and Alg + HA 70:30 (~70 layers)	Bioprinting	✓	✓			Janarthanan et al. (2022) [138]
Osteochondral	Col + PLGA and Ti/Col + PLGA	3D printing/freeze-drying	✓				Yang et al. (2021) [135]
Osteochondral	Ti/PLA/ Col + PLGA and Col + HA	3D printing	✓				Tamaddon et al. (2022) [139]
Osteochondral	PCS/Col + HA	Freeze-drying	✓				Rashidi et al. (2021) [140]
Osteochondral	PCL + PEO + PES and PCL + PEO/rGO + HA-Sr + PES-$BH_{0.5\%}$	Electrospinning	✓				Dargoush et al. (2022) [134]
Osteochondral	Silk and BMP-2 + Silk and Silk/Silk (bilayer) and BMP-2 + bilayer and BMP-2 + bilayer/TGF-β1 + SilMA	Hydrogel solution	✓	✓	✓		Wu et al. (2021) [133]
Osteochondral	GelMA + PEO + HA/GelMA + PEO + HA/GelMA + PEO	Hydrogel solution	✓			✓	Li et al. (2022) [141]
Periodontal	PCL/BG and PCL/HyA	3D printing	✓	✓			Nejad et al. (2021) [142]

Table 4. *Cont.*

Tissue Type	Material per Layer	Fabrication Method	Mechanical Properties Modified and Assessed				Author, Year, Reference
			Tensile/ Compressive Modulus	Failure Strain	Ultimate Tensile Strength	Storage/Loss Modulus	
Vascular	PLGA/PCL	Electrospinning	✓				Bazgir et al. (2021) [143]
Vascular	PCL/PCL	Electrospinning	✓	✓	✓		Li et al. (2021) [144]
Vascular	RHC + PCL/PEO + PCL	Electrospinning	✓	✓	✓		Do et al. (2021) [145]
Vascular	dECM/GAGF/dECM/dECM	Decellurizing and gas foaming	✓	✓	✓		Smith et al. (2022) [146]

Ti: titanium, PLGA: Poly-lactic-co-glycolic acid: PLA: Polylactic acid, PCS: polymerised chondroitin sulphate, PEO: Polyethylene oxide, PES: Polyethersulfone, HA-Sr: Strontium-doped hydroxyapatite, BH: Benzyl hyaluronan, BMP-2: bone morphological protein-2, TGF-β3: transforming growth factor-β3, SilMA: Methacrylated silk, BG: bioglass, HyA: hyaluronic acid, RHC: recombinant human collagen, GAGF: Glycosaminoglycan foam, PLLA: Poly(L-lactic acid).

Table 5. An overview of Young's modulus value ranges attained during studies presented in Table 4.

Tissue Type	Material per Layer	Fabrication Method	Elastic Modulus Range Attained in Tension (T) or Compression (C)						Author, Year, Reference
			<1 kPa	1–100 kPa	100–1000 kPa	1–100 MPa	100–1000 MPa	>1 GPa	
Bone	PLLA/Cell sheet	Electrospinning				✓ (T)			Tevlek et al. (2021) [136]
Cartilage	PCL + gelatine/Gelatine + Alginate	Electrospinning			✓ (T)	✓ (T)			Semitela et al. (2021) [137]
Cartilage	Alg + HA 50:50 (~70 layers) and Alg + HA 70:30 (~70 layers)	Bioprinting		✓ (C)					Janarthanan et al. (2022) [138]
Osteochondral	Col + PLGA and Ti/Col + PLGA	3D printing/freeze-drying				✓ (C)			Yang et al. (2021) [135]
Osteochondral	Ti/PLA/ Col + PLGA and Col + HA	3D printing				✓ (C)	✓ (C)	✓ (C)	Tamaddon et al. (2022) [139]

Table 5. *Cont.*

Tissue Type	Material per Layer	Fabrication Method	Elastic Modulus Range Attained in Tension (T) or Compression (C)						Author, Year, Reference
			<1 kPa	1–100 kPa	100–1000 kPa	1–100 MPa	100–1000 MPa	>1 GPa	
Osteochondral	PCS/Col + HA	Freeze-drying		✓ (C)					Rashidi et al. (2021) [140]
Osteochondral	PCL + PEO + PES and PCL + PEO/rGO + HA-Sr + PES-$BH_{0.5\%}$	Electrospinning		✓ (T)					Dargoush et al. (2022) [134]
Osteochondral	Silk and BMP-2 + Silk and silk/silk (bilayer) and BMP-2 + bilayer and BMP-2 + bilayer/TGF-β1 + SilMA	Hydrogel solution			✓ (C)	✓ (C)			Wu et al. (2021) [133]
Osteochondral	GelMA + PEO + HA/GelMA + PEO + HA/GelMA + PEO	Hydrogel solution		✓ (C)					Li et al. (2022) [141]
Periodontal	PCL/BG and PCL/HyA	3D printing			✓ (C)	✓ (C)			Nejad et al. (2021) [142]
Vascular	PLGA/PCL	Electrospinning				✓ (T)			Bazgir et al. (2021) [143]
Vascular	PCL/PCL	Electrospinning				✓ (T)			Li et al. (2021) [144]
Vascular	RHC + PCL/PEO + PCL	Electrospinning				✓ (T)			Do et al. (2021) [145]
Vascular	dECM/GAGF/dECM/dECM	Decellurizing and gas foaming				✓ (T)			Smith et al. (2022) [146]

While the properties of multi-layered scaffolds are largely determined by the material selection, several groups have investigated the potential of applying novel fabrication methods in order to construct a multi-layered construct which can function within the physiological load range. Killian et al. studied the combination of calcium phosphate cement (CPC), fabricated via 3D printing, and PCL, drawn using a melting electrowriting (MEW) process for the purpose of bone tissue engineering [147]. These two materials were printed upon one another up to 10 layers in height, creating a grid-like pattern of CPC which contained multiple interspatial variations, and from which the strands of PCL fibre could intersect. Interestingly, while the addition of the PCL fibres slightly diminished the yield strength and elastic modulus of the scaffolds (which were approximately 4–10 and 40–85 MPa, respectively), the inclusion of these fibres would occasionally cause the construct to mechanically yield twice under load: once for the PCL strands, and again as the CPC failed, offering an element of redundancy to the design. Liu et al. fabricated a bilayered construct by combining electrospinning and 3D printing processes, with the intention of promoting guided bone regeneration [148]. The electrospun layer consisted of a PCL/gelatine membrane, while the 3D-printed scaffold consisted of a PCL/gelatine/HA mixture, which were combined by dissolving the electrospun membrane and adding it to the scaffold. It was found that the compressive strength of the resulting scaffold, at 13.86 MPa, closely resembled that of cancellous bone (up to ~13 MPa [149]). The similarity of these compressive stress values suggests that a greatly diminished stress-shielding effect would be observed in clinical trials, in comparison to existing bone implant devices formed from a Ti-6Al-4V alloy with a Young's modulus of 110 GPa [150].

In addition, Thompson et al. employed a process known as embedded 3D printing to fabricate a multi-layer scaffold in an effort to replicate human vocal cord tissue [151]. This method utilises a cavity which has pre-printed ink filaments embedded within a given medium, allowing for the fabrication of a component without interference from gravity or requiring additional supports (Figure 3) [152]. For this particular study, various silicone elastomers and thinners were used to construct each layer, and then they were cured via a 700 W microwave. The elastic modulus varied depending on the layer assessed, with the superficial lamina propria (SLP) providing a 0.91 kPa tensile modulus, and the epithelium increasing this value to 39.74 kPa, which lies within the range of human vocal cords, at approximately 30 kPa [153]. The group were also successful in functionalising the construct by designing it to exhibit flow-induced vibration, representing human phonation.

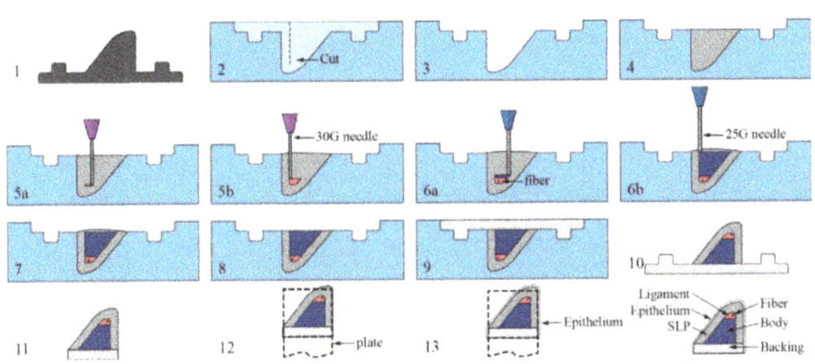

Figure 3. An illustration of the embedded 3D printing process of the vocal cords. Steps as follows: (**1**) Create positive mold of SLP layer. (**2**) Create reservoir with cut for fiber insertion. (**3**) Coat reservoir with release agent. (**4**) Fill reservoir with SLP support matrix silicone. (**5a**,**b**) Print ligament layer within reservoir, beginning near reservoir bottom, and then insert fiber. (**6a**,**b**) Print body layer beginning at interface of ligament layer. (**7**–**8**) Remove overflow material and then cure in microwave. (**9**) Pour backing into remaining cavity and let model cure and cool completely. (**10**) Remove VF model from reservoir. (**11**) Trim excess backing material and gently clean model with acetone. (**12**) Attached model to mounting plate. (**13**) Pour epithelial layer and cure. Reproduced with the publisher's permission [152].

5. Surface Modification

The mechanical properties of engineered tissue are not necessarily set in stone. By design, a substance such as a polymeric compound may have its constituents altered such that it may better serve its intended function [154,155]. In a similar manner, a tissue-engineered graft, being typically composed of one or more complex substances, may be modified in such a way that a desired property, such as increased mechanical strength, can be achieved without having to resort to less suitable or more expensive material alternatives, via processes such as annealing or salt leaching [156,157]. Recent studies have considered a broad range of surface modifications, depending on the desired functionality of the design. These range from topographical nano-modification of the scaffold [158] to grafting post-fabrication and polymer coatings [159,160]. Table 6 offers an overview of the materials, fabrication methods, and modification methods employed in recent studies involving surface modification of tissue-engineered grafts. Table 7 expands on this by showing the Young's modulus value ranges achieved during each of the studies featured in this section.

Table 6. Summarisation of materials, modification methods, intended tissue type, and which mechanical properties were altered and assessed during studies which utilised surface modification techniques.

Tissue Type	Materials	Fabrication Method	Modification Method	Mechanical Properties Modified and Assessed				Author, Year, Reference
				Tensile/ Compressive Modulus	Failure Strain	Ultimate Tensile strength	Storage/Loss Modulus	
Bone	PLA/HA/PDA	3D printing	PDA coating	✓				Chi et al. (2022) [160]
Bone	SF/OcPh/PDA	Freeze-drying	PDA coating	✓				Peng et al. (2022) [161]
Bone	Forsterite/Copper ferrite/P3HB	Sol–gel combustion	P3HB coating	✓				Aghajanian et al. (2022) [64]
Neural	PCL + Chitosan and PCL + Chitosan + Alginate	Electrospinning	Alginate coating	✓	✓	✓		Habibizadeh et al. (2022) [162]
Osteochondral	AC-dECM/ Bone-dECM/	Freeze-drying	Annealing	✓				Browe et al. (2022) [156]
Osteochondral	PCL/GelMA	PDMS mould UV curing	Salt leaching	✓				DiCerbo et al. (2022) [157]
Skin	PCL/PEG/ PCEC/Hydrogel	Electrospinning	Copolymer/hydrogel adsorption	✓			✓	Zhong et al. (2022) [163]
Skin	PCL/Chitosan/Gelatine	Electrospinning	Collagen grafting	✓	✓	✓		Sheikhi et al. (2022) [159]
Vascular	PCL/ ePTFE/RGD	Material/solvent solution	Induced crystallisation and RGD coating	✓	✓	✓		Wang et al. (2022) [164]
Vascular	PCL + PGS + DTβ4	Electrospinning and 3D printing	Molecular modification	✓	✓			Xiao et al. (2022) [165]

PDA: Polydopamine, OcPh: Octacalcium phosphate, P3HB: Poly-3-hydroxybutyrate, AC-dECM: articular cartilage decellurised extra-cellular matrix, ePTFE: expanded Polytetrafluoroethylene, RGD: Arginine-glycine-aspartate, PGS: Poly(glycerol sebacate), DTβ4: Dimeric thymosin β4, PEG: Poly(ethylene glycol), PCEC: PCL-b-PEG-b-PCL.

Table 7. An overview of Young's modulus value ranges attained during studies presented in Table 6.

Tissue Type	Materials	Fabrication Method	Modification Method	Elastic Modulus Range Attained in Tension (T) or Compression (C)						Author, Year, Reference
				<1 kPa	1–100 kPa	100–1000 kPa	1–100 MPa	100–1000 MPa	>1 GPa	
Bone	PLA/HA/PDA	3D printing	PDA coating				✓ (C)	✓ (T)		Chi et al. (2022) [160]
Bone	SF/OcPh/PDA	Freeze-drying	PDA coating		✓ (C)					Peng et al. (2022) [161]
Bone	Forsterite/Copper ferrite/P3HB	Sol–gel combustion	P3HB coating				✓ (C)			Aghajanian et al. (2022) [64]
Neural	PCL + Chitosan and PCL + Chitosan + Alginate	Electrospinning	Alginate coating				✓ (T)			Habibizadeh et al. (2022) [162]
Osteochondral	PCL/ GelMA	PDMS mould UV curing	Salt leaching		✓ (C)	✓ (C)				Browe et al. (2022) [156]
Osteochondral	AC-dECM/Bone-dECM/	Freeze-drying	Annealing				✓ (C)			DiCerbo et al. (2022) [157]
Skin	PCL/PEG/ PCEC/Hydrogel	Electrospinning	Copolymer/hydrogel adsorption		✓ (C)					Zhong et al. (2022) [163]
Skin	PCL/Chitosan/Gelatine	Electrospinning	Collagen grafting				✓ (T)			Sheikhi et al. (2022) [159]
Vascular	PCL/ ePTFE/RGD	Material/solvent solution	Induced crystallisation and RGD coating				✓ (T)	✓ (T)		Wang et al. (2022) [164]
Vascular	PCL + PGS + DTβ4	Electrospinning and 3D printing	Molecular modification					✓ (C)		Xiao et al. (2022) [165]

Surface modification also has the capacity to facilitate the functionalisation of more unlikely material candidates for this research into more suitable substances. An example of this is offered by Mahendiran et al., who examined the potential of cellulose scaffolds derived from *Borassus flabellifer* for the purpose of bone tissue regeneration [166]. The scaffolds were derived from the plant's immature endosperm which was then, after washing and oxidising the scaffolds, modified by two organosilanes; amino-terminated aminopropyltriethoxysilane (APTES) and methyl-terminated octadecyltrichlorosilane (OTS). The subsequent scaffolds formed a foamy architecture, and, in comparison to the unmodified constructs employed as a control during the study, both the APTES and OTS treatments enhanced the compressive modulus of the material, from approximately 0.3 MPa to 0.9 MPa and 1.2 MPa, respectively. Crucially, both treatment methods also bore osteoinductive properties, demonstrating the potential of such a design in a clinical context.

Nanotopographical roughness is a feature which may also be employed to modify the properties of a given scaffold. The influence of surface roughness on cell behaviour is well documented, with several groups reporting relative increases in cell proliferation, adhesion, and desired protein expression when seeded on irregular scaffold surface morphologies [167–169]. One method of creating roughness on engineered scaffolds is alkaline hydrolysis, which was employed by Meng et al. to alter an existing PLLA design for the purposes of bone tissue engineering [170]. For this particular study, the scaffolds, formed via MEW, were immersed in alkali solutions consisting of concentrations of 0.25 M and 0.5 M sodium hydroxide (NaOH) in ethanol (1:1 ratio) for 1, 2, 3, and 4 h at a time. While the effect of this process was perhaps most notable in terms of cell count, which, after 15 days, was largely increased in comparison to the control scaffold, perhaps the most surprising result was the improved tensile modulus of the 0.5 M NaOH scaffold. After 1, 2, and 3 h periods of immersion, a maximum modulus of 5 GPa was achieved, in comparison to 2.5 GPa of the control scaffold. It was theorised that the alkaline treatment process increased the crystallinity of the individual PLLA fibres, thus providing additional tensile strength for the scaffold overall.

6. In Vitro Limitations and Animal Research

The mechanical and biological performance of a tissue-engineered scaffold can be analysed in a number of ways: mechanically, via the methods described in Section 1 such as compressive, shear, and dynamic testing [61,171,172], while biological viability of the scaffold can be studied in vitro via cell seeding of primary or clonal cell lines, from which parameters such as cell adhesion, differentiation, and viability can be derived [173,174]. While these methods may suggest how such a design would perform in a clinical context, there are several clear limitations to such an evaluation, which exist primarily due to the complexity of physiological in vivo conditions. Mechanically, the human body is in a constant state of flux; internal and external loads are met with biomechanical responses as a product of the mechanotransducive process outlined in Section 2. As an illustration of these mechanical stresses, Figure 4 features an arterial section cutaway diagram, where the biological tissue is subjected to physiological stress conditions as a result of hemodynamic flow.

A construct designed to operate in such an environment, such as a vascular graft, would be expected to function under these load conditions; mechanical testing is therefore key, prior to the further development of such a design. However, classical mechanical assessment methods such as those outlined earlier in this section, while perhaps indicative of a particular design's fundamental mechanical properties, are often incapable of evaluating the graft's performance under true physiological conditions, as these are typically non-linear, heterogeneous loads which are nearly impossible to replicate via contemporary testing equipment [175].

The complex nature of these forces has led scientists to explore several avenues of research in an effort to better comprehend, and design for, their impact on tissue-engineered designs. Recent advances in finite element analysis (FEA) and computational fluid dynamics (CFD) analyses have allowed researchers to model tissues such as pulmonary arteries [176], airways [177], ventricles [178], and eye lenses [179], which carries benefits ranging from a greater understanding of the morphological and physiological characteristics of these tissue types, to assisting surgeons during complex surgical procedures [180,181]. A brief overview of additional studies such as these is given in Table 8.

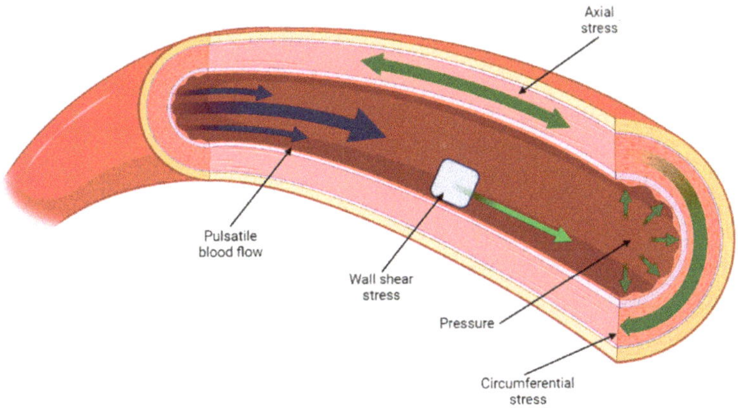

Figure 4. A cross-sectional diagram of arterial flow, illustrating the various load conditions present at any one time in the cardiovascular system. Created using BioRender.com, 6 March 2023.

Table 8. Synopsis of studies which utilised FEA analyses to model various tissues, including model type and whether the model was based on experimental or simulation data.

Tissue Type	Model Type	Based on Native Tissue Assessment?	Author, Year, Reference
Bone	Linear	✓	Irarrázaval et al. (2021) [182]
Cartilage	Non-linear	×	Jahangir et al. (2022) [183]
Kidney	Linear	×	Jing et al. (2021) [184]
Liver	Linear	✓	Fujimoto (2021) [185]
Neural	Non-linear	✓	Peshin (2023) [186]
Osteochondral	Non-linear	×	Hislop et al. (2021) [187]
Skin	Linear	✓	Jobanputra et al. (2021) [188]
Vascular	Linear	✓	Helou et al. (2023) [189]

Despite the promising outlook of this research, a true estimation of the in vivo performance of a tissue-engineered scaffold is still beyond the grasp of contemporary computational or in vitro methods.

The next logical step, therefore, is to assess the viability of tissue-engineered implants in vivo. This will allow researchers to understand how viable their scaffold's design is once implanted into biological tissue, and will also validate the use of particular materials and construction methods within living organisms. Table 9 subcategorises the previously described studies thus far into those which assessed the performance of their designs in an in vivo setting, and provides an overview of the study parameters, as well as a brief summary of each respective group's findings.

Table 9. Summary of materials and intended tissue type, as well as animal species study length, whether the proposed implant fully integrated into the animal, and a brief synopsis of all studies considered thus far, where animal testing was employed.

Tissue Type	Material Type	Fabrication Method	Translation Species	Length of Study	Minimised Immune Response?	Study Summary	Author, Year, Reference
Bone	Gelatine + HA and gelatine + HA + hPE	3D printing	Sprague Dawley rats	12 weeks	✓	The loading of hPE into the gelatine/HA scaffold induced a superior osteogenic response compared to that of the unmodified scaffold.	Lee et al. (2022) [117]
Bone	Col and Col + PGA and Col + PGA/HA + PGA and Col + PGA/HA + PGA (2-layer membrane)	3D printing	Sprague Dawley rats/Nude mice	4 weeks/1, 2, and 4 weeks	✓	The Col + PGA/HA + PGA scaffold indicated the highest cell proliferation and osteogenesis. The next highest cell viability was found in the Col + PGA scaffold. For the membrane-based scaffold, the cell core appeared on the surface of the membrane, with the ECM inside it. The Col scaffold showed the lowest cell viability.	Nguyen et al. (2022) [115]
Bone	PCL + Gel + HA and PCL + Gel + Hep/PCL + Gel + HA	Electrospinning and 3D printing	New Zealand white rabbits	5 and 20 weeks	✓	The composite scaffold showed good integrative and regenerative properties, and displayed no cytotoxicity, while also acting as a barrier to prevent infiltration by fibrous connective tissue. The bilayer scaffold demonstrated much greater new bone tissue formation, however.	Liu et al. (2021) [148]
Bone	Forsterite + copper ferrite and Forsterite + copper ferrite/P3HB	Sol–gel combustion	Wistar rats	8 weeks	✓	Both scaffolds induced a positive response in terms of new tissue formation and trabecular thickness when compared to the control study. However, the P3HB was observed to slightly increase these properties further. Additionally, neither specimen seemed to induce an immune response.	Aghajanian et al. (2022) [64]
Bone	Palm dECM + silicon (OTS-modified) and Palm dECM + silicon (APTES-modified)	Decellurizing	Wistar rats	2 and 4 weeks	✓	Neither scaffold presented signs of inflammation nor infection. Both scaffolds exhibited neovascularisation, the presence of endothelial cells, and collagen network fibres. The quantification of these differences was not presented.	Mahendiran et al. (2022) [166]
Cartilage	HA-NFs + GelMA	Hydrogel	Sprague Dawley rats	12 weeks	✓	Increasing quantities of HA-NFs in the GelMA promoted a stronger osteogenic response, with 15 and 25 wt/wt% HA-NFs showing new bone deposition and blood tissue formation, compared to 0 and 5 wt/wt% which showed little new tissue formation.	Wang et al. (2022) [118]
Cartilage	GelMA + HAMA + chondrospheroids	Hydrogel	Nude mice	1 and 2 months	✓	The chondro-spheroids maintained their morphology during the study. Genes COL 2, SOX 9, and HIF-1a were upregulated in comparison to the positive control (natural cartilage), while COL 10 was downregulated in comparison.	Wang et al. (2021) [113]

Table 9. *Cont.*

Tissue Type	Material Type	Fabrication Method	Translation Species	Length of Study	Minimised Immune Response?	Study Summary	Author, Year, Reference
Cartilage	Alg + HA 50:50 (~70 layers) and Alg + HA 70:30 (~70 layers)	Bioprinting	C57BL/6 mice	1 and 4 weeks	✓	No significant differences were found, in terms of integration, between 50:50 and 70:30 ratios of Alg + HA; however, both scaffolds demonstrated high expression rates of macrophage F4/80 and angiogenesis protein CD31 compared to the control solution.	Janarthanan et al. (2022) [138]
Neural	PCL + Chitosan and PCL + Chitosan + Alginate	Electrospinning	Wistar rats	2, 4, and 8 weeks	✓	The PCL + Chitosan sheet showed a moderate inflammatory response and rapid degradation in comparison to the PCL + Chitosan + Alginate construct, which induced a mild inflammatory response and featured a slower degradation rate.	Habibizadeh et al. (2022) [162]
Osteochondral	Gellan gum + Alginate sodium and Gellan gum + Alginate sodium + TMP-BG	3D printing	New Zealand white rabbits	6 and 12 weeks	✓	After 6 weeks of implantation, subchondral bone growth was slightly diminished in the alginate + gellan growth, and significantly higher in the TMP-BG group, compared to the control. By week 12, the alginate + gellan group showed improved subchondral bone growth compared to the control, and the TMP group showed further enhanced proliferation.	Chen et al. (2022) [122]
Osteochondral	Col + PLGA and Ti/Col + PLGA	3D printing/freeze-drying	New Zealand white rabbits	4, 12, and 24 weeks	✓	For the Col-PLGA group, while it indicated superior cell proliferation at the defect site after 24 weeks, this was mostly just fibrous tissue. In comparison, the bilayered scaffold showed more new bone tissue and better integration with the host. The defect did not fully heal after 24 weeks for either construct.	Yang et al. (2021) [135]
Osteochondral	Ti/PLA/Col + PLGA and Col + HA	3D printing	Ovine condyle model	12 weeks	✓	The multi-layer scaffold provided a more homogenous response in terms of 'filling in' the defect. In addition, the Col + HA scaffold was rougher than the multi-layer, featuring cracks and fissures. In general, the multi-layer scaffold offered a more complete repair response.	Tamaddon et al. (2022) [139]
Osteochondral	PCL + PEO + PES and PCL + PEO/rGO + HAP-Sr + PES-$BH_{0.5\%}$	Electrospinning	Wistar rats	2 months	✓	The nanocomposite scaffold showed larger upregulation in the COL II, COL X, SOX 9, ALP, and Osteocalcin genes and protein compared to the hybrid scaffold. The hybrid scaffold may have induced an immune response due to degradation.	Dargoush et al. (2022) [134]
Osteochondral	AC-dECM/ Bone-dECM/	Freeze-drying	Caprine models	24 weeks	✓	Broad variation in defect repair quality was found. Generally, however, the bilayered scaffold promoted zonally defined tissue, and was able to return the mechanical properties of the region close to that of the surrounding osseous region. Bone repair was more consistent than that of the natural healing process.	Browe et al. (2022) [156]

Table 9. *Cont.*

Tissue Type	Material Type	Fabrication Method	Translation Species	Length of Study	Minimised Immune Response?	Study Summary	Author, Year, Reference
Osteochondral	Silk and BMP-2 + silk and silk/silk (bilayer) and BMP-2 + bilayer and BMP-2 + bilayer/TGF-β1 + SilMA	Hydrogel solution	New Zealand white rabbits	0, 3, and 8 weeks	✓	The silk and BMP-2 + silk integrated poorly; however, the latter of these did promote large volumes of new bone tissue. Comparatively, the bilayer scaffold alone showed very little new tissue formation. The BMP-2 + bilayer integrated and promoted new tissue growth well, superseded only by the silk + SilMA composite scaffold.	Wu et al. (2021) [133]
Osteochondral	GelMA + PEO + HA/GelMA + PEO + HA/GelMA + PEO	Hydrogel solution	New Zealand white rabbits	12 weeks	✓	It was stated and illustrated that the tri-layer scaffold demonstrated good capacity for regenerating cartilage, subchondral bone, and trabecular bone. These properties were not quantified or examined further, however.	Li et al. (2022) [141]
Skin	PCL+ Col and PCL + Col + ZnO and PCL + Col + ZnO + VEGF	Electrospinning	Sprague Dawley rats	6 and 12 days	✓	Gross imaging and linked diagrams indicated that the wound healing rate was enhanced by the use of PCL + Col, and further enhanced by the inclusion of ZnO, before reaching its highest rate with the inclusion of Zno + VEGF. A similar trend was noted during Col and TGF-β1 expression analysis.	Li et al. (2021) [126]
Vascular	PCL + PGS + DTβ4	Electrospinning and 3D printing	New Zealand white rabbits	2 and 12 weeks	×	Patency rate for scaffold was maintained at 80% across test animals, showing no signs of dilation or thrombosis. However, the grafts degraded before native tissue could remodel around the grafts. The slight generation of cross-linked elastin was noted, as well as rapid endothelialisation.	Xiao et al. (2022) [165]

The assessment of tissue-engineered designs in vivo can also offer insight into cell behaviour when uncommon scaffold morphologies are implanted within the animal. Feng et al. examined the effects of implanting a conch-like scaffold, featuring a helical inner structure, into the upper femoral region of New Zealand rabbits, with the aim of assessing such a design to promote guided bone regeneration [190]. Formed via 3D printing with β-tricalcium phosphate, various diameters, pitches, and pore sizes of the scaffold were examined, with a maximum compressive modulus of 1.75 GPa achieved with a 9.8 mm scaffold diameter, 1 mm pore diameter, and 2.4 mm pitch. It was found that the distinctive design of the scaffold encouraged a capillary action effect when placed within a cell media solution; this phenomenon was fully demonstrated when the scaffold was assessed in vivo, as the cells rapidly proliferated up the helical section, illustrating the directional osteoinductive benefits of such a scaffold morphology.

Despite developments in this area, in vivo testing in animals, especially in the current exploratory phase of tissue engineering research, suffers from a widely variable animal-to-human translational success rate [191]. This is primarily due to differing physiology between animals and humans, which also extends to genetics and epigenetics, as well as the low reproducibility of such experiments and lack of use of prospective systematic reviews [191,192]. As a potential solution to this, a rapidly emerging technology termed organ-on-chip removes the in vivo aspect of organic tissue development by providing a biomimetic substrate upon which living tissue can develop. This tissue can progress to function as an organoid, with models such as liver- and kidney-on-chip providing valuable insights as to how such tissue may be fully vascularised and function prior to use in vivo [193,194]. A brief overview of which organ-on-chip models are currently in use is provided in Table 10.

Table 10. An overview of which organ-on-chip models are currently in use, illustrating from this subset which of these studies have considered the mechanical properties of the tissue in question.

Cell/Molecule Type	Chip Material Type/System	Fabrication Method	Study Purpose	Mechanical Properties Considered?	Author, Year, Reference
RAW264.7 macrophages and NIH-3T3 fibroblasts	Polydimethylsiloxane (PDMS)	Soft lithography	Skin wound healing	×	Li et al. (2023) [195]
Chinese hamster ovary cells	Silicon	Micro-electro-mechanical system	Single cell analysis	×	Xu et al. (2023) [196]
Caco-2 and HepG2 cells	PDMS	Soft lithography	Disease modelling (non-alcoholic fatty liver disease)	×	Yang et al. (2023) [197]
Keratinocytes	PDMS	Soft lithography	Skin reconstruction	×	Ahn et al. (2023) [198]
Hepatocytes, stellate cells, Kupffer-like macrophages, and endothelial cells	PDMS	Soft lithography	Studying effects of inflammation and cirrhosis on drug metabolism during hepatocellular carcinoma	✓	Özkan et al. (2023) [199]
Cardiomyocytes and cardiac fibroblasts	PDMS	Casting	Examining effects on cardiac tissue mechanical response following infarction	✓	Das et al. (2022) [200]
Glucose molecules	Polymethyl methacrylate	Stereolithography	Glucose sensing	×	Podunavac et al. (2023) [201]
H9c2 rat cardiac myoblasts and adult human dermal fibroblasts	Agarose	Casting	Cardiac remodelling following arteriovenous fistula	×	Waldrop et al. (2023) [202]
Human embryonic kidney 293 cells, NIH3T3 embryonic mouse fibroblasts, and human mammary MCF10A cells	PDMS	Soft lithography	Cell spheroid viscoelasticity quantification	✓	Boot et al. (2023) [203]

Table 10. *Cont.*

Cell/Molecule Type	Chip Material Type/System	Fabrication Method	Study Purpose	Mechanical Properties Considered?	Author, Year, Reference
Human umbilical vein endothelial cells	PDMS	Stereolithography	Effect of biomechanical/biochemical stimuli on angio- and vasculogenesis	×	Ferrari et al. (2023) [204]
Mouse podocytes and mouse glomerular endothelial cells	PDMS	Casting	Investigating crosstalk between glomerular endothelial cells and glomerular epithelial cells	×	Hart et al. (2023) [205]
Taste receptor cells	MEA2100-System	Unstated	Ex vivo sense of taste simulation	×	Wu et al. (2023) [206]
Bone marrow mesenchymal stem cells	PDMS	Casting	Cross-cellular interactions in osseous tissue	✓	Erbay et al. (2022) [207]
Human lung fibroblasts and human umbilical vein endothelial cells	PDMS	Casting	Controlling angiogenesis in lung cancer spheroids	×	Kim et al. (2022) [208]
Bovine aortic endothelial cells and human pulmonary artery endothelial cells	PDMS	Spin coating and dual-layer lithography	Cell shear stress analysis and imaging	✓	Sinclair et al. (2023) [209]

Organ-on-chip technology bears significant promise, as it may provide a more agile and less ethically challenging means of assessing tissue-engineered constructs. However, one can observe from Table 10 that consideration of the mechanical properties of the engineered tissue examined in these designs is often not assessed in these designs.

7. Clinical Translation, Challenges, and Future Outlook

The use of synthetic designs to supplant existing damaged native tissue is a long-established practice, with the earliest total hip arthroplasty, made with ivory, being recorded in Germany in 1891 [210]. Scientific advancement since then has led to several crucial developments in the field of tissue engineering, such as the creation of various polymers, the advents of 3D printing and electrospinning, and a broader understanding of the biomechanical underpinnings of the human body. However, as illustrated in this article thus far, tissue engineering is still a developing field. Further progression in key areas such as the mechanical characteristics, biocompatibility, and fabrication methods involved in proposed designs are required before the use of tissue-engineered constructs in clinical settings is likely to become commonplace.

A further hurdle is that tissue-engineered designs are subject to various regulatory frameworks depending on the jurisdiction in which the design is to be marketed and sold within. While several jurisdictions in the European Union, United States of America, Canada, Australia, Japan, and South Korea have various classifications for tissue-based products, many others do not [211]; difficulties in introducing or adapting legislation for such complex products will likely involve protracted legal processes, potentially limiting the use of tissue-engineered designs within these regions for a significant period.

Regardless, progress in this field is clear, with several novel tissue-engineered implants having recently been successfully trialled. Table 11 provides an overview of these trials, including the material type, fabrication method, intended tissue type or region, the procedure method and study length, total study size, and a brief summary of the results of each trial.

Table 11. Summary of clinical trials with results published within the last two years, including material type, fabrication method and intended tissue type, procedure undergone by patient(s), length of subsequent study, post-operative outcome of patient(s), and a brief summary of each study.

Tissue Type (Region)	Material Type	Fabrication Method	Procedure	Study Length	Post-Operative Positive Outcome (Patient Number/Total Cases)	Study Summary	Author, Year, Reference
Bone (hand)	PLA	3D printing	Splint fitting	4–6 weeks	10/10	Compared to the control (thermoplastic) splint, patients reported a similar level of comfort, with the cost-effectiveness of the design potentially outperforming current designs. Two patients reported splint breakage after heavy use.	Waldburger et al. (2021) [212]
Bone (jaw)	PCL + β-TCP	3D printing	Maxillary reconstruction	6 months	8/8	While one patient suffered post-operative wound dehiscence (which was subsequently covered with a local flap), all cases showed signs of bone regeneration and scaffold integration. However, the group concluded that definitive parameters, such as implant efficacy and degradation, could not be accurately assessed.	Jeong et al. (2022) [31]
Bone (leg)	PCL + TCP	3D printing	Scaffold implantation	9–23 months	4/4	The scaffold, designed as a mesh which wraps around the bone defect site, was tested in four cases. In all cases, load-bearing functionality of the affected bone area was eventually restored, with bone development into the scaffold noted in three of the four cases.	Laubach et al. (2022) [213]
Bone (skull)	Surgical guide resin	3D printing	Cranioplasty	3 days	0/1	Patient mobility and cognition improved during the period of study. Patient symptoms began to redevelop, but not to the same extent as prior to the procedure, before passing away due to complications as a result of their initial condition.	Mee et al. (2021) [214]
Bone (skull)	PMMA	3D printing	Cranioplasty	10 days	3/3	The first case initially showed signs of fluid collection between the implant and dura; however, this was resolved shortly thereafter. The other two cases progressed well as expected.	Dabadi et al. (2021) [215]
Bone (sternum)	Aluminium oxide	Unspecified—company trademarked	Sternal replacement	28 months	1/1	The patient did not present any physiological complaints at the 28 month follow-up appointment, and was able to take part in physical activities without displacement or irritation of the prosthesis. This particular construct has a history of prior use with paediatric patients.	Mainard et al. (2022) [216]
Cardiac	UPy	Supramolecular chemistry	RVOT reconstruction	12 months	12/18	Overall, the conduit performed well across patients; however, several outlier complicated cases developed. These were either a result of immune response and thrombi formation, failure of the valve leaflets, or poor integration times. These were resolved using alternative measures.	Morales et al. (2021) [217]
Dental	Undefined resin—company trademarked	3D printing	Aligner fitting	76 weeks	120/120	Patients presented dental problems such as crowding, over-, under-, and cross-bite. All 120 cases were corrected using 3D-printed clear aligners and resulted in well-aligned teeth.	Yu et al. (2022) [218]

Table 11. *Cont.*

Tissue Type (Region)	Material Type	Fabrication Method	Procedure	Study Length	Post-Operative Positive Outcome (Patient Number/Total Cases)	Study Summary	Author, Year, Reference
Dermal (face)	Unspecified thermoplastic	3D printing	Volumetric modulated arc therapy	12 weeks	23/35	Twelve of the patients treated with the 3D-printed bolus suffered from grade ≥ 2 radiation dermatitis, while seventeen did with the conventional bolus. Additionally, none of the patients suffered from radiation pneumonitis, which is significant compared to the three which did using the conventional bolus.	Zhang et al. (2021) [219]
Oral tissue	Acrylic	CAD/Casting	Impression of cleft palate tray	Procedure length (~16 min)	Not stated	In comparison to the standard 'finger technique' of the impression of cleft palate trays, the 3D-modelled acrylic tray recorded greater detail, showed zero distortions, took less time to obtain the impression, and induced a lower heart rate in the infants who were tested.	Kalaskar et al. (2021) [220]
Oral tissue/Limb	PLA, PLU-branded elastomer, TPU	3D printing	Splint fitting	-	1/18	Three sets of constructs were tested: oral, upper-extremity, and lower-extremity splints. The oral splints could not be tested fully as patients resisted their use. Only one patient tolerated the upper-extremity splint, which did remove the risk of contracture during its use. The lower-extremity splints caused discomfort due to tightness and incongruity. The group recommend the usage of dynamic splints as a remedial measure.	Şenayli et al. (2021) [221]
Renal	PLA	3D printing	Renal autotransplantation	12 months	1/1	This construct took the form of a 3D-printed cold jacket, for use during kidney transplantation. One year after the operation, an ultrasound scan indicated normal kidney size and shape, and that ordinary blood flow had resumed. However, random comparison to existing methods and/or extending this procedure to other patients were not considered.	Cui et al. (2022) [222]
Soft tissue	dECM	Cell sheet	Flap reconstruction	6 months	7/9	Integration of the latter cases was complicated by wound dehiscence. This was later resolved via secondary healing. Additionally, five patients received incisional negative pressure wound therapy, but the impact of which could not be fully assessed.	Desvigne et al. (2021) [223]

PMMA: Poly(methyl methacrylate), UPy; Supramolecular 2-ureido-4[1H]-pyrimidone, PLU: Polyurethane, TPU: thermoplastic polyurethane, TCP: Tricalcium phosphate, β-TCP: β-Tricalcium phosphate.

A clear trend from Table 11 is that, by and large, 3D printing techniques have been the dominant fabrication method for the fabrication of clinically translated implants thus far. The logic for this is relatively self-evident: human soft tissue and cardiovascular and musculoskeletal systems are unique as a result of age, health, and genetics; injuries or pathologies pertaining to these systems increase this distinctiveness further, often requiring an equally bespoke treatment method that a technique such as additive manufacturing is readily able to provide. Additionally, the short lead times associated with 3D printing as a result of rapid prototyping, manufacturing, and delivery times are a boon to often time-sensitive clinical circumstances [224].

Whilst the continuous utilisation of 3D printing technology as the basis for the clinical translation of tissue-engineered designs remains to be seen, alternative fabrication methods such as electrospun grafts, hydrogel systems, and bioprinting continue to be developed in parallel, bringing tissue-engineered constructs ever closer to clinical and commercial viability.

8. Conclusions

The limitations of contemporary surgical and therapeutic techniques to restore the original functionality of damaged organic tissue have driven the development of various tissue-engineered platforms, which aim to support and encourage the regeneration of healthy autologous tissue. One of the key challenges in this area is establishing a high degree of similarity in the mechanical properties between the engineered scaffold and the surrounding biological tissue. This is due to the typical inability of manufactured materials to replicate the complex biomechanics of organic tissue. This capability, however, is crucial for a successful tissue-engineered design.

In response to this issue, current tissue engineering research is examining a broad spectrum of potential designs aimed at enhancing the mechanical and biological compatibility of tissue-engineered constructs. From the studies described in this work, it is clear that relatively well-documented fabrication methods such as electrospinning and 3D printing can produce designs with, in some cases, a high degree of mechanical equivalence to native tissues. Techniques such as bioprinting and MEW, while promising, will likely require further validation before widespread use is achieved. In terms of materials, the usage of particular polymers such as PCL and PLA, ceramics such as HA and β-TCP, and hydrogels consisting of gelatine and alginate have seen widespread use, while the effects of more novel materials such as CNTs and quantum dots are yet to be fully examined in a tissue engineering context.

A notable observation from these studies is that accounting for the complex range of mechanical properties present in biological tissues is generally beyond the remit of a single-material-based design. Rather, a multifaceted approach consisting of multiple material types, layers, or surface treatment methods working in tandem is likely to yield the most successful designs. In terms of translation to clinical use, the time-sensitive nature of surgical treatment suggests that 3D printing, as a fabrication method, is a strong contender for the successful realisation of these designs. Regardless of approach, it is clear that with further development, tissue engineering is set to act as a transformative treatment method.

Supplementary Materials: The following supporting information can be downloaded at https://www.mdpi.com/article/10.3390/biomimetics8020205/s1: Table S1: A synopsis of each study described within this article, including the authors and publication year, material and tissue type, and testing and mechanical property information, as well as a brief summary, where appropriate, of each research paper reported thus far [29,64,113,115–129,133–148,152,156,157,159–166,170,190].

Author Contributions: Conceptualization, A.J. and A.C.; methodology, A.J. and A.C.; software, A.J.; validation, A.J. and A.C.; formal analysis, A.J.; investigation, A.J.; resources, A.J.; data curation, A.J.; writing—original draft preparation, A.J.; writing—review and editing, A.J. and A.C.; visualization, A.J.; supervision, A.C.; project administration, A.C.; funding acquisition, A.C. All authors have read and agreed to the published version of the manuscript.

Funding: This research was funded by Engineering and Physical Sciences Research Council (EPSRC), grant number EP/W524384/1.

Data Availability Statement: The authors confirm that the data supporting the findings of this study are available within the article and its Supplementary Materials.

Conflicts of Interest: The authors declare no conflict of interest.

References

1. Sarasa-Renedo, A.; Chiquet, M. Mechanical signals regulating extracellular matrix gene expression in fibroblasts. *Scand. J. Med. Sci. Sports* **2005**, *15*, 223–230. [CrossRef] [PubMed]
2. Papanicolaou, M.; Cox, T.R. Extracellular Matrix (ECM). In *Encyclopedia of Molecular Pharmacology*; Springer: Cham, Switzerland, 2022; pp. 643–650.
3. Yue, B. Biology of the extracellular matrix: An overview. *J. Glaucoma* **2014**, *23* (Suppl. 1), S20–S23. [CrossRef] [PubMed]
4. Chow, M.-J.; Turcotte, R.; Lin, C.P.; Zhang, Y. Arterial Extracellular Matrix: A Mechanobiological Study of the Contributions and Interactions of Elastin and Collagen. *Biophys. J.* **2014**, *106*, 2684–2692. [CrossRef] [PubMed]
5. Baumann, L.; Bernstein, E.F.; Weiss, A.S.; Bates, D.; Humphrey, S.; Silberberg, M.; Daniels, R. Clinical Relevance of Elastin in the Structure and Function of Skin. *Aesthetic Surg. J. Open Forum* **2021**, *3*, ojab019. [CrossRef]
6. Sinnott, M.D.; Cleary, P.W.; Harrison, S.M. Peristaltic transport of a particulate suspension in the small intestine. *Appl. Math. Model.* **2017**, *44*, 143–159. [CrossRef]
7. Mecham, R. Elastin in lung development and disease pathogenesis. *Matrix Biol.* **2018**, *73*, 6–20. [CrossRef]
8. Faury, G.; Ristori, M.; Verdetti, J.; Jacob, M.; Robert, L. Effect of Elastin Peptides on Vascular Tone. *J. Vasc. Res.* **1995**, *32*, 112–119. [CrossRef]
9. Godinho, M.S.C.; Thorpe, C.T.; Greenwald, S.E.; Screen, H.R.C. Elastin is Localised to the Interfascicular Matrix of Energy Storing Tendons and Becomes Increasingly Disorganised with Ageing. *Sci. Rep.* **2017**, *7*, 9713. [CrossRef]
10. De Brouwer, B.; Drent, M.; van den Ouweland, J.M.; Wijnen, P.A.; van Moorsel, C.H.; Bekers, O.; Grutters, J.C.; White, E.S.; Janssen, R. Increased circulating desmosine and age-dependent elastinolysis in idiopathic pulmonary fibrosis. *Respir. Res.* **2018**, *19*, 45. [CrossRef]
11. Merla, G.; Brunetti-Pierri, N.; Piccolo, P.; Micale, L.; Loviglio, M.N. Supravalvular aortic stenosis: Elastin arteriopathy. *Circ. Cardiovasc. Genet.* **2012**, *5*, 692–696. [CrossRef]
12. Heinz, A. Elastic fibers during aging and disease. *Ageing Res. Rev.* **2021**, *66*, 101255. [CrossRef] [PubMed]
13. Rouchi, A.H.; Mahdavi-Mazdeh, M. Regenerative Medicine in Organ and Tissue Transplantation: Shortly and Practically Achievable? *Int. J. Organ Transplant. Med.* **2015**, *6*, 93–98.
14. Chiarello, E.; Cadossi, M.; Tedesco, G.; Capra, P.; Calamelli, C.; Shehu, A.; Giannini, S. Autograft, allograft and bone substitutes in reconstructive orthopedic surgery. *Aging Clin. Exp. Res.* **2013**, *25*, 101–103. [CrossRef] [PubMed]
15. Tonk, G.; Yadav, P.K.; Agarwal, S.; Jamoh, K. Donor site morbidity in autologous bone grafting—A comparison between different techniques of anterior iliac crest bone harvesting: A prospective study. *J. Orthop. Trauma Rehabil.* **2022**, *29*, 22104917221092163. [CrossRef]
16. Batten, P.; Rosenthal, N.; Yacoub, M.H. Immune response to stem cells and strategies to induce tolerance. *Philos. Trans. R. Soc. B: Biol. Sci.* **2007**, *362*, 1343–1356. [CrossRef] [PubMed]
17. Zakrzewski, W.; Dobrzyński, M.; Szymonowicz, M.; Rybak, Z. Stem cells: Past, present, and future. *Stem Cell Res. Ther.* **2019**, *10*, 68. [CrossRef] [PubMed]
18. Heim, M.; Nixon, I.J.; Emmerson, E.; Callanan, A. From hormone replacement therapy to regenerative scaffolds: A review of current and novel primary hypothyroidism therapeutics. *Front. Endocrinol.* **2022**, *13*, 2392. [CrossRef]
19. Handley, E.; Callanan, A. Modulation of Tissue Microenvironment following Myocardial Infarction. *Adv. NanoBiomed Res.* **2022**, *2*, 2200005. [CrossRef]
20. Sturtivant, A.; Callanan, A. The use of antifreeze proteins to modify pore structure in directionally frozen alginate sponges for cartilage tissue engineering. *Biomed. Phys. Eng. Express* **2020**, *6*, 055016. [CrossRef]
21. Sofokleous, P.; Chin, M.H.; Day, R. Phase-separation technologies for 3D scaffold engineering. In *Functional 3D Tissue Engineering Scaffolds—Materials, Technologies and Applications*; Elsevier: Amsterdam, The Netherlands, 2018; pp. 101–126.
22. Xie, Z.; Gao, M.; Lobo, A.O.; Webster, T.J. 3D Bioprinting in Tissue Engineering for Medical Applications: The Classic and the Hybrid. *Polymers* **2020**, *12*, 1717. [CrossRef]
23. Costantini, M.; Barbetta, A. Gas foaming technologies for 3D scaffold engineering. In *Functional 3D Tissue Engineering Scaffolds*; Woodhead Publishing: Sawston, UK, 2018; pp. 127–149.
24. Chan, B.P.; Leong, K.W. Scaffolding in tissue engineering: General approaches and tissue-specific considerations. *Eur. Spine J.* **2008**, *17* (Suppl. 4), 467–479. [CrossRef] [PubMed]
25. Prakoso, A.T.; Basri, H.; Adanta, D.; Yani, I.; Ammarullah, M.I.; Akbar, I.; Ghazali, F.A.; Syahrom, A.; Kamarul, T. The Effect of Tortuosity on Permeability of Porous Scaffold. *Biomedicines* **2023**, *11*, 427. [CrossRef] [PubMed]

26. Putra, R.U.; Basri, H.; Prakoso, A.T.; Chandra, H.; Ammarullah, M.I.; Akbar, I.; Syahrom, A.; Kamarul, T. Level of Activity Changes Increases the Fatigue Life of the Porous Magnesium Scaffold, as Observed in Dynamic Immersion Tests, over Time. *Sustainability* **2023**, *15*, 823. [CrossRef]
27. Hosseinkhani, M.; Mehrabani, D.; Karimfar, M.H.; Bakhtiyari, S.; Manafi, A.; Shirazi, R. Tissue Engineered Scaffolds in Regenerative Medicine. *World J. Plast. Surg.* **2014**, *3*, 3–7. [PubMed]
28. Qiu, Y.; Chen, X.; Hou, Y.; Hou, Y.; Tian, S.; Chen, Y.; Yu, L.; Nie, M.; Liu, X. Characterization of different biodegradable scaffolds in tissue engineering. *Mol. Med. Rep.* **2019**, *19*, 4043–4056. [CrossRef] [PubMed]
29. Rezania, N.; Asadi-Eydivand, M.; Abolfathi, N.; Bonakdar, S.; Mehrjoo, M.; Solati-Hashjin, M. Three-dimensional printing of polycaprolactone/hydroxyapatite bone tissue engineering scaffolds mechanical properties and biological behavior. *J. Mater. Sci. Mater. Med.* **2022**, *33*, 31. [CrossRef]
30. De Pieri, A.; Rochev, Y.; Zeugolis, D.I. Scaffold-free cell-based tissue engineering therapies: Advances, shortfalls and forecast. *Npj Regen. Med.* **2021**, *6*, 18. [CrossRef]
31. Jeong, W.-S.; Kim, Y.-C.; Min, J.-C.; Park, H.-J.; Lee, E.-J.; Shim, J.-H.; Choi, J.-W. Clinical Application of 3D-Printed Patient-Specific Polycaprolactone/Beta Tricalcium Phosphate Scaffold for Complex Zygomatico-Maxillary Defects. *Polymers* **2022**, *14*, 740. [CrossRef]
32. Borschel, G.H.; Kia, K.F.; Kuzon, W.M.; Dennis, R.G. Mechanical properties of acellular peripheral nerve. *J. Surg. Res.* **2003**, *114*, 133–139. [CrossRef]
33. Tomlins, P. *1—Material Types for Tissue Scaffolds, in Characterisation and Design of Tissue Scaffolds*; Tomlins, P., Ed.; Woodhead Publishing: Sawston, UK, 2016; pp. 1–21.
34. Perez-Puyana, V.; Jiménez-Rosado, M.; Romero, A.; Guerrero, A. Polymer-Based Scaffolds for Soft-Tissue Engineering. *Polymers* **2020**, *12*, 1566. [CrossRef]
35. Baino, F.; Novajra, G.; Vitale-Brovarone, C. Bioceramics and Scaffolds: A Winning Combination for Tissue Engineering. *Front. Bioeng. Biotechnol.* **2015**, *3*, 202. [CrossRef] [PubMed]
36. Zhang, X.; Chen, X.; Hong, H.; Hu, R.; Liu, J.; Liu, C. Decellularized extracellular matrix scaffolds: Recent trends and emerging strategies in tissue engineering. *Bioact. Mater.* **2021**, *10*, 15–31. [CrossRef] [PubMed]
37. Radulescu, D.M.; Neacsu, I.A.; Grumezescu, A.-M.; Andronescu, E. New Insights of Scaffolds Based on Hydrogels in Tissue Engineering. *Polymers* **2022**, *14*, 799. [CrossRef]
38. Koushik, T.M.; Miller, C.M.; Antunes, E. Bone Tissue Engineering Scaffolds: Function of Multi-Material Hierarchically Structured Scaffolds. *Adv. Healthc. Mater.* **2023**, *12*, 2202766. [CrossRef]
39. Salati, M.A.; Khazai, J.; Tahmuri, A.M.; Samadi, A.; Taghizadeh, A.; Taghizadeh, M.; Zarrintaj, P.; Ramsey, J.D.; Habibzadeh, S.; Seidi, F.; et al. Agarose-Based Biomaterials: Opportunities and Challenges in Cartilage Tissue Engineering. *Polymers* **2020**, *12*, 1150. [CrossRef] [PubMed]
40. Raus, R.A.; Nawawi, W.M.F.W.; Nasaruddin, R.R. Alginate and alginate composites for biomedical applications. *Asian J. Pharm. Sci.* **2020**, *16*, 280–306. [CrossRef] [PubMed]
41. Tian, Y.; Wu, D.; Wu, D.; Cui, Y.; Ren, G.; Wang, Y.; Wang, J.; Peng, C. Chitosan-Based Biomaterial Scaffolds for the Repair of Infected Bone Defects. *Front. Bioeng. Biotechnol.* **2022**, *10*, 755. [CrossRef]
42. Gao, Y.; Callanan, A. Influence of surface topography on PCL electrospun scaffolds for liver tissue engineering. *J. Mater. Chem. B* **2021**, *9*, 8081–8093. [CrossRef]
43. Dai, Y.; Lu, T.; Shao, M.; Lyu, F. Recent advances in PLLA-based biomaterial scaffolds for neural tissue engineering: Fabrication, modification, and applications. *Front. Bioeng. Biotechnol.* **2022**, *10*, 1011783. [CrossRef]
44. Kotturi, H.; Abuabed, A.; Zafar, H.; Sawyer, E.; Pallipparambil, B.; Jamadagni, H.; Khandaker, M. Evaluation of Polyethylene Glycol Diacrylate-Polycaprolactone Scaffolds for Tissue Engineering Applications. *J. Funct. Biomater.* **2017**, *8*, 39. [CrossRef]
45. Ammarullah, M.I.; Hartono, R.; Supriyono, T.; Santoso, G.; Sugiharto, S.; Permana, M.S. Polycrystalline Diamond as a Potential Material for the Hard-on-Hard Bearing of Total Hip Prosthesis: Von Mises Stress Analysis. *Biomedicines* **2023**, *11*, 951. [CrossRef] [PubMed]
46. Mazzoni, E.; Iaquinta, M.R.; Lanzillotti, C.; Mazziotta, C.; Maritati, M.; Montesi, M.; Sprio, S.; Tampieri, A.; Tognon, M.; Martini, F. Bioactive Materials for Soft Tissue Repair. *Front. Bioeng. Biotechnol.* **2021**, *9*, 61378. [CrossRef] [PubMed]
47. Bernardo, M.P.; da Silva, B.C.R.; Hamouda, A.E.I.; de Toledo, M.A.S.; Schalla, C.; Rütten, S.; Goetzke, R.; Mattoso, L.H.C.; Zenke, M.; Sechi, A. PLA/Hydroxyapatite scaffolds exhibit in vitro immunological inertness and promote robust osteogenic differentiation of human mesenchymal stem cells without osteogenic stimuli. *Sci. Rep.* **2022**, *12*, 2333. [CrossRef] [PubMed]
48. Amirazad, H.; Dadashpour, M.; Zarghami, N. Application of decellularized bone matrix as a bioscaffold in bone tissue engineering. *J. Biol. Eng.* **2022**, *16*, 1. [CrossRef] [PubMed]
49. Sarrigiannidis, S.O.; Rey, J.; Dobre, O.; González-García, C.; Dalby, M.; Salmeron-Sanchez, M. A tough act to follow: Collagen hydrogel modifications to improve mechanical and growth factor loading capabilities. *Mater. Today Bio* **2021**, *10*, 100098. [CrossRef]
50. Andreazza, R.; Morales, A.; Pieniz, S.; Labidi, J. Gelatin-Based Hydrogels: Potential Biomaterials for Remediation. *Polymers* **2023**, *15*, 1026. [CrossRef]
51. Chircov, C.; Grumezescu, A.M.; Bejenaru, L.E. Hyaluronic acid-based scaffolds for tissue engineering. *Rom. J. Morphol. Embryol.* **2018**, *59*, 71–76.

52. Gao, H.; Yang, J.; Jin, X.; Qu, X.; Zhang, F.; Zhang, D.; Chen, H.; Wei, H.; Zhang, S.; Jia, W.; et al. Porous tantalum scaffolds: Fabrication, structure, properties, and orthopedic applications. *Mater. Des.* **2021**, *210*, 110095. [CrossRef]
53. Hanawa, T. Titanium–Tissue Interface Reaction and Its Control With Surface Treatment. *Front. Bioeng. Biotechnol.* **2019**, *7*, 170. [CrossRef]
54. Sprague, A.L.; Awokuse, D.; Pohlig, R.T.; Cortes, D.H.; Silbernagel, K.G. Relationship between mechanical properties (shear modulus and viscosity), age, and sex in uninjured Achilles tendons. *Transl. Sports Med.* **2020**, *3*, 321–327. [CrossRef]
55. LaCroix, A.S.; Duenwald-Kuehl, S.E.; Lakes, R.S.; Vanderby, R. Relationship between tendon stiffness and failure: A metaanalysis. *J. Appl. Physiol.* **2013**, *115*, 43–51. [CrossRef] [PubMed]
56. Maganaris, C.N.; Paul, J.P. In vivo human tendon mechanical properties. *J. Physiol.* **1999**, *521 Pt 1*, 307–313. [CrossRef]
57. Urban, M.W.; Rule, A.D.; Atwell, T.D.; Chen, S. Novel Uses of Ultrasound to Assess Kidney Mechanical Properties. *Kidney360* **2021**, *2*, 1531–1539. [CrossRef] [PubMed]
58. Karimi, A.; Shojaei, A. Measurement of the Mechanical Properties of the Human Kidney. *IRBM* **2017**, *38*, 292–297. [CrossRef]
59. Palmeri, M.L.; Wang, M.; Dahl, J.; Frinkley, K.; Nightingale, K. Quantifying Hepatic Shear Modulus In Vivo Using Acoustic Radiation Force. *Ultrasound Med. Biol.* **2008**, *34*, 546–558. [CrossRef] [PubMed]
60. Kemper, A.R.; Santago, A.C.; Stitzel, J.D.; Sparks, J.L.; Duma, S.M. Biomechanical response of human liver in tensile loading. *Ann. Adv. Automot. Med.* **2010**, *54*, 15–26.
61. Yeh, W.-C.; Li, P.-C.; Jeng, Y.-M.; Hsu, H.-C.; Kuo, P.-L.; Li, M.-L.; Yang, P.-M.; Lee, P.H. Elastic modulus measurements of human liver and correlation with pathology. *Ultrasound Med. Biol.* **2002**, *28*, 467–474. [CrossRef]
62. Arikawa, H. Dynamic Shear Modulus in Torsion of Human Dentin and Enamel. *Dent. Mater. J.* **1989**, *8*, 223–235,287. [CrossRef]
63. Chun, K.; Choi, H.; Lee, J. Comparison of mechanical property and role between enamel and dentin in the human teeth. *J. Dent. Biomech.* **2014**, *5*, 1758736014520809. [CrossRef]
64. Aghajanian, A.H.; Bigham, A.; Sanati, A.; Kefayat, A.; Salamat, M.R.; Sattary, M.; Rafienia, M. A 3D macroporous and magnetic Mg_2SiO_4-$CuFe_2O_4$ scaffold for bone tissue regeneration: Surface modification, in vitro and in vivo studies. *Biomater. Adv.* **2022**, *137*, 212809.
65. Park, S.; Tao, J.; Sun, L.; Fan, C.-M.; Chen, Y. An Economic, Modular, and Portable Skin Viscoelasticity Measurement Device for In Situ Longitudinal Studies. *Molecules* **2019**, *24*, 907. [CrossRef] [PubMed]
66. Gauthier, R.; Attik, N.; Chevalier, C.; Salles, V.; Grosgogeat, B.; Gritsch, K.; Trunfio-Sfarghiu, A.-M. 3D Electrospun Polycaprolactone Scaffolds to Assess Human Periodontal Ligament Cells Mechanobiological Behaviour. *Biomimetics* **2023**, *8*, 108. [CrossRef] [PubMed]
67. Mirzaali, M.J.; Schwiedrzik, J.J.; Thaiwichai, S.; Best, J.P.; Michler, J.; Zysset, P.K.; Wolfram, U. Mechanical properties of cortical bone and their relationships with age, gender, composition and microindentation properties in the elderly. *Bone* **2016**, *93*, 196–211. [CrossRef] [PubMed]
68. Post, A.; Diaz-Rodriguez, P.; Balouch, B.; Paulsen, S.; Wu, S.; Miller, J.; Hahn, M.; Cosgriff-Hernandez, E. Elucidating the role of graft compliance mismatch on intimal hyperplasia using an ex vivo organ culture model. *Acta Biomater.* **2019**, *89*, 84–94. [CrossRef]
69. Cao, T.; Jiang, Z.; Zhao, H.; Zhang, K.-Q.; Meng, K. Numerical simulation to study the impact of compliance mismatch between artificial and host blood vessel on hemodynamics. *Med. Nov. Technol. Devices* **2022**, *15*, 100152. [CrossRef]
70. Teichtahl, A.J.; Wluka, A.E.; Wijethilake, P.; Wang, Y.; Ghasem-Zadeh, A.; Cicuttini, F.M. Wolff's law in action: A mechanism for early knee osteoarthritis. *Thromb. Haemost.* **2015**, *17*, 207. [CrossRef]
71. Ambrosi, D.; Ben Amar, M.; Cyron, C.J.; DeSimone, A.; Goriely, A.; Humphrey, J.D.; Kuhl, E. Growth and remodelling of living tissues: Perspectives, challenges and opportunities. *J. R. Soc. Interface* **2019**, *16*, 2019023. [CrossRef]
72. Frost, H.M. A 2003 Update of Bone Physiology and Wolff's Law for Clinicians. *Angle Orthod.* **2004**, *74*, 3–15.
73. Raffa, M.L.; Nguyen, V.H.; Hernigou, P.; Flouzat-Lachaniette, C.H.; Haiat, G. Stress shielding at the bone-implant interface: Influence of surface roughness and of the bone-implant contact ratio. *J. Orthop. Res.* **2021**, *39*, 1174–1183. [CrossRef]
74. Munir, N.; Callanan, A. Novel phase separated polycaprolactone/collagen scaffolds for cartilage tissue engineering. *Biomed. Mater.* **2018**, *13*, 051001. [CrossRef]
75. Garrison, C.M.; Singh-Varma, A.; Pastino, A.K.; Steele, J.A.; Kohn, J.; Murthy, N.S.; Schwarzbauer, J.E. A multilayered scaffold for regeneration of smooth muscle and connective tissue layers. *J. Biomed. Mater. Res. Part A* **2021**, *109*, 733–744. [CrossRef] [PubMed]
76. Biazar, E.; Kamalvand, M.; Avani, F. Recent advances in surface modification of biopolymeric nanofibrous scaffolds. *Int. J. Polym. Mater. Polym. Biomater.* **2021**, *71*, 493–512. [CrossRef]
77. Bailey, S.; Vashishth, D. Mechanical Characterization of Bone: State of the Art in Experimental Approaches-What Types of Experiments Do People Do and How Does One Interpret the Results? *Curr. Osteoporos. Rep.* **2018**, *16*, 423–433. [CrossRef] [PubMed]
78. Gu, M.; Fan, S.; Zhou, G.; Ma, K.; Yao, X.; Zhang, Y. Effects of dynamic mechanical stimulations on the regeneration of in vitro and in vivo cartilage tissue based on silk fibroin scaffold. *Compos. Part B Eng.* **2022**, *235*, 109764. [CrossRef]
79. Yang, Y.; Li, K.; Sommer, G.; Yung, K.-L.; A Holzapfel, G. Mechanical characterization of porcine liver properties for computational simulation of indentation on cancerous tissue. *Math. Med. Biol. J. IMA* **2020**, *37*, 469–490. [CrossRef]
80. Wei, W.; Dai, H. Articular cartilage and osteochondral tissue engineering techniques: Recent advances and challenges. *Bioact. Mater.* **2021**, *6*, 4830–4855. [CrossRef]

81. Rayudu, N.M.; Subburaj, K.; Mei, K.; Dieckmeyer, M.; Kirschke, J.S.; Noël, P.B.; Baum, T. Finite Element Analysis-Based Vertebral Bone Strength Prediction Using MDCT Data: How Low Can We Go? *Front. Endocrinol.* **2020**, *11*, 442. [CrossRef]
82. Kurtaliaj, I.; Golman, M.; Abraham, A.C.; Thomopoulos, S. Biomechanical Testing of Murine Tendons. *J. Vis. Exp.* **2019**, *152*, e60280.
83. Gough, A.; Stern, A.M.; Maier, J.; Lezon, T.; Shun, T.-Y.; Chennubhotla, C.; Schurdak, M.E.; Haney, S.A.; Taylor, D.L. Biologically Relevant Heterogeneity: Metrics and Practical Insights. *SLAS Discov. Adv. Sci. Drug Discov.* **2017**, *22*, 213–237. [CrossRef]
84. Zhang, K.; Zhu, M.; Thomas, E.; Hopyan, S.; Sun, Y. Existing and Potential Applications of Elastography for Measuring the Viscoelasticity of Biological Tissues In Vivo. *Front. Phys.* **2021**, *9*, 670571. [CrossRef]
85. Mitchell, G.R.; Tojeira, A. Role of Anisotropy in Tissue Engineering. *Procedia Eng.* **2013**, *59*, 117–125. [CrossRef]
86. Karimi, A.; Navidbakhsh, M.; Shojaei, A.; Faghihi, S. Measurement of the uniaxial mechanical properties of healthy and atherosclerotic human coronary arteries. *Mater. Sci. Eng. C* **2013**, *33*, 2550–2554. [CrossRef] [PubMed]
87. Claes, E.; Atienza, J.; Guinea, G.; Rojo, F.; Bernal, J.; Revuelta, J.; Elices, M. Mechanical properties of human coronary arteries. 2010, 2010, 3792–3795. In Proceedings of the 2010 Annual International Conference of the IEEE Engineering in Medicine and Biology, Buenos Aires, Argentina, 31 August–4 September 2010; pp. 3792–3795.
88. A Vorp, D.; Schiro, B.J.; Ehrlich, M.P.; Juvonen, T.S.; Ergin, M.; Griffith, B.P. Effect of aneurysm on the tensile strength and biomechanical behavior of the ascending thoracic aorta. *Ann. Thorac. Surg.* **2003**, *75*, 1210–1214. [CrossRef]
89. Gijsen, F.J.H.; Wentzel, J.J.; Thury, A.; Mastik, F.; Schaar, J.A.; Schuurbiers, J.C.H.; Slager, C.J.; Van Der Giessen, W.J.; De Feyter, P.J.; Van Der Steen, A.F.W.; et al. Strain distribution over plaques in human coronary arteries relates to shear stress. *Am. J. Physiol. Circ. Physiol.* **2008**, *295*, H1608–H1614. [CrossRef] [PubMed]
90. Lu, X.; Yang, J.; Zhao, J.; Gregersen, H.; Kassab, G.S. Shear modulus of porcine coronary artery: Contributions of media and adventitia. *Am. J. Physiol. Circ. Physiol.* **2003**, *285*, H1966–H1975. [CrossRef] [PubMed]
91. Eschweiler, J.; Horn, N.; Rath, B.; Betsch, M.; Baroncini, A.; Tingart, M.; Migliorini, F. The Biomechanics of Cartilage—An Overview. *Life* **2021**, *11*, 302. [CrossRef]
92. Chizhik, S.A.; Wierzcholski, K.; Trushko, A.; Zhytkova, M.A.; Miszczak, A. Properties of Cartilage on Micro- and Nanolevel. *Adv. Tribol.* **2010**, *2010*, 8. [CrossRef]
93. Wang, S.; Bao, Y.; Guan, Y.; Zhang, C.; Liu, H.; Yang, X.; Gao, L.; Guo, T.; Chen, Q. Strain distribution of repaired articular cartilage defects by tissue engineering under compression loading. *J. Orthop. Surg. Res.* **2018**, *13*, 19. [CrossRef]
94. Wong, B.L.; Bae, W.C.; Chun, J.; Gratz, K.R.; Lotz, M.; Robert, L.S. Biomechanics of Cartilage Articulation. *Arthritis Rheum.* **2008**, *58*, 2065–2074.
95. Morgan, E.F.; Unnikrisnan, G.U.; Hussein, A.I. Bone Mechanical Properties in Healthy and Diseased States. *Annu. Rev. Biomed. Eng.* **2018**, *20*, 119–143. [CrossRef]
96. Ma, Z.; Qiang, Z.; Zhao, H.; Piao, H.; Ren, L. Mechanical properties of cortical bones related to temperature and orientation of Haversian canals. *Mater. Res. Express* **2020**, *7*, 015408. [CrossRef]
97. Spatz, H.C.; Vincent, J.F. Young's moduli and shear moduli in cortical bone. *Proc. R. Soc. Lond. B* **1996**, *263*, 287–294.
98. Pawlaczyk, M.; Lelonkiewicz, M.; Wieczorowski, M. Age-dependent biomechanical properties of the skin. *Adv. Dermatol. Allergol.* **2013**, *30*, 302–306. [CrossRef] [PubMed]
99. Ottenio, M.; Tran, D.; Annaidh, A.N.; Gilchrist, M.D.; Bruyère, K. Strain rate and anisotropy effects on the tensile failure characteristics of human skin. *J. Mech. Behav. Biomed. Mater.* **2015**, *41*, 241–250. [CrossRef]
100. Kwan, M.K.; Wall, E.J.; Massie, J.; Garfin, S.R. Strain, stress and stretch of peripheral nerve Rabbit experiments in vitro and in vivo. *Acta Orthop.* **1992**, *63*, 267–272. [CrossRef]
101. Nicholson, K.J.; Winkelstein, B.A. Nerve and Nerve Root Biomechanics. *Neural Tissue Biomech.* **2010**, *3*, 203–229.
102. Singh, A.; Magee, R. Mechanical Properties of Cervical Spinal Cord in Neonatal Piglet: In Vitro. *Neurol. Neurobiol.* **2020**, *2*, 3. [CrossRef]
103. Durand, S.; Raffoul, W.; Christen, T.; Pedrazzi, N. Post-Operative Assessment of Ulnar Nerve Tension Using Shear-Wave Elastography. *Neurol. Int.* **2021**, *13*, 469–476. [CrossRef]
104. Nicolle, S.; Palierne, J.-F. Dehydration effect on the mechanical behaviour of biological soft tissues: Observations on kidney tissues. *J. Mech. Behav. Biomed. Mater.* **2010**, *3*, 630–635. [CrossRef]
105. Phillip, J.M.; Aifuwa, I.; Walston, J.; Wirtz, D. The Mechanobiology of Aging. *Annu. Rev. Biomed. Eng.* **2016**, *17*, 113–141. [CrossRef]
106. Martino, F.; Perestrelo, A.R.; Vinarsky, V.; Pagliari, S.; Forte, G. Cellular Mechanotransduction: From Tension to Function. *Front. Physiol.* **2018**, *9*, 824. [CrossRef] [PubMed]
107. Mierke, C.T.; Kollmannsberger, P.; Zitterbart, D.P.; Smith, J.; Fabry, B.; Goldmann, W.H. Mechano-Coupling and Regulation of Contractility by the Vinculin Tail Domain. *Biophys. J.* **2008**, *94*, 661–670. [CrossRef] [PubMed]
108. Khan, K.M.; Scott, A. Mechanotherapy: How physical therapists' prescription of exercise promotes tissue repair. *Br. J. Sports Med.* **2009**, *43*, 247–252. [CrossRef]
109. Charkoudian, N. Mechanisms and modifiers of reflex induced cutaneous vasodilation and vasoconstriction in humans. *J. Appl. Physiol.* **2010**, *109*, 1221–1228. [CrossRef] [PubMed]

110. Yamako, G.; Janssen, D.; Hanada, S.; Anijs, T.; Ochiai, K.; Totoribe, K.; Chosa, E.; Verdonschot, N. Improving stress shielding following total hip arthroplasty by using a femoral stem made of β type Ti-33.6Nb-4Sn with a Young's modulus gradation. *J. Biomech.* **2017**, *63*, 135–143. [CrossRef]
111. Be'Ery-Lipperman, M.; Gefen, A. A method of quantification of stress shielding in the proximal femur using hierarchical computational modeling. *Comput. Methods Biomech. Biomed. Eng.* **2007**, *9*, 35–44. [CrossRef]
112. Lei, B.; Guo, B.; Rambhia, K.J.; Ma, P.X. Hybrid polymer biomaterials for bone tissue regeneration. *Front. Med.* **2018**, *13*, 189–201. [CrossRef]
113. Wang, X.; Li, S.; Yu, H.; Lv, J.; Fan, M.; Wang, X.; Wang, X.; Liang, Y.; Mao, L.; Zhao, Z. The Biocompatibility of Multi-Source Stem Cells and Gelatin-Carboxymethyl Chitosan-Sodium Alginate Hybrid Biomaterials. *Tissue Eng. Regen. Med.* **2022**, *19*, 491–503. [CrossRef]
114. Nokhbatolfoghahaei, H.; Paknejad, Z.; Bohlouli, M.; Rad, M.R.; Aminishakib, P.; Derakhshan, S.; Amirabad, L.M.; Nadjmi, N.; Khojasteh, A. Fabrication of Decellularized Engineered Extracellular Matrix through Bioreactor-Based Environment for Bone Tissue Engineering. *ACS Omega* **2020**, *5*, 31943–31956. [CrossRef]
115. Nguyen, T.T.; Hu, C.C.; Sakthivel, R.; Nabilla, S.C.; Huang, Y.W.; Yu, J.; Cheng, N.C.; Kuo, Y.J.; Chung, R.J. Preparation of gamma poly-glutamic acid/hydroxyapatite/collagen composite as the 3D-printing scaffold for bone tissue engineering. *Biomater. Res.* **2022**, *26*, 21. [CrossRef]
116. Singh, B.N.; Nallakumarasamy, A.; Sinha, S.; Rastogi, A.; Mallick, S.P.; Divakar, S.; Srivastava, P. Generation of hybrid tissue engineered construct through embedding autologous chondrocyte loaded platelet rich plasma/alginate based hydrogel in porous scaffold for cartilage regeneration. *Int. J. Biol. Macromol.* **2022**, *203*, 389–405. [CrossRef] [PubMed]
117. Lee, J.U.; Kim, D.; Jang, C.H.; Kim, G.H. Highly elastic 3D-printed gelatin/HA/placental-extract scaffolds for bone tissue engineering. *Theranostics* **2022**, *12*, 4051–4066. [CrossRef] [PubMed]
118. Wang, H.; Hu, B.; Li, H.; Feng, G.; Pan, S.; Chen, Z.; Li, B.; Song, J. Biomimetic Mineralized Hydroxyapatite Nanofiber-Incorporated Methacrylated Gelatin Hydrogel with Improved Mechanical and Osteoinductive Performances for Bone Regeneration. *Int. J. Nanomed.* **2022**, *17*, 1511–1529. [CrossRef] [PubMed]
119. Tabatabaee, S.; Baheiraei, N.; Salehnia, M. Fabrication and characterization of PHEMA–gelatin scaffold enriched with graphene oxide for bone tissue engineering. *J. Orthop. Surg. Res.* **2022**, *17*, 216. [CrossRef]
120. Min, Q.; Tian, D.; Zhang, Y.; Wang, C.; Wan, Y.; Wu, J. Strong and Elastic Chitosan/Silk Fibroin Hydrogels Incorporated with Growth-Factor-Loaded Microspheres for Cartilage Tissue Engineering. *Biomimetics* **2022**, *7*, 41. [CrossRef]
121. Dorishetty, P.; Balu, R.; Gelmi, A.; Mata, J.P.; Quigley, A.; Dutta, N.K.; Choudhury, N.R. Microporosity engineered printable silk/graphene hydrogels and their cytocompatibility evaluations. *Mater. Today Adv.* **2022**, *14*, 100233. [CrossRef]
122. Chen, Y.; Chen, Y.; Xiong, X.; Cui, R.; Zhang, G.; Wang, C.; Xiao, D.; Qu, S.; Weng, J. Hybridizing gellan/alginate and thixotropic magnesium phosphate-based hydrogel scaffolds for enhanced osteochondral repair. *Mater. Today Bio* **2022**, *14*, 100261. [CrossRef]
123. Baskapan, B.; Callanan, A. Electrospinning Fabrication Methods to Incorporate Laminin in Polycaprolactone for Kidney Tissue Engineering. *Tissue Eng. Regen. Med.* **2022**, *19*, 73–82. [CrossRef]
124. Xu, T.; Zhang, X.; Dai, X. Properties of Electrospun Aligned Poly(lactic acid)/Collagen Fibers With Nanoporous Surface for Peripheral Nerve Tissue Engineering. *Macromol. Mater. Eng.* **2022**, *307*, 2200256. [CrossRef]
125. Salehi, R.; Mohammadzadeh, L.; Mahkam, M.; Jafarizad, A.; Rahbarghazi, R. Electrospun gelatin/methylcellulose hybrid nanofibers promoted the maturation of human cutaneous tissue progenitor cells toward keratinocyte-like cells. *Cellulose* **2022**, *29*, 7837–7848. [CrossRef]
126. Li, P.; Ruan, L.; Wang, R.; Liu, T.; Song, G.; Gao, X.; Jiang, G.; Liu, X. Electrospun Scaffold of Collagen and Polycaprolactone Containing ZnO Quantum Dots for Skin Wound Regeneration. *J. Bionic Eng.* **2021**, *18*, 1378–1390. [CrossRef] [PubMed]
127. Sang, S.; Cheng, R.; Cao, Y.; Yan, Y.; Shen, Z.; Zhao, Y.; Han, Y. Biocompatible chitosan/polyethylene glycol/multi-walled carbon nanotube composite scaffolds for neural tissue engineering. *J. Zhejiang Univ. B* **2022**, *23*, 58–73. [CrossRef] [PubMed]
128. Ramasamy, M.S.; Kaliannagounder, V.K.; Rahaman, A.; Park, C.H.; Kim, C.S.; Kim, B. Synergistic Effect of Reinforced Multiwalled Carbon Nanotubes and Boron Nitride Nanosheet-Based Hybrid Piezoelectric PLLA Scaffold for Efficient Bone Tissue Regeneration. *ACS Biomater. Sci. Eng.* **2022**, *8*, 3542–3556. [CrossRef] [PubMed]
129. Jiawei, S.; Xuemeng, B.; Yahui, Z.; Jianfei, C.; Yinghong, X. Sugarcane Stem Derived Hybrid Scaffold for Bone Tissue Engineering via Top-down Approach. *Compos. Interfaces* **2022**, *30*, 323–340. [CrossRef]
130. Karšaj, I.; Humphrey, J.D. A multilayered wall model of arterial growth and remodeling. *Mech. Mater.* **2013**, *44*, 110–119. [CrossRef]
131. Liu, Y.; Luo, D.; Wang, T. Hierarchical Structures of Bone and Bioinspired Bone Tissue Engineering. *Small* **2016**, *12*, 4611–4632. [CrossRef]
132. Totonelli, G.; Maghsoudlou, P.; Fishman, J.M.; Orlando, G.; Ansari, T.; Sibbons, P.; A Birchall, M.; Pierro, A.; Eaton, S.; De Coppi, P. Esophageal tissue engineering: A new approach for esophageal replacement. *World J. Gastroenterol.* **2012**, *18*, 6900–6907. [CrossRef]
133. Wu, X.; Zhou, M.; Jiang, F.; Yin, S.; Lin, S.; Yang, G.; Lu, Y.; Zhang, W.; Jiang, X. Marginal sealing around integral bilayer scaffolds for repairing osteochondral defects based on photocurable silk hydrogels. *Bioact. Mater.* **2021**, *6*, 3976–3986. [CrossRef]
134. Dargoush, S.A.; Hanaee-Ahvaz, H.; Irani, S.; Soleimani, M.; Khatami, S.M.; Sohi, A.N. A composite bilayer scaffold functionalized for osteochondral tissue regeneration in rat animal model. *J. Tissue Eng. Regen. Med.* **2022**, *16*, 559–574. [CrossRef]

135. Yang, T.; Tamaddon, M.; Jiang, L.; Wang, J.; Liu, Z.; Liu, Z.; Meng, H.; Hu, Y.; Gao, J.; Yang, X.; et al. Bilayered scaffold with 3D printed stiff subchondral bony compartment to provide constant mechanical support for long-term cartilage regeneration. *J. Orthop. Transl.* **2021**, *30*, 112–121. [CrossRef]
136. Tevlek, A.; Aydin, H.M. Multi-layered in vitro 3D-bone model via combination of osteogenic cell sheets with electrospun membrane interlayer. *J. Biomater. Appl.* **2021**, *36*, 818–833. [CrossRef] [PubMed]
137. Semitela, Â.; Leal Pereira, A.; Sousa, C.; Mendes, A.F.; Marques, P.A.; Completo, A. Multi-layered electrospinning and electrospraying approach: Effect of polymeric supplements on chondrocyte suspension. *J. Biomater. Appl.* **2021**, *36*, 1629–1640. [CrossRef] [PubMed]
138. Janarthanan, G.; Kim, J.H.; Kim, I.; Lee, C.; Chung, E.-J.; Noh, I. Manufacturing of self-standing multi-layered 3D-bioprinted alginate-hyaluronate constructs by controlling the cross-linking mechanisms for tissue engineering applications. *Biofabrication* **2022**, *14*, 035013. [CrossRef]
139. Tamaddon, M.; Blunn, G.; Tan, R.; Yang, P.; Sun, X.; Chen, S.-M.; Luo, J.; Liu, Z.; Wang, L.; Li, D.; et al. In vivo evaluation of additively manufactured multi-layered scaffold for the repair of large osteochondral defects. *Bio-Design Manuf.* **2022**, *5*, 481–496. [CrossRef]
140. Rashidi, N.; Tamaddon, M.; Liu, C.; Brand, D.D.; Czernuszka, J. A Bilayer Osteochondral Scaffold with Self-Assembled Monomeric Collagen Type-I, Type-II, and Polymerized Chondroitin Sulfate Promotes Chondrogenic and Osteogenic Differentiation of Mesenchymal Stem Cells. *Adv. NanoBiomed Res.* **2021**, *2*, 2100089. [CrossRef]
141. Li, M.; Song, P.; Wang, W.; Xu, Y.; Li, J.; Wu, L.; Gui, X.; Zeng, Z.; Zhou, Z.; Liu, M.; et al. Preparation and characterization of biomimetic gradient multi-layer cell-laden scaffolds for osteochondral integrated repair. *J. Mater. Chem. B* **2022**, *10*, 4172–4188. [CrossRef]
142. Nejad, Z.M.; Zamanian, A.; Saeidifar, M.; Vanaei, H.R.; Amoli, M.S. 3D Bioprinting of Polycaprolactone-Based Scaffolds for Pulp-Dentin Regeneration: Investigation of Physicochemical and Biological Behavior. *Polymers* **2021**, *13*, 4442. [CrossRef]
143. Bazgir, M.; Zhang, W.; Zhang, X.; Elies, J.; Saeinasab, M.; Coates, P.; Youseffi, M.; Sefat, F. Fabrication and Characterization of PCL/PLGA Coaxial and Bilayer Fibrous Scaffolds for Tissue Engineering. *Materials* **2021**, *14*, 6295. [CrossRef]
144. Li, M.X.; Li, L.; Zhou, S.Y.; Cao, J.H.; Liang, W.H.; Tian, Y.; Shi, X.T.; Yang, X.B.; Wu, D.Y. A biomimetic orthogonal-bilayer tubular scaffold for the co-culture of endothelial cells and smooth muscle cells. *RSC Adv.* **2021**, *11*, 31783–31790. [CrossRef]
145. Do, T.M.; Yang, Y.; Deng, A. Porous Bilayer Vascular Grafts Fabricated from Electrospinning of the Recombinant Human Collagen (RHC) Peptide-Based Blend. *Polymers* **2021**, *13*, 4042. [CrossRef]
146. Smith, M.J.; Dempsey, S.G.; Veale, R.W.; Duston-Fursman, C.G.; Rayner, C.A.F.; Javanapong, C.; Gerneke, D.; Dowling, S.G.; A Bosque, B.; Karnik, T.; et al. Further structural characterization of ovine forestomach matrix and multi-layered extracellular matrix composites for soft tissue repair. *J. Biomater. Appl.* **2021**, *36*, 996–1010. [CrossRef] [PubMed]
147. Kilian, D.; von Witzleben, M.; Lanaro, M.; Wong, C.S.; Vater, C.; Lode, A.; Allenby, M.C.; Woodruff, M.A.; Gelinsky, M. 3D Plotting of Calcium Phosphate Cement and Melt Electrowriting of Polycaprolactone Microfibers in One Scaffold: A Hybrid Additive Manufacturing Process. *J. Funct. Biomater.* **2022**, *13*, 75. [CrossRef] [PubMed]
148. Liu, J.; Zou, Q.; Wang, C.; Lin, M.; Li, Y.; Zhang, R.; Li, Y. Electrospinning and 3D printed hybrid bi-layer scaffold for guided bone regeneration. *Mater. Des.* **2021**, *210*, 110047. [CrossRef]
149. Schoenfeld, C.M.; Lautenschlager, E.P.; Meyer, P.R. Mechanical properties of human cancellous bone in the femoral head. *Med. Biol. Eng. Comput.* **1974**, *12*, 313–317. [CrossRef] [PubMed]
150. Chiba, D.; Yamada, N.; Mori, Y.; Oyama, M.; Ohtsu, S.; Kuwahara, Y.; Baba, K.; Tanaka, H.; Aizawa, T.; Hanada, S.; et al. Mid-term results of a new femoral prosthesis using Ti-Nb-Sn alloy with low Young's modulus. *BMC Musculoskelet. Disord.* **2021**, *22*, 987. [CrossRef] [PubMed]
151. Zhao, J.; He, N. A mini-review of embedded 3D printing: Supporting media and strategies. *J. Mater. Chem. B* **2020**, *8*, 10474–10486. [CrossRef] [PubMed]
152. Greenwood, T.E.; Thomson, S.L. Embedded 3D printing of multi-layer, self-oscillating vocal fold models. *J. Biomech.* **2021**, *121*, 110388. [CrossRef]
153. Alipour, F.; Vigmostad, S. Measurement of Vocal Folds Elastic Properties for Continuum Modeling. *J. Voice* **2012**, *26*, 816.e21–816.e29. [CrossRef]
154. Peng, Y.-Y.; Srinivas, S.; Narain, R. *Chapter 5—Modification of Polymers, in Polymer Science and Nanotechnology*; Narain, R., Ed.; Elsevier: Amsterdam, The Netherlands, 2020; pp. 95–104.
155. Sengupta, P.; Prasad, B.L.V. Surface Modification of Polymers for Tissue Engineering Applications: Arginine Acts as a Sticky Protein Equivalent for Viable Cell Accommodation. *ACS Omega* **2018**, *3*, 4242–4251. [CrossRef]
156. Browe, D.C.; Díaz-Payno, P.J.; Freeman, F.E.; Schipani, R.; Burdis, R.; Ahern, D.P.; Nulty, J.M.; Guler, S.; Randall, L.D.; Buckley, C.T.; et al. Bilayered extracellular matrix derived scaffolds with anisotropic pore architecture guide tissue organization during osteochondral defect repair. *Acta Biomater.* **2022**, *143*, 266–281. [CrossRef]
157. DiCerbo, M.; Benmassaoud, M.M.; Vega, S.L. Porous Scaffold-Hydrogel Composites Spatially Regulate 3D Cellular Mechanosensing. *Front. Med. Technol.* **2022**, *4*, 884314. [CrossRef] [PubMed]
158. Zhan, L.; Wang, L.; Deng, J.; Zheng, Y.; Ke, Q.; Yang, X.; Zhang, X.; Jia, W.; Huang, C. Enhanced cellular infiltration of tissue-engineered scaffolds fabricated by PLLA nanogrooved microfibers. *Nano Res.* **2022**, *16*, 1614–1625. [CrossRef]

159. Sheikhi, F.; Khorram, M.; Hashemi, S.-S.; Mohammadi, A.; Peyrovedin, H. Preparation, Characterization, and Surface Modification of Polycaprolactone-Based Nanofibrous Scaffold by Grafting with Collagen for Skin Tissue Engineering. *Regen. Eng. Transl. Med.* **2022**, *8*, 545–562. [CrossRef]
160. Chi, M.; Li, N.; Cui, J.; Karlin, S.; Rohr, N.; Sharma, N.; Thieringer, F.M. Biomimetic, mussel-inspired surface modification of 3D-printed biodegradable polylactic acid scaffolds with nano-hydroxyapatite for bone tissue engineering. *Front. Bioeng. Biotechnol.* **2022**, *10*, 989729. [CrossRef] [PubMed]
161. Peng, C.; Shu, Z.; Zhang, C.; Chen, X.; Wang, M.; Fan, L. Surface modification of silk fibroin composite bone scaffold with polydopamine coating to enhance mineralization ability and biological activity for bone tissue engineering. *J. Appl. Polym. Sci.* **2022**, *139*, e52900. [CrossRef]
162. Habibizadeh, M.; Nadri, S.; Fattahi, A.; Rostamizadeh, K.; Mohammadi, P.; Andalib, S.; Hamidi, M.; Forouzideh, N. Surface modification of neurotrophin-3 loaded PCL/chitosan nanofiber/net by alginate hydrogel microlayer for enhanced biocompatibility in neural tissue engineering. *J. Biomed. Mater. Res. Part A* **2021**, *109*, 2237–2254. [CrossRef]
163. Zhong, H.; Li, Z.; Zhao, T.; Chen, Y. Surface Modification of Nanofibers by Physical Adsorption of Fiber-Homologous Amphiphilic Copolymers and Nanofiber-Reinforced Hydrogels with Excellent Tissue Adhesion. *ACS Biomater. Sci. Eng.* **2021**, *7*, 4828–4837. [CrossRef]
164. Wang, D.; Yu, X.; Xu, Y.; Wang, X.; Wang, H.; Zhang, Y.; Li, Q.; Turng, L.-S. Physical shish-kebab modification vs. chemical surface coating on expanded polytetrafluoroethylene vascular grafts for enhanced endothelial cell adhesion. *Mater. Des.* **2022**, *220*, 110889. [CrossRef]
165. Xiao, W.; Chen, W.; Wang, Y.; Zhang, C.; Zhang, X.; Zhang, S.; Wu, W. Recombinant DTβ4-inspired porous 3D vascular graft enhanced antithrombogenicity and recruited circulating CD93+/CD34+ cells for endothelialization. *Sci. Adv.* **2022**, *8*, eabn1958. [CrossRef]
166. Mahendiran, B.; Muthusamy, S.; Janani, G.; Mandal, B.B.; Rajendran, S.; Krishnakumar, G.S. Surface Modification of Decellularized Natural Cellulose Scaffolds with Organosilanes for Bone Tissue Regeneration. *ACS Biomater. Sci. Eng.* **2022**, *8*, 2000–2015. [CrossRef]
167. Zareidoost, A.; Yousefpour, M.; Ghaseme, B.; Amanzadeh, A. The relationship of surface roughness and cell response of chemical surface modification of titanium. *J. Mater. Sci. Mater. Med.* **2012**, *23*, 1479–1488. [CrossRef]
168. Wang, T.; Feng, Z.-Q.; Leach, M.K.; Wu, J.; Jiang, Q. Nanoporous fibers of type-I collagen coated poly(l-lactic acid) for enhancing primary hepatocyte growth and function. *J. Mater. Chem. B* **2012**, *1*, 339–346. [CrossRef] [PubMed]
169. Zamani, F.; Tehran, M.A.; Latifi, M.; Shokrgozar, M.A. The influence of surface nanoroughness of electrospun PLGA nanofibrous scaffold on nerve cell adhesion and proliferation. *J. Mater. Sci. Mater. Med.* **2013**, *24*, 1551–1560. [CrossRef] [PubMed]
170. Meng, J.; Boschetto, F.; Yagi, S.; Marin, E.; Adachi, T.; Chen, X.; Pezzotti, G.; Sakurai, S.; Sasaki, S.; Aoki, T.; et al. Enhancing the bioactivity of melt electrowritten PLLA scaffold by convenient, green, and effective hydrophilic surface modification. *Biomater. Adv.* **2022**, *135*, 112686. [CrossRef] [PubMed]
171. Dokos, S.; LeGrice, I.J.; Smaill, B.H.; Kar, J.; Young, A. A Triaxial-Measurement Shear-Test Device for Soft Biological Tissues. *J. Biomech. Eng.* **2000**, *122*, 471–478. [CrossRef] [PubMed]
172. Patra, S.; Ajayan, P.M.; Narayanan, T.N. Dynamic mechanical analysis in materials science: The Novice's Tale. *Oxf. Open Mater. Sci.* **2020**, *1*, itaa001. [CrossRef]
173. Villalona, G.A.; Udelsman, B.; Duncan, D.R.; McGillicuddy, E.; Sawh-Martinez, R.F.; Hibino, N.; Painter, C.; Mirensky, T.; Erickson, B.; Shinoka, T.; et al. Cell-Seeding Techniques in Vascular Tissue Engineering. *Tissue Eng. Part B Rev.* **2010**, *16*, 341–350. [CrossRef]
174. Keong, L.C.; Halim, A.S. In Vitro Models in Biocompatibility Assessment for Biomedical-Grade Chitosan Derivatives in Wound Management. *Int. J. Mol. Sci.* **2009**, *10*, 1300–1313. [CrossRef]
175. Buffinton, C.M.; Tong, K.J.; Blaho, R.A.; Buffinton, E.M.; Ebenstein, D.M. Comparison of mechanical testing methods for biomaterials: Pipette aspiration, nanoindentation, and macroscale testing. *J. Mech. Behav. Biomed. Mater.* **2015**, *51*, 367–379. [CrossRef]
176. Wałdoch, A.; Sabiniewicz, R.; Meyer-Szary, J. Interventional treatment using a 3D model of a right pulmonary artery to left atrial fistula in an infant. *Adv. Interv. Cardiol.* **2022**, *18*, 170–172. [CrossRef]
177. Chen, W.; Ma, L.; Shao, J.; Bi, C.; Xie, Y.; Zhao, S. Morphological specificity analysis of an image-based 3D model of airway filling in a difficult airway. *BMC Anesthesiol.* **2022**, *22*, 336. [CrossRef] [PubMed]
178. Hameed, M.; Prather, R.; Divo, E.; Kassab, A.; Nykanen, D.; Farias, M.; DeCampli, W.M. Computational fluid dynamics investigation of the novel hybrid comprehensive stage II operation. *JTCVS Open* **2021**, *7*, 308–323. [CrossRef] [PubMed]
179. Rossi, T.; Querzoli, G.; Badas, M.G.; Angius, F.; Telani, S.; Ripandelli, G. Computational Fluid Dynamics of Intraocular Silicone Oil Tamponade. *Transl. Vis. Sci. Technol.* **2021**, *10*, 22. [CrossRef] [PubMed]
180. Schollenberger, J.; Osborne, N.H.; Hernandez-Garcia, L.; Figueroa, C.A. A Combined Computational Fluid Dynamics and Arterial Spin Labeling MRI Modeling Strategy to Quantify Patient-Specific Cerebral Hemodynamics in Cerebrovascular Occlusive Disease. *Front. Bioeng. Biotechnol.* **2021**, *9*, 722445. [CrossRef] [PubMed]
181. Eltes, P.E.; Bartos, M.; Hajnal, B.; Pokorni, A.J.; Kiss, L.; Lacroix, D.; Varga, P.P.; Lazary, A. Development of a Computer-Aided Design and Finite Element Analysis Combined Method for Affordable Spine Surgical Navigation with 3D-Printed Customized Template. *Front. Surg.* **2021**, *7*, 583386. [CrossRef]

182. Irarrázaval, S.; Ramos-Grez, J.A.; Pérez, L.I.; Besa, P.; Ibáñez, A. Finite element modeling of multiple density materials of bone specimens for biomechanical behavior evaluation. *SN Appl. Sci.* **2021**, *3*, 776. [CrossRef]
183. Jahangir, S.; Mohammadi, A.; Mononen, M.E.; Hirvasniemi, J.; Suomalainen, J.-S.; Saarakkala, S.; Korhonen, R.K.; Tanska, P. Rapid X-ray-Based 3-D Finite Element Modeling of Medial Knee Joint Cartilage Biomechanics during Walking. *Ann. Biomed. Eng.* **2022**, *50*, 666–679. [CrossRef]
184. Jing, M.; Cui, Z.; Fu, H.; Chen, X. Real-Time Deformation Simulation of Kidney Surgery Based on Virtual Reality. *J. Shanghai Jiaotong Univ. (Sci.)* **2021**, *26*, 290–297. [CrossRef]
185. Fujimoto, K.; Shiinoki, T.; Yuasa, Y.; Tanaka, H. Estimation of liver elasticity using the finite element method and four-dimensional computed tomography images as a biomarker of liver fibrosis. *Med. Phys.* **2021**, *48*, 1286–1298. [CrossRef]
186. Peshin, S.; Karakulova, Y.; Kuchumov, A.G. Finite Element Modeling of the Fingers and Wrist Flexion/Extension Effect on Median Nerve Compression. *Appl. Sci.* **2023**, *13*, 1219. [CrossRef]
187. Hislop, B.D.; Heveran, C.M.; June, R.K. Development and analytical validation of a finite element model of fluid transport through osteochondral tissue. *J. Biomech.* **2021**, *123*, 110497. [CrossRef] [PubMed]
188. Jobanputra, R.D.; Hayes, J.; Royyuru, S.; Masen, M.A. A numerical analysis of skin–PPE interaction to prevent facial tissue injury. *Sci. Rep.* **2021**, *11*, 16248. [CrossRef]
189. Helou, B.; Bel-Brunon, A.; Dupont, C.; Ye, W.; Silvestro, C.; Rochette, M.; Lucas, A.; Kaladji, A.; Haigron, P. Patient-specific finite element simulation of peripheral artery percutaneous transluminal angioplasty to evaluate the procedure outcome without stent implantation. *Int. J. Numer. Methods Biomed. Eng.* **2023**, *39*, e3685. [CrossRef] [PubMed]
190. Feng, B.; Zhang, M.; Qin, C.; Zhai, D.; Wang, Y.; Zhou, Y.; Chang, J.; Zhu, Y.; Wu, C. 3D printing of conch-like scaffolds for guiding cell migration and directional bone growth. *Bioact. Mater.* **2022**, *22*, 127–140. [CrossRef] [PubMed]
191. Leenaars, C.H.C.; Kouwenaar, C.; Stafleu, F.R.; Bleich, A.; Ritskes-Hoitinga, M.; De Vries, R.B.M.; Meijboom, F.L.B. Animal to human translation: A systematic scoping review of reported concordance rates. *J. Transl. Med.* **2019**, *17*, 223. [CrossRef] [PubMed]
192. Pound, P.; Ritskes-Hoitinga, M. Can prospective systematic reviews of animal studies improve clinical translation? *J. Transl. Med.* **2020**, *18*, 15. [CrossRef]
193. Bas-Cristóbal Menéndez, A.; Du, Z.; van den Bosch, T.P.; Othman, A.; Gaio, N.; Silvestri, C.; Quirós, W.; Lin, H.; Korevaar, S.; Merino, A.; et al. Creating a kidney organoid-vasculature interaction model using a novel organ-on-chip system. *Sci. Rep.* **2022**, *12*, 20699. [CrossRef]
194. Dhwaj, A.; Roy, N.; Jaiswar, A.; Prabhakar, A.; Verma, D. 3D-Printed Impedance Micropump for Continuous Perfusion of the Sample and Nutrient Medium Integrated with a Liver-On-Chip Prototype. *ACS Omega* **2022**, *7*, 40900–40910. [CrossRef]
195. Li, P.; Cui, F.; Chen, H.; Yang, Y.; Li, G.; Mao, H.; Lyu, X. A Microfluidic Cell Co-Culture Chip for the Monitoring of Interactions between Macrophages and Fibroblasts. *Biosensors* **2022**, *13*, 70. [CrossRef]
196. Xu, C.; Wang, K.; Huang, P.; Liu, D.; Guan, Y. Single-Cell Isolation Microfluidic Chip Based on Thermal Bubble Micropump Technology. *Sensors* **2023**, *23*, 3623. [CrossRef]
197. Yang, J.; Hirai, Y.; Iida, K.; Ito, S.; Trumm, M.; Terada, S.; Sakai, R.; Tsuchiya, T.; Tabata, O.; Kamei, K.-I. Integrated-gut-liver-on-a-chip platform as an in vitro human model of non-alcoholic fatty liver disease. *Commun. Biol.* **2023**, *6*, 310. [CrossRef] [PubMed]
198. Ahn, J.; Ohk, K.; Won, J.; Choi, D.-H.; Jung, Y.H.; Yang, J.H.; Jun, Y.; Kim, J.-A.; Chung, S.; Lee, S.-H. Modeling of three-dimensional innervated epidermal like-layer in a microfluidic chip-based coculture system. *Nat. Commun.* **2023**, *14*, 1488. [CrossRef] [PubMed]
199. Özkan, A.; Stolley, D.L.; Cressman, E.N.K.; McMillin, M.; Yankeelov, T.E.; Rylander, M.N. Vascularized Hepatocellular Carcinoma on a Chip to Control Chemoresistance through Cirrhosis, Inflammation and Metabolic Activity. *Small Struct.* **2023**, 2200403. [CrossRef]
200. Das, S.L.; Sutherland, B.P.; Lejeune, E.; Eyckmans, J.; Chen, C.S. Mechanical response of cardiac microtissues to acute localized injury. *Am. J. Physiol. Circ. Physiol.* **2022**, *323*, H738–H748. [CrossRef]
201. Podunavac, I.; Djocos, M.; Vejin, M.; Birgermajer, S.; Pavlovic, Z.; Kojic, S.; Petrovic, B.; Radonic, V. 3D-Printed Microfluidic Chip for Real-Time Glucose Monitoring in Liquid Analytes. *Micromachines* **2023**, *14*, 503. [CrossRef]
202. Waldrop, T.I.; Graham, C.; Gard, W.; Ingle, K.; Ptacek, T.; Nguyen, N.; Lose, B.; Sethu, P.; Lee, T. Biomimetic cardiac tissue chip and murine arteriovenous fistula models for recapitulating clinically relevant cardiac remodeling under volume overload conditions. *Front. Bioeng. Biotechnol.* **2023**, *11*, 1101622. [CrossRef]
203. Boot, R.C.; Roscani, A.; van Buren, L.; Maity, S.; Koenderink, G.H.; Boukany, P.E. High-throughput mechanophenotyping of multicellular spheroids using a microfluidic micropipette aspiration chip. *Lab Chip* **2023**, *23*, 1768–1778. [CrossRef]
204. Ferrari, D.; Sengupta, A.; Heo, L.; Pethö, L.; Michler, J.; Geiser, T.; Perez, V.A.D.J.; Kuebler, W.M.; Zeinali, S.; Guenat, O.T. Effects of biomechanical and biochemical stimuli on angio- and vasculogenesis in a complex microvasculature-on-chip. *iScience* **2023**, *26*, 106198. [CrossRef]
205. Hart, D.C.t.; Yildiz, D.; Palacio-Castañeda, V.; Li, L.; Gumuscu, B.; Brock, R.; Verdurmen, W.P.R.; van der Vlag, J.; Nijenhuis, T. Co-Culture of Glomerular Endothelial Cells and Podocytes in a Custom-Designed Glomerulus-on-a-Chip Model Improves the Filtration Barrier Integrity and Affects the Glomerular Cell Phenotype. *Biosensors* **2023**, *13*, 339. [CrossRef]
206. Wu, J.; Chen, C.; Qin, C.; Li, Y.; Jiang, N.; Yuan, Q.; Duan, Y.; Liu, M.; Wei, X.; Yu, Y.; et al. Mimicking the Biological Sense of Taste In Vitro Using a Taste Organoids-on-a-Chip System. *Adv. Sci.* **2023**, *10*, e2206101. [CrossRef]

207. Erbay, I.H.; Polatli, E.; Koc, A.C.; Özbilgiç, R.; Karaman, O.; Güven, S. Bioengineering Bone-on-a-Chip Model Harnessing Osteoblastic and Osteoclastic Resolution. *Adv. Eng. Mater.* **2023**, *25*, 2201063. [CrossRef]
208. Kim, D.; Hwang, K.S.; Seo, E.U.; Seo, S.; Lee, B.C.; Choi, N.; Choi, J.; Kim, H.N. Vascularized Lung Cancer Model for Evaluating the Promoted Transport of Anticancer Drugs and Immune Cells in an Engineered Tumor Microenvironment. *Adv. Healthc. Mater.* **2022**, *11*, 2102581. [CrossRef] [PubMed]
209. Sinclair, W.E.; Pawate, A.S.; Larry, T.A.; Schieferstein, J.M.; Whittenberg, J.J.; Leckband, D.E.; Kenis, P.J.A. Development of microfluidic platform that enables 'on-chip' imaging of cells exposed to shear stress and cyclic stretch. *Microfluid. Nanofluidics* **2023**, *27*, 11. [CrossRef]
210. Knight, S.R.; Aujla, R.; Biswas, S.P. Total Hip Arthroplasty—Over 100 years of operative history. *Orthop. Rev.* **2011**, *3*, e16.
211. Oberweis, C.V.; Marchal, J.A.; López-Ruiz, E.; Galvez-Martin, M.P. A Worldwide Overview of Regulatory Frameworks for Tissue-Based Products. *Tissue Eng. Part B Rev.* **2020**, *26*, 181–196. [CrossRef]
212. Waldburger, L.; Schaller, R.; Furthmüller, C.; Schrepfer, L.; Schaefer, D.J.; Kaempfen, A. 3D-Printed Hand Splints versus Thermoplastic Splints: A Randomized Controlled Pilot Feasibility Trial. *Int. J. Bioprinting* **2021**, *8*, 474. [CrossRef]
213. Laubach, M.; Suresh, S.; Herath, B.; Wille, M.-L.; Delbrück, H.; Alabdulrahman, H.; Hutmacher, D.W.; Hildebrand, F. Clinical translation of a patient-specific scaffold-guided bone regeneration concept in four cases with large long bone defects. *J. Orthop. Transl.* **2022**, *34*, 73–84. [CrossRef]
214. Hee, M.; Greasley, S.; Whiting, G.; Harkin, C.; Oliver, G.; Marsden, D.; Andrews, R.; Sireau, S.; Price, R.D.; Anwar, F.; et al. 3D printed customised external cranial plate in a patient with syndrome of trephined: 'A case report'. *3D Print Med.* **2021**, *7*, 35.
215. Sharma, U.; Dabadi, S.; Dhungel, R.R.; Shrestha, D.; Gurung, P.; Shrestha, R.; Pant, B. Customized cost-effective polymethylmethacrylate cranioplasty implant using three-dimensional printer. *Asian J. Neurosurg.* **2021**, *16*, 150–154.
216. Mainard, N.; Sharma, D.; Fron, D.; Mezel, A.; Canavese, F.; Bonnevalle, M.; Nectoux, E. Porous Ceramic Sternal Prosthesis Implantation in a 13-Year-Old Patient Presenting with Metastatic Ewing's Sarcoma. *Eur. J. Pediatr. Surg. Rep.* **2022**, *10*, e1–e5. [CrossRef]
217. Morales, D.L.; Herrington, C.; Bacha, E.A.; Morell, V.O.; Prodán, Z.; Mroczek, T.; Sivalingam, S.; Cox, M.; Bennink, G.; Asch, F.M. A Novel Restorative Pulmonary Valve Conduit: Early Outcomes of Two Clinical Trials. *Front. Cardiovasc. Med.* **2021**, *7*, 583360. [CrossRef] [PubMed]
218. Yu, X.; Li, G.; Zheng, Y.; Gao, J.; Fu, Y.; Wang, Q.; Huang, L.; Pan, X.; Ding, J. 'Invisible' orthodontics by polymeric 'clear' aligners molded on 3D-printed personalized dental models. *Regen. Biomater.* **2022**, *9*, rbac007. [CrossRef] [PubMed]
219. Zhang, Y.; Huang, Y.; Ding, S.; Liang, J.; Kuang, J.; Mao, Q.; Ying, W.; Shu, Y.; Li, J.; Jiang, C. A clinical trial to compare a 3D-printed bolus with a conventional bolus with the aim of reducing cardiopulmonary exposure in postmastectomy patients with volumetric modulated arc therapy. *Cancer Med.* **2022**, *11*, 1037–1047. [CrossRef]
220. Kalaskar, R.; Bhaje, P.; Balasubramanian, S.; Kalaskar, A. Effectiveness of the novel impression tray "cleftray" for infants with cleft lip and palate: A randomized controlled clinical trial. *J. Korean Assoc. Oral Maxillofac. Surg.* **2021**, *47*, 82–90. [CrossRef] [PubMed]
221. Şenayli, A.; Çankaya, G.; Öztorun, C.I.; Oflaz, H.; Şenel, E. Clinical trials of 3D printing splints to avoid contracture development in burned children. *Turk. J. Med. Sci.* **2021**, *51*, 2543–2553. [PubMed]
222. Cui, D.; Wu, B.; He, D.; Wang, Y.; Jiao, Y.; Zhang, B. 3D-Printed Cold Preservation Device in Renal Autotransplantation for the Treatment of a Patient with Renal Artery Stenosis. *Front. Bioeng. Biotechnol.* **2022**, *9*, 1312. [CrossRef] [PubMed]
223. Desvigne, M.N.; Bauer, K.; Holifield, K.; Day, K.; Gilmore, D.; Wardman, A.L. Case Report: Surgical Closure of Chronic Soft Tissue Defects Using Extracellular Matrix Graft Augmented Tissue Flaps. *Front. Surg.* **2021**, *7*, 559450. [CrossRef]
224. Aimar, A.; Palermo, A.; Innocenti, B. The Role of 3D Printing in Medical Applications: A State of the Art. *J. Healthc. Eng.* **2019**, 5340616. [CrossRef]

biomimetics

MDPI

Article

Designing Biomimetic Conductive Gelatin-Chitosan–Carbon Black Nanocomposite Hydrogels for Tissue Engineering

Kamol Dey [1,*], Emanuel Sandrini [2], Anna Gobetti [2], Giorgio Ramorino [2], Nicola Francesco Lopomo [3], Sarah Tonello [4], Emilio Sardini [3] and Luciana Sartore [2]

1 Bio-Nanomaterials and Tissue Engineering Laboratory (BNTELab), Department of Applied Chemistry and Chemical Engineering, Faculty of Science, University of Chittagong, Chittagong 4331, Bangladesh
2 Department of Mechanical and Industrial Engineering, Materials Science and Technology Laboratory, University of Brescia, Via Branze 38, 25123 Brescia, Italy; e.sandrini005@studenti.unibs.it (E.S.); a.gobetti@unibs.it (A.G.); giorgio.ramorino@unibs.it (G.R.); luciana.sartore@unibs.it (L.S.)
3 Department of Information Engineering, University of Brescia, Via Branze 38, 25123 Brescia, Italy; nicola.lopomo@unibs.it (N.F.L.); emilio.sardini@unibs.it (E.S.)
4 Department of Information Engineering, University of Padova, 35131 Padua, Italy; sarah.tonello@unipd.it
* Correspondence: kamoldey@cu.ac.bd

Abstract: Conductive nanocomposites play a significant role in tissue engineering by providing a platform to support cell growth, tissue regeneration, and electrical stimulation. In the present study, a set of electroconductive nanocomposite hydrogels based on gelatin (G), chitosan (CH), and conductive carbon black (CB) was synthesized with the aim of developing novel biomaterials for tissue regeneration application. The incorporation of conductive carbon black (10, 15 and 20 wt.%) significantly improved electrical conductivity and enhanced mechanical properties with the increased CB content. We employed an oversimplified unidirectional freezing technique to impart anisotropic morphology with interconnected porous architecture. An investigation into whether any anisotropic morphology affects the mechanical properties of hydrogel was conducted by performing compression and cyclic compression tests in each direction parallel and perpendicular to macroporous channels. Interestingly, the nanocomposite with 10% CB produced both anisotropic morphology and mechanical properties, whereas anisotropic pore morphology diminished at higher CB concentrations (15 and 20%), imparting a denser texture. Collectively, the nanocomposite hydrogels showed great structural stability as well as good mechanical stability and reversibility. Under repeated compressive cyclic at 50% deformation, the nanocomposite hydrogels showed preconditioning, characteristic hysteresis, nonlinear elasticity, and toughness. Overall, the collective mechanical behavior resembled the mechanics of soft tissues. The electrical impedance associated with the hydrogels was studied in terms of the magnitude and phase angle in dry and wet conditions. The electrical properties of the nanocomposite hydrogels conducted in wet conditions, which is more physiologically relevant, showed a decreasing magnitude with increased CB concentrations, with a resistive-like behavior in the range 1 kHz–1 MHz and a capacitive-like behavior for frequencies <1 kHz and >1 MHz. Overall, the impedance of the nanocomposite hydrogels decreased with increased CB concentrations. Together, these nanocomposite hydrogels are compositionally, morphologically, mechanically, and electrically similar to native ECMs of many tissues. These gelatin-chitosan–carbon black nanocomposite hydrogels show great promise for use as conducting substrates for the growth of electro-responsive cells in tissue engineering.

Keywords: hydrogel; gelatin; chitosan; conductive carbon black; nanocomposite; cyclic compression; dissipation energy; anisotropy; tissue engineering

Citation: Dey, K.; Sandrini, E.; Gobetti, A.; Ramorino, G.; Lopomo, N.F.; Tonello, S.; Sardini, E.; Sartore, L. Designing Biomimetic Conductive Gelatin-Chitosan–Carbon Black Nanocomposite Hydrogels for Tissue Engineering. *Biomimetics* **2023**, *8*, 473. https://doi.org/10.3390/biomimetics8060473

Academic Editors: Ryszard Uklejewski and Mariusz Winiecki

Received: 21 August 2023
Revised: 17 September 2023
Accepted: 26 September 2023
Published: 3 October 2023

1. Introduction

The natural extracellular matrix (ECM) consists of dynamic, highly complex, and hierarchically organized nanocomposites that govern and regulate cells' fate and functions [1].

Moreover, the composition, density, nanostructure, and microstructure of the ECM quite differently vary between different tissue types [2]. As such, adopting a nanotechnological strategy for designing advanced nanocomposites to both better architecturally and functionally emulate the ECM, has gained intense interest [3,4]. Tissue engineering has emerged as a promising field for the regeneration and repair of damaged tissues and organs by copying the architectural and functional features of the native ECM. However, traditional tissue engineering approaches have limitations in mimicking the complex electrical properties of native ECM and tissues. Conductive nanocomposites offer an innovative solution to this challenge by combining conductive nanoparticles with biocompatible matrices to create a conducive microenvironment for cell growth and tissue regeneration [5,6]. The conductive properties of these nanocomposites facilitate better cell adhesion to the scaffold material [5]. Conductive nanocomposites can be used to deliver electrical stimulation to cells and tissues, which can promote cell proliferation, migration, and differentiation [5,7]. Electrical stimulation has been shown to accelerate tissue regeneration and improve the functionality of engineered tissues [6].

Therefore, conductive nanocomposites have emerged as a promising class of materials for tissue engineering, particularly for neural tissue engineering [8–10], cardiac tissue engineering [6,7,11,12], bone tissue engineering [13–15], cartilage tissue engineering [16–18], muscle tissue engineering [19,20], and skin tissue engineering [21–23]. They can be used to create neural scaffolds that support the growth and differentiation of neurons [10]. The electrical stimulation provided by the nanocomposite can also facilitate neural network formation and functional integration with host tissues [11]. In cardiac tissue engineering, conductive nanocomposites can be used to create scaffolds for heart muscle regeneration [11]. Electrical stimulation can help promote cardiomyocyte alignment and contractile behavior, leading to improved tissue functionality [12]. In bone tissue engineering, conductive nanocomposites can be integrated into scaffolds to support the growth and differentiation of bone-forming cells (osteoblasts) [13]. Electrical stimulation can also promote bone mineralization and enhance the overall healing process [14]. In addition to scaffolds, conductive nanocomposites can also be used as bioelectrodes for electrical stimulation applications as well as biosensors [24–27].

Additionally, the mechanical properties and degradation rates of the nanocomposites should be tailored to match the specific tissue engineering application [1,5,28]. It has been shown that conducting polymers, such as polyaniline and polypyrrole, can stimulate the attachment and proliferation of a variety of mammalian cell lines, including myoblasts, fibroblasts, and endothelial cells; however, these conducting polymers fail to mimic the physiological moduli of the ECM of various tissues [29]. This is why the incorporation of conductive nanomaterials, including metallic, inorganic, organic, or polymeric nanomaterials, into a suitable biopolymer matrix that can mimic a wide range of native ECM properties has become an alternative strategy to create multifunctional scaffolds that provide the appropriate structural, mechanical, electrical, and biological properties to foster healthy cell function and tissue formation [30,31].

Notably, there is a wide variety of carbon-based nanomaterials, such as carbon nanotubes, nanodiamonds, graphene oxide, and reduced graphene oxide, which can be used to develop electroconductive nanocomposite hydrogels [32–34]. However, compared to the abovementioned carbon-based nanomaterials, nanosized carbon black (CB) is an ill-explored nanofiller for biomedical applications [35]. As a sensing conductive filler, CB has a low cost, low density, and excellent intrinsic conductivity [36]. A CB-filled polymer composite has many advantages, including easy fabrication and superior environmental stability [36]. The electrical conductivity of CB is influenced by its particle size, aggregate shape and structure, porosity, and surface chemistry. The electrical conductivity of carbon black/polymer mixtures also depends on polymer characteristics such as the chemical structure, porosity, and processing conditions. The improved conductivity of polymer/carbon black mixtures is achieved by using CB of smaller particle size (larger surface area), lower particle density (higher particle porosity), and higher structure (better aggregation).

Here, we present a series of nanocomposite hydrogels, which are composed of CB nanoparticles homogeneously dispersed into a gelatin-based hydrogel system. In tissue engineering, gelatin (G), a denatured collagen product, is widely used because the material is biocompatible, biodegradable, inexpensive, and easy to use [37]. However, G alone is structurally and mechanically weak and dissolves in the physiological environment, which limits its application as a scaffold material in tissue engineering [37,38]. To overcome these limitations, a mild crosslinking strategy is applied where G acts as a backbone polymer and poly(ethylene glycol)diglycidyl ether (PEGDGE) acts as the crosslinking agent. By utilizing end epoxide groups, PEGDGE can react with multiple functional groups of G and act as a spacer between two natural macromolecules, reducing steric hindrance and facilitating cell adhesion [39]. Chitosan (CH), a partially deacetylated derivative of chitin, is used to further enhance the biomimetic properties of the hydrogel [40,41]. Most importantly, in the context of tissue engineering, CH is structurally similar to the native ECM, which facilitates cell-chitosan interaction [42–44]. Furthermore, CH possesses excellent biological properties such as biodegradability, anti-bacterial activity, and biocompatibility [42,44]. Moreover, many biological tissues including skeletal muscle, bone, cartilage, and the heart exhibit orientation-dependent anisotropic structures, and the structural, mechanical, and functional anisotropies of these tissues are critical for maintaining healthy physiological activities [45–47]. In this regard, anisotropic hydrogel scaffolds with geometrical resemblance to the ECM of anisotropic tissues are promising solutions for restoring their structural and functional integrity [48–50]. To achieve structurally and mechanically anisotropic porous hydrogel, we employed a unidirectional freezing technique using liquid nitrogen. We forced the degree of crosslinking through the concerted effect of chemical reactions in the solution and post-curing treatment. Together, the preparation of electroconductive and structurally stable macroporous nanocomposite hydrogels using gelatin, chitosan, and conductive carbon black was performed in this study in four consecutive steps: liquid-phase pre-crosslinking/grafting, oversimplified unidirectional freezing, lyophilization, and post-curing—this is the first report of this kind.

The overall aim of this study is to develop gelatin-based conductive hydrogel scaffolds via a mild processing condition, which includes aqueous media, various polymer assembly, and crosslinking chemistry facilitating gelation with CB nanomaterial, and to characterize the structural property, mechanical property, thermal property, morphology, and electrical conductivity of the synthesized nanocomposite hydrogels to find their potential applications as tissue engineering scaffolds.

2. Materials and Methods

2.1. Materials

Type A gelatin (pharmaceutical grade, 280 bloom, viscosity 4.30 mPs), produced from pig skin, was purchased from ITALGELATINE, Santa Vittoria d'Alba (CN), Italy. CH (molecular weight between 50,000–190,000 Da and degree of deacetylation 75–85%) was obtained from Fluka, Milano, Italy. PEGDGE (molecular weight 526 Da) was supplied by Sigma-Aldrich Co Milano, Italy. Conductive nanosized carbon black (Printex XE2B) was purchased from Degussa Huls Chemicals S.p.A. with an average particle size of 30 nm and BET surface area of 1000 m^2/g. Ethylene diamine (EDA) and acetic acid were provided by Fluka, Milano, Italy.

2.2. Methods

2.2.1. Synthesis of G/PEG/CH (CB) Nanocomposite Hydrogels

The G/PEG/CH (CB) nanocomposite hydrogel was prepared in different CB compositions, specifically 10%, 15%, and 20% of CB, following a slightly modified previously mentioned method [37]. For preparing G/PEG/CH (CB) nanocomposites with 10% CB, an amount of 0.80 g pulverized CB was dissolved in 60 mL water and treated with magnetic stirring and ultrasonication until completely dispersed. Then, gelatin granules (5.90 g) were added into the CB solution and magnetically stirred for 2 h at 45 °C to homogeneously dissolve the gelatin. After that, PEGDGE (1.4 g) was introduced into the mixture followed by EDA (70 mg) under continuous magnetic stirring at 45 °C. The chitosan solution of 2 wt.% was prepared by dissolving 0.7g CH in 1% aqueous acetic acid solution (35 g) and was subsequently stirred overnight. Then, the previously prepared chitosan solution (33 g) was added into a gelatin-PEG-CB reaction mixture and continuously stirred for another 1 h (with 5 min sonication after an interval of 20 min magnetic stirring) under the same temperature to obtain the homogeneous reaction mixture. Finally, the reaction mixture was poured into a plastic box, cooled to room temperature to form a gel, frozen with liquid nitrogen, lyophilized, and post cured at 45 °C for 2 h. The same sequences were applied for fabricating G/PEG/CH (CB) nanocomposites with 15 and 20% CB with the required amount of CB for each formulation. Three different G/PEG/CH (CB) nanocomposites were prepared; the formulations are tabulated in Table 1.

Table 1. Composition, density, and porosity of the G/PEG/CH (control sample) and G/PEG/CH (CB) nanocomposite hydrogels.

	Composition (*w/w* %)					Physical Properties	
Hydrogels	**G**	**PEG**	**CH**	**CB ***	**CB ɸ**	**Apparent Density (g/cm³)**	**Porosity (%)**
G/PEG/CH	74	18	8	-	-	0.11 ± 0.03	77 ± 2.0
G/PEG/CH (CB 1)	67	16	7	10	6.95	0.12 ± 0.03	65 ± 10
G/PEG/CH (CB 2)	63	15	7	15	13.6	0.13 ± 0.01	64 ± 3.0
G/PEG/CH (CB 3)	60	14	6	20	18.2	0.13 ± 0.01	59 ± 10

* Denotes initial CB content and ɸ denotes CB content in the final product after washing and calculated using TGA analysis.

2.2.2. Density and Porosity Measurement

The ethanol displacement method was used to measure the apparent density and porosity of the dry gel by soaking the pre-massed sample in a defined ethanol volume. The sample was previously placed under vacuum to remove the entrapped air, and the mass of the dry gel (W) was measured. Later, the sample was immersed into a graded cylinder containing a known volume (V_1) of ethanol, and the total volume (V_2) of ethanol and gel was recorded. The gel was carefully removed from the ethanol 5 h later, and the residual volume (V_3) of the ethanol was measured. Finally, the total volume of gel was calculated as $V = V_2 - V_3$.

The apparent density (ρ) of hydrogel was calculated using the following equation:

$$\rho = W/(V_2 - V_3)$$

The porosity (ϵ) of hydrogel was measured using the following equation:

$$\epsilon = (V_1 - V_3) \times 100/(V_2 - V_3)$$

2.2.3. Swelling Ratio (%) Measurement

The swelling ratio of the hydrogel was estimated by soaking the previously weighed dry sample in the distilled water in a thermostatic bath at a temperature of 37 °C. At predetermined time intervals, the sample was removed from the distilled water, the excess surface water was removed by gently pressing it with absorbent paper, and the weight was accurately measured using an electronic analytical balance. The percentage swelling ratio (%) was determined as:

$$S_R\,(\%) = (W_s - W_d) \times 100/W_d$$

where W_s and W_d indicate the wet and initial dry weight of the sample, respectively. Each experiment was conducted using five samples, and the average value was exposed with standard deviation.

2.2.4. Chemical Structure Analysis

The chemical characterization of nanocomposite hydrogels was performed using Fourier transform infrared (FTIR) spectroscopy on dry hydrogels using a Thermo Scientific, Waltham, MA, USA, Nicolet iS50 FTIR spectrophotometer equipped with a PIKE MIRacle attenuated total reflectance attachment and recorded over a range of 400 to 4000 cm^{-1} at a resolution of 4 cm^{-1}.

2.2.5. Thermogravimetric Analysis

The exact amounts of CB in the nanocomposite hydrogels were evaluated by thermogravimetric analysis (TGA) run by a TGA 500 equipped using the Hi-Res-Dynamic method with a Mettler TG50 microbalance heating a sample of about 5 mg into an alumina crucible from room temperature to 700 °C at 50 °C/min under nitrogen flow (40 mL/min).

2.2.6. Morphology Analysis

The morphology of the nanocomposite hydrogels was analyzed using a stereomicroscope (LEICA DMS 300) with reflected light. The dried samples were cut in parallel (direction of ice crystal growth) and perpendicular directions to the macroporous channels, and the morphologies of the surfaces and textures were observed using the stereomicroscope.

2.2.7. Compression and Cyclic Compression Tests

The mechanical properties of the nanocomposite hydrogels were measured using a universal testing system (INSTRON series 3366) equipped with a 50 N load cell, in unconfined compression mode between two impermeable parallel plates. Specimens were cut from the sample bars into cuboid-shaped samples using a precision rotary saw. Before the tests, the as-prepared samples were soaked in distilled water at 37 °C for 2 h. The actual height (H), width (W), and thickness (T) of the specimens were measured using an optical traveling microscope. Prior to initiating the compression test, a pre-load of 0.005 N was applied in order to reduce the influence of surface artifacts. For the compression test, cuboid-shaped samples were compressed at a strain rate of 10 mm/min up to a maximum of 50% strain of the original heights, either along the parallel or perpendicular directions to the macroporous channels. The cyclic consecutive loading-unloading test was continuously performed for 10 cycles, as long as no significant change in curve shape was observed. At least three specimens were tested for each direction. For unconfined compression testing, the load–displacement (F–x) data are converted to stress–strain ($\sigma - \varepsilon$) data through simple geometrical relationships. Engineering stress was calculated by dividing the recorded force by the initial cross-section area. The engineering strain under compression was defined as the change in height relative to the original height of the individual specimen. The initial elastic modulus (stiffness) was calculated from the first compression cycle and determined as the linear segment slope of the compressive stress-strain within the range of 5–10% strain. Compressive strength was defined as the maximum stress at 50% strain.

The successive cyclic compression experiment involved a loading-unloading pattern to reach different maximum strains (%) at each cycle. The specimen was first compressed (loaded) to a maximum strain of 20% and then relaxed (unloaded). Sequentially, the specimen was compressed to 30% maximum strain and relaxed again, repeating the operation increasing the maximum strain to 40%, 50%, and 60%. The energy absorbed by the nanocomposite hydrogels was calculated by the cyclic compression stress-deformation curves. The total energy applied to the hydrogel during the loading, defined as compression energy (kJ/m^3), was derived from the area included by the loading curve and horizontal axis, while the energy released by the hydrogel during the unloading, defined as relaxation energy (kJ/m^3), was the area bounded between the unloading curve and horizontal axis.

The hysteresis loop area indicates the dissipated energy due to the viscous nature of the hydrogels, so the dissipation energy (kJ/m^3) loss during the hysteresis cycle was calculated as the difference between the area under the load curve (compression energy) and the discharge (relaxation energy) curve from the stress-deformation curve. The percentage of dissipation energy (%) was determined by dividing the dissipation energy by the compression energy. The same computations were performed for the cyclic compression test with gradually increasing maximum compressive strain from 20% to 60%, calculating the dissipation energy (kJ/m^3) and percentage dissipation energy (%) at each maximum strain.

2.2.8. Electrical Impedance Measurement

Preliminary electrical impedance analysis was carried out considering an alternate current (AC) regimen in the frequency range of 10^2–10^7 Hz at room temperature by using a commercial impedance analyzer (HP4194A). Nanocomposite hydrogels filled with different weight percentages of carbon black nanoparticles were tested in both dry and hydrated states. In order to highlight the resistive and capacitive behavior, the complex impedance associated with the samples was studied in terms of the magnitude and phase angle. Tests were performed in triplicate.

3. Results

3.1. *Preparation of G/PEG/CH (CB) Nanocomposite Hydrogels and Their Physical Properties*

Conductive nanocomposites have emerged as promising materials in the field of tissue engineering due to their unique properties that can enhance tissue regeneration and repair. These materials combine conductive elements such as conductive nanoparticles or conductive polymers with biocompatible polymers, providing electrical conductivity while supporting cell growth and tissue integration. Motivated by this, we synthesized G/PEG/CH (CB) nanocomposites using a versatile approach consisting of liquid-phase pre-crosslinking/grafting, a unidirectional freezing process carried out by liquid nitrogen subsequent freeze-drying, and a post-curing process, as shown in Figure 1.

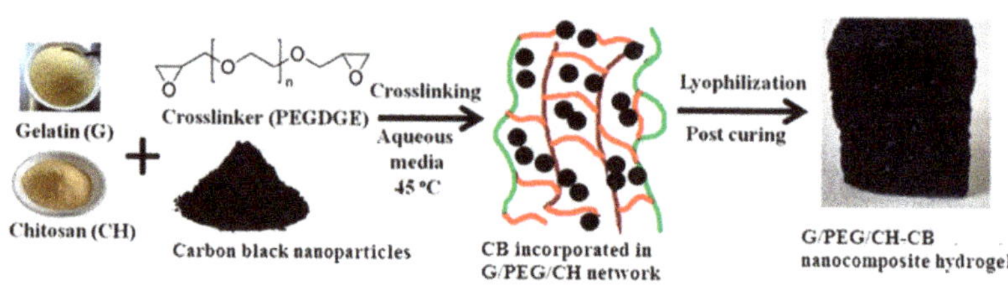

Figure 1. Schematic diagram illustrating the synthesis process of nanocomposite hydrogels.

The composition of the G/PEG/CH (CB) nanocomposite hydrogels is presented in Table 1. We calculated the exact amount of CB present in the final nanocomposite hydrogels with the help of TGA analysis (Figure 2a) and the results are illustrated in Table 1. TGA was performed with a heating rate of 50 °C/min to 700 °C under a nitrogen atmosphere. Under this operating condition, pristine conductive carbon black showed no weight loss, indicating indifference to thermal decomposition. However, G/PEG/CH (control sample) hydrogel showed significant weight loss and retained 19.80% of its original weight under the same conditions. Figure 2a showed the increased percentage residue with the higher CB content in the sample, and we applied the 'subtraction' approach to determine the final CB content, assuming that the increased percentage residue was attributed to thermally resistant CB. We calculated the final CB content in the nanocomposite hydrogel by subtracting the individual residue content from the control residue. The homogeneous dispersion of CB into the polymeric network is important to obtain the most beneficial contribution of CB incorporation. On visual inspection during synthesis, we could not observe any agglomeration of CB into the polymeric system. Furthermore, to confirm the uniform dispersion, we carried out TGA analysis of the G/PEG/CH (CB-1) nanocomposite using different sections of the same sample; namely, the top, middle, and bottom parts. Figure 2b exhibits no difference in residue for all sections, confirming uniform CB dispersion into nanocomposite hydrogels.

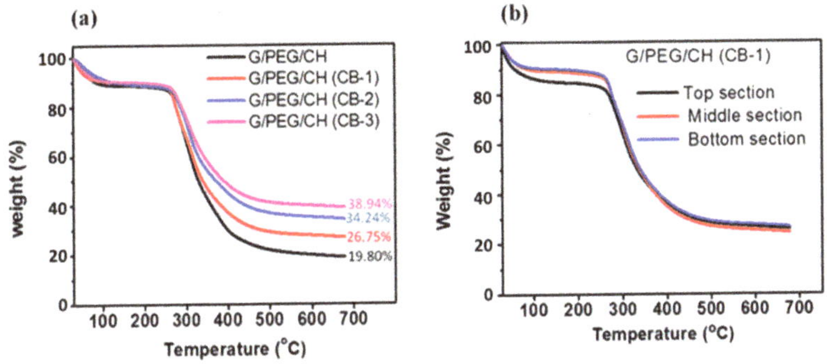

Figure 2. TGA curves of control G/PEG/CH and G/PEG/CH (CB) nanocomposite hydrogels. (**a**) Thermograms showing the residue at 700 °C under a nitrogen atmosphere with a purge rate of 50 mL/min and (**b**) showing the trend of weight loss and residue of different layers of the same G/PEG/CH (CB-1) nanocomposite hydrogel.

3.2. Chemical Structure Characterization

The chemical structure (functional groups) of the nanocomposite hydrogels was analyzed by FTIR using the washed samples; Figure 3 displays the IR spectrum of the control and nanocomposite hydrogels. The physico-mechanical properties of the nanocomposite hydrogels depend on the molecular level interaction of G-CB, CH-CB and PEG-CB. Gelatin showed prominent amine peaks around 1540 cm^{-1} (due to -NH bending vibrations and C-N stretching vibrations) and 1650 cm^{-1} (due to C=O stretching vibrations) [29]. The representative intense peak presented around 1096 cm^{-1}, which was assigned to C-O stretching (ether bond C-O-C), validated the presence of PEG in the final hydrogels. The addition of CB neither split nor shifted any peaks or did not induce new peaks, indicating no chemical interaction with polymers at the molecular level.

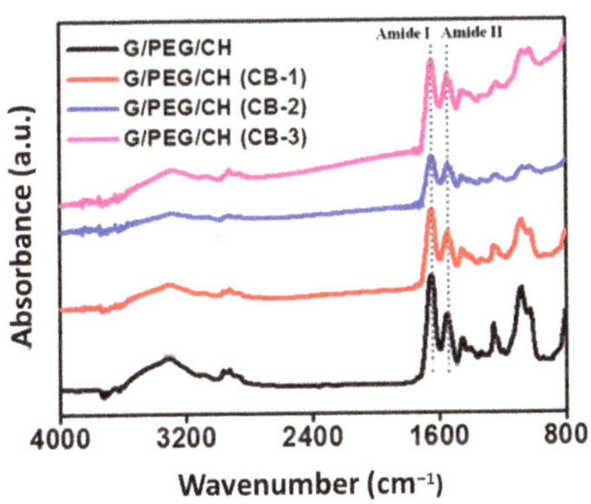

Figure 3. FTIR spectra of control G/PEG/CH and G/PEG/CH (CB) nanocomposite hydrogels.

3.3. Apparent Density, Porosity, and Swelling Ratio

The apparent density and total average porosity of the nanocomposite hydrogels were investigated using the ethanol displacement method and the results are shown in Table 1. The apparent density of G/PEG/CH (CB) nanocomposite hydrogels showed higher values than that of the G/PEG/CH control sample. However, the different amount of CB in these hydrogels appeared to slightly influence the density among the nanocomposites, showing similar density values around 0.13 g/cm^3. At the same time, the decrease in hydrogel porosity was observed with the increase in the nanofiller CB amount. The increase in the carbon black amount as filler actually decreased the porosity and favored the formation of a hydrogel with a wider distribution of micropores. The control G/PEG/CH hydrogel showed a percentage porosity of 77%, whereas the G/PEG/CH (CB) nanocomposite hydrogels showed an average percentage porosity of 62.66%. We also investigated the effect of CB incorporation on the swelling ratio of the nanocomposite hydrogels. As shown in Figure 4, the swelling ratio of nanocomposites did not seem to be significantly influenced by the addition of CB. All nanocomposite hydrogels showed a higher swelling ratio (~730%) suitable for tissue engineering applications. Using a freeze-drying process resulted in a net-like porous structure, which enhanced the hydrogel's ability to absorb large amounts of water.

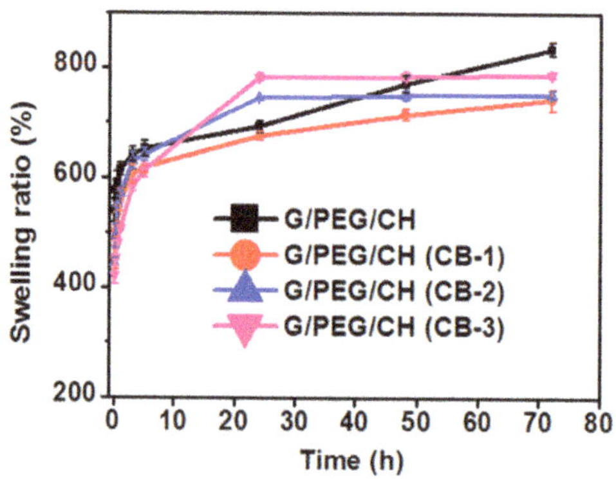

Figure 4. Swelling ratio of control G/PEG/CH and G/PEG/CH (CB) nanocomposite hydrogels.

3.4. Morphological Evaluation

Scaffolds for tissue engineering applications should have an interconnected porous structure for cell infiltration and the effective transport of nutrients and oxygen. The freeze-drying technique produced a thin skin with a dense texture (Figure 5a,b). However, beneath the skin, the hydrogel was highly porous (Figure 5c,d). A closer examination revealed this skin was composed of micropores (inset). However, removing the skin disclosed interconnected macropores of thin walls, which were themselves composed of micropores (Figure 5c,d). Overall, the freeze-drying technique used in our protocol produced highly porous hydrogels below a thin skin with a dense texture.

We also investigated the effect of CB incorporation into the pore morphology, and we analyzed the pore morphology in the parallel (side view) and perpendicular (cross-section view) directions of the ice crystal growth, as shown in Figure 6a–f. With 10% CB content, the hydrogel showed an anisotropic pore morphology: larger macroporous channels and random pores formation along and normal to the ice crystal growth, respectively (Figure 6a,c). The unidirectional freezing process allowed water molecules to become preferentially crystallized (ice crystals) along the thermal gradient (bottom to top) induced by bathing in liquid nitrogen. The lyophilization process sublimated ice crystals leaving behind the interconnected macroporous channels along the thermal gradient (Scheme 1a). Surprisingly, a higher CB content compromised the anisotropic nature of the hydrogels as well as the pore sizes. A cross-section view of the G/PEG/CH (CB-2) hydrogel revealed a denser texture with smaller-sized pores (Figure 6b). Smaller channels, both in terms of length and diameter, were observed with a 15% CB content along the direction of ice crystal growth (white color arrows indicate the smaller channels) (Figure 6e). A further increment of CB concentration to 20% (G/PEG/CH (CB-2)) imparted more compact structures in both directions, completely nullifying the anisotropic pore morphology and introducing an isotropic structure (Figure 6e). The most plausible explanation behind the transition of anisotropic to isotropic pore morphology with the increased CB content might be that the higher CB concentration restricted the directional ice crystal growth. Excitingly, this CB concentration-imposed pore morphology accordingly triggered a mechanical response (discussed below).

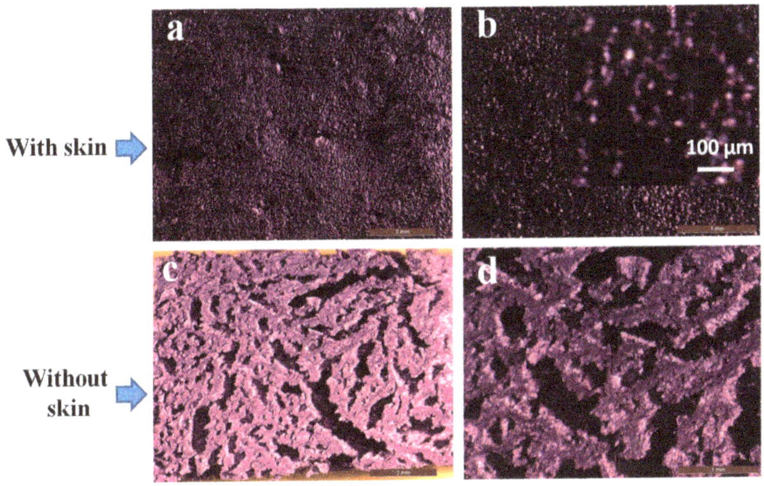

Figure 5. Optical microscopic visualization of nanocomposite hydrogels. (**a**,**b**) with skin (inset shows magnification with scale bar of 100 μm and dark parts resemble the pores) and (**c**,**d**) without skin.

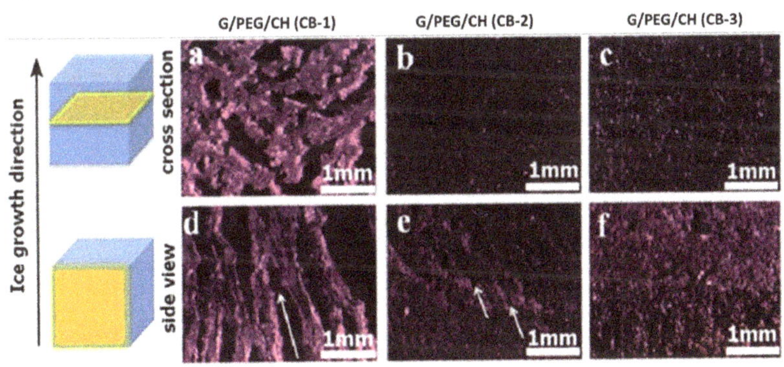

Figure 6. Optical images displaying the cross-section and side view of the nanocomposite hydrogels for (**a**,**d**) G/PEG/CH (CB-1), (**b**,**e**) G/PEG/CH (CB-2), and (**c**,**f**) G/PEG/CH (CB-3). The arrows indicate the macroporous channeling. Scale bars are 1 mm (**a**–**f**).

3.5. Compressive Mechanical Properties

To explore the effect of CB incorporation on the mechanical properties, a uniaxial unconfined compression test was carried out on swollen samples in distilled water at 50% strain (Figure 7). It is obvious from Figure 7 that the nanocomposite hydrogels showed reversible behavior (returned to their original position) during the compression test. The initial elastic modulus (stiffness) was calculated from the initial linear regions of the stress-strain curves (3 to 10% strain) and the stress at 50% strain was considered at strength. All the hydrogels were compressed in two directions (parallel and perpendicular) to examine the effect of the pore morphology on the mechanical stiffness and strength; the average values of the results are summarized in Table 2. Figure 8a–c shows the stress-strain curves of nanocomposite hydrogels obtained from mechanical data when samples were subjected to compression in both parallel and perpendicular directions to the ice crystal growth. It was noticed that the anisotropic mechanical phenomenon became negligible, increasing the CB amount in the nanocomposite hydrogels. Clearly, the G/PEG/CH (CB-1) nanocomposite hydrogel showed a significant anisotropic mechanical property, displaying an anisotropic ratio-ratio of the modulus in the parallel direction to the perpendicular direction-of 3.70 (Figure 8a). Notably, the anisotropic mechanical response significantly reduced with a 15% CB content (G/PEG/CH (CB-2), Figure 8b), while it completely diminished with a 20% CB content (G/PEG/CH (CB-3), Figure 8c). This behavior might be explained by the more compact structure due to the larger amount of conductive carbon black aggregated with increasing CB content in the gel structure, which did not allow water molecules to accrete in larger crystals during the freeze-casting process. This reduced the gel matrix voids and made the macroporous channels less aligned with the temperature gradient generated by nitrogen freezing, as shown in Scheme 1b.

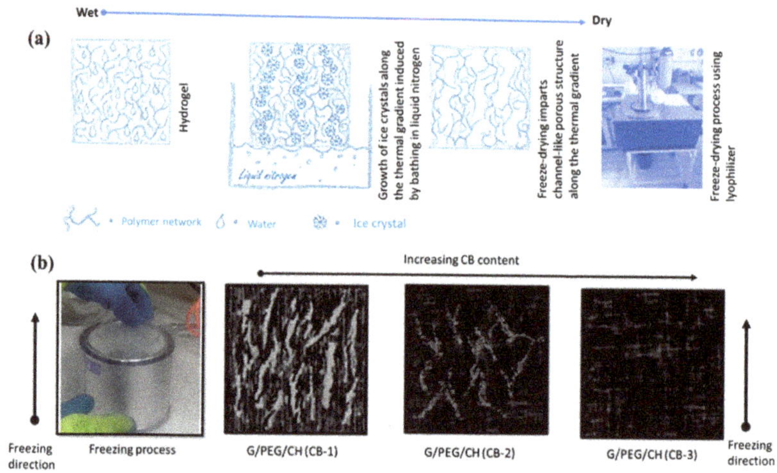

Scheme 1. (**a**) The scheme showing the formation of macroporous interconnected channels along the direction of freezing and (**b**) illustration of the different macroporous structures in the hydrogels depending on the CB content: larger macroporous structure in G/PEG/CH (CB-1), aligned with the temperature gradient during the freeze casting procedure; smaller and less aligned macroporous structure in G/PEG/CH (CB-2); and no observable macroporous channels in G/PEG/CH (CB-3).

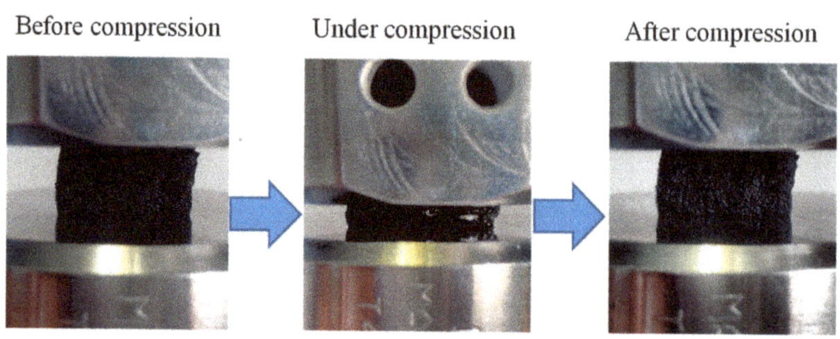

Figure 7. Optical images showing the compression test with up to 50% strain and mechanical reversibility after deformation.

Table 2. Compressive mechanical properties of the G/PEG/CH control and nanocomposite hydrogels.

Sample Name	Modulus (MPa)	Stress (MPa) at 50% Strain	Compression Direction	Pore Morphology
G/PEG/CH	0.247 ± 0.03	0.0462 ± 0.01	Parallel to smaller macropore channels	Anisotropic
G/PEG/CH	0.042 ± 0.01	0.0534 ± 0.01	Perpendicular to smaller macropore channels	
G/PEG/CH (CB-1)	0.230 ± 0.05	0.053 ± 0.01	Parallel to smaller macropore channels	Anisotropic
G/PEG/CH (CB-1)	0.063 ± 0.02	0.057 ± 0.02	Perpendicular to smaller macropore channels	
G/PEG/CH (CB-2)	0.128 ± 0.01	0.132 ± 0.03	Parallel to smaller macropore channels	Intermediate behavior
G/PEG/CH (CB-2)	0.112 ± 0.01	0.106 ± 0.04	Perpendicular to smaller macropore channels	
G/PEG/CH (CB-3)	0.173 ± 0.01	0.146 ± 0.01	No observable macropore channels	Isotropic

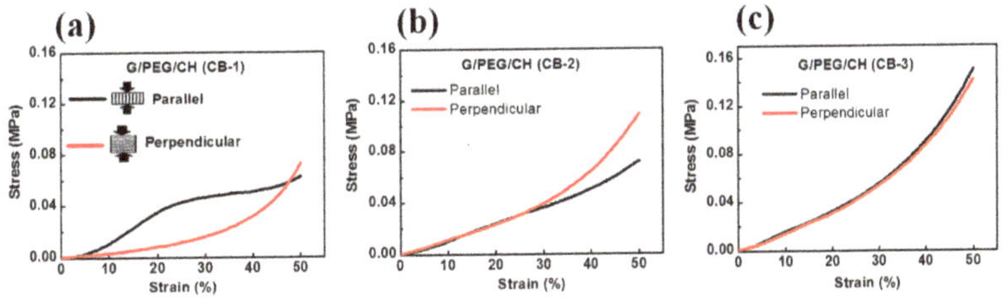

Figure 8. Stress-strain curves obtained from compression tests of fully swollen nanocomposite hydrogels. (**a**) G/PEG/CH (CB-1), (**b**) G/PEG/CH (CB-2), and (**c**) G/PEG/CH (CB-3).

The pore morphology of the nanocomposite hydrogels imparted these direction-dependent mechanical responses. Notably, G/PEG/CH (CB-1) nanocomposite hydrogel showed around a four-fold (0.063 MPa to 0.230 MPa) increase in stiffness when compressed in the parallel direction to the macropore channels compared to the perpendicular direction (Table 2). The two best plausible explanations for the increased stiffness when compressed in the parallel direction in contrast to the perpendicular compression for G/PEG/CH (CB-1) nanocomposite hydrogel might be due to: (i) the large lamellae, which act as pillars reinforcing the hydrogel, and (ii) the presence of more entrapped pressurized water between two compressive plates so that water could not be easily squeezed out [37].

The direction-dependent stiffness of anisotropic hydrogel was correlated with their anisotropic pore structures. As a possible confirmation of this, we could notice that a greater extent of macroporous channels along the direction of freezing in the G/PEG/CH (CB-1) nanocomposite hydrogel made it an anisotropic hydrogel. On the other hand, a dimensional reduction of the walls of macrochannels in the hydrogel with 15% of CB corresponded to a loss of the anisotropic behavior. In the G/PEG/CH (CB-2) nanocomposite hydrogel, the difference between the directional-dependent stiffness and strength was so little that we could treat it as an intermediate hydrogel in terms of mechanical anisotropy. The typical channel-like morphology disappeared in the G/PEG/CH (CB-3) nanocomposite hydrogel, and the more compact pore structure probably allowed the water to squeeze out in a similar way in both directions, making it a perfectly isotropic hydrogel. Figure 9 demonstrates the trend of compressive stiffness and strength variation with CB content. Figure 9 and Table 2 show a significant improvement in stiffness and strength upon the addition of CB into the hydrogel system. For example, a three-fold increase in both stiffness and strength was observed with 20% CB content. However, the nanocomposite hydrogel containing 10% CB displayed similar stiffness and strength values of G/PEG/CH alone in both directions.

In addition to investigating the effect of CB content on the mechanical stability of the nanocomposite hydrogels, cyclic compression tests with ten loading-unloading cycles were also performed. Figure 10 shows the cyclic compressive stress-strain curves for up to 10 consecutive cycles at 50% maximum strain without waiting time in the parallel and perpendicular directions to the macroporous channels of the G/PEG/CH (CB-1) nanocomposite hydrogel in wet conditions. Figure 11 presents the cyclic compressive stress-strain curves of the G/PEG/CH (CB-2) and G/PEG/CH (CB-3) nanocomposite hydrogels, as well as the corresponding stress-time curves.

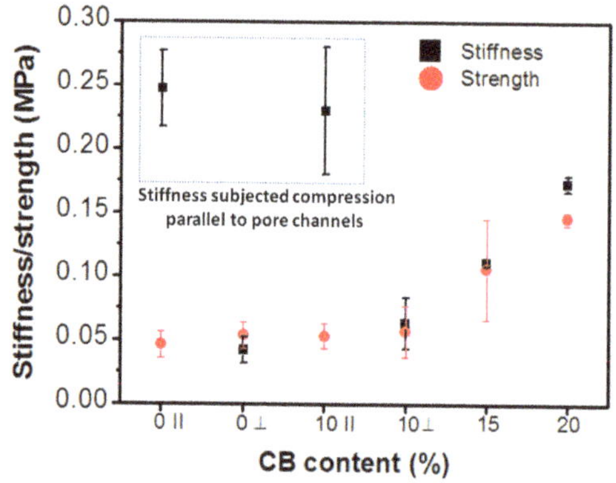

Figure 9. Stiffness and strength of the nanocomposite hydrogels as a function of CB content (|| indicates compression parallel to smaller macropore channels and ⊥ indicates compression perpendicular to smaller macropore channels).

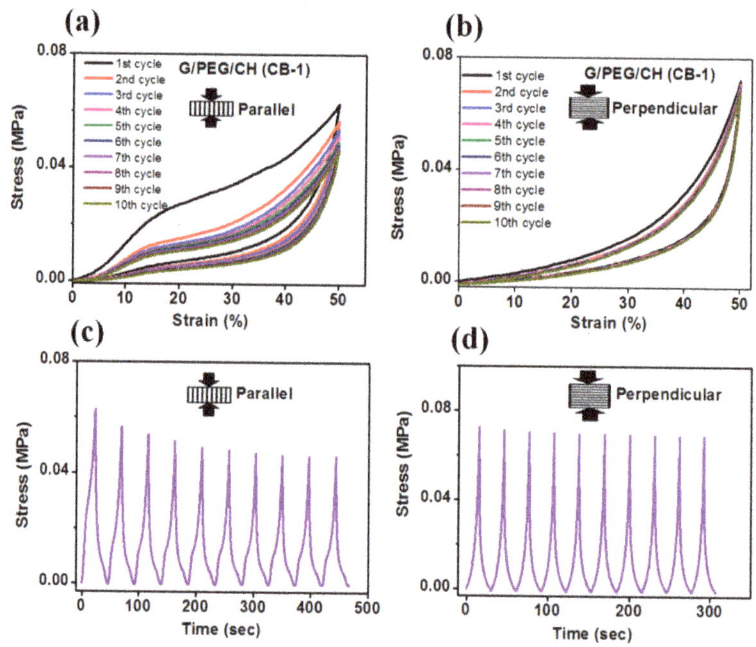

Figure 10. Cyclic compressive stress-strain curves of G/PEG/CH (CB-1) nanocomposite hydrogel for up to 10 consecutive cycles at 50% maximum strain along the (**a**) parallel and (**b**) perpendicular directions to the macroporous channels. Corresponding stress-time curves in the (**c**) parallel and (**d**) perpendicular directions.

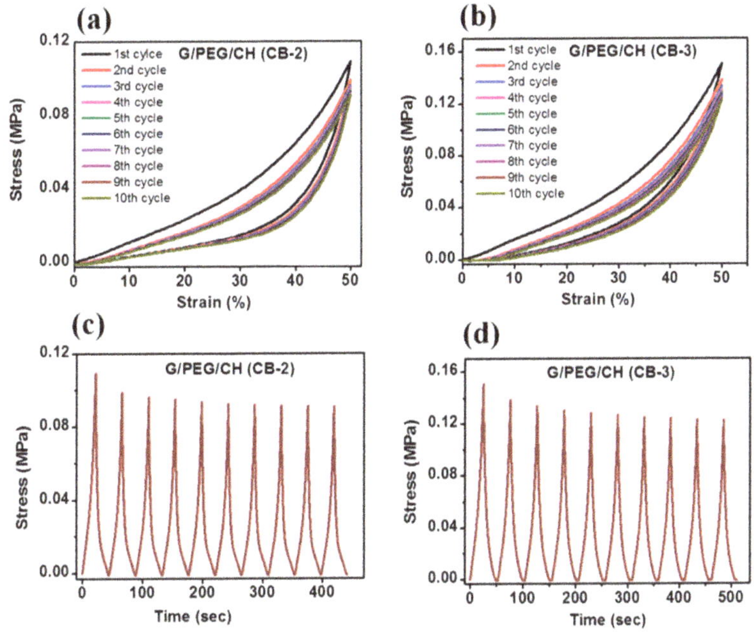

Figure 11. Cyclic compressive stress-strain curves of hydrogels for up to 10 consecutive cycles at 50% maximum strain without waiting time. (**a**) G/PEG/CH (CB-2); (**b**) G/PEG/CH (CB-3); and (**c**,**d**) corresponding stress-time curves, respectively.

As displayed in Figure 10a, the stress-strain curves of the G/PEG/CH (CB-1) nanocomposite hydrogel clearly showed two different pathways during the loading and unloading cycles, resulting in hysteresis loops. Regarding the cyclic compression in a direction parallel to the macroporous channels, a pronounced deviation in the stress-strain loading-unloading curves was shown between the first and second cycles, which indicated the occurrence of irreversible damage during the first cycle. It has been hypothesized that those smaller lamellae bridges connecting larger lamellae might be fractured during the first cycle; furthermore, the characteristic plateau region was affected by these major micro fractures in the hydrogel, involving a more collapsed structure and reduced buckling resistance of the lamellae for the consecutive cycles [37,38]. However, a slighter reduction in the slope and maximal stress at 50% strain was observed at each compressive cycle from the second to fifth consecutive loading-unloading cycle in the material, due to the occurrence of other minor micro fractures. After the fifth cycle, all following hysteresis loops were closely overlapped, indicating that the hydrogel achieved a mechanically stable structure. It was noticed that all unloading curves returned to 0% strain, indicating full shape recovery of the hydrogels even at 50% deformation. The stress responses (Figure 10c) of the G/PEG/CH (CB-1) nanocomposite hydrogel compressed at a constant strain level of 50% were shown as a function of time during the 10 cycles. It was observed that the induced stress exhibited a transition phenomenon: during the first cycle, the induced stress was 0.064 MPa, but during subsequent cycles, it gradually decreased and became stationary, reaching a constant value of 0.056 MPa after multiple cycles. This mechanical response was the outcome of a stress-softening tendency, described as Mullins' effect, which is characterized by a lower resulting stress for the same applied strain. It is often reported in filled and non-filled rubber-like materials, and such behavior is known as "preconditioning" [51].

Figure 10b shows the stress-strain curves resulting from the cyclic loading-unloading curves compressed in the perpendicular direction to the macroporous channels. In this case, the stress-strain curves indicated a very small variation between the first and second cycle curves, indicating some minor irreversible micro fractures during the first loading. However, the second cycle and all subsequent hysteresis loops were overlapped, suggesting that the hydrogel had a good reversible behavior. The corresponding stress-time plot (Figure 10d) also confirmed the mechanical stability. The maximal stress slightly reduced during the second cycle; however, it rapidly reached a stationary value after subsequent cycles. The mechanical responses observed in the G/PEG/CH (CB-2) and G/PEG/CH (CB-3) nanocomposite hydrogels during compression testing (Figure 11) were similar. During the first loading, some sort of micro fracture occurred, however, from the second cycle onward, and after subsequent cycles, nearly identical hysteresis loops were observed. They exhibited a preconditioning behavior up to the fifth cycle, as a result of the stress-softening effect explained in the previous paragraph.

Cyclic stress softening can be characterized by the amplitude of a normalized stress decrease and by the number of cycles needed to reach a stabilized state. This effect was also evaluated by calculating the ratio of the maximum stress of every cycle to the maximum stress of the first cycle (normalized stress). As observed in Figure 12, the normalized stress over 10 cycles preserved at least 85% of the maximal stress reached in the first cycle for G/PEG/CH (CB-1) in the parallel direction, G/PEG/CH (CB-2) and G/PEG/CH (CB-3). Lower stress-softening was observed particularly for G/PEG/CH (CB-1) in the perpendicular direction, showing an amplitude of a normalized stress decrease of 5%, confirming that discussed above.

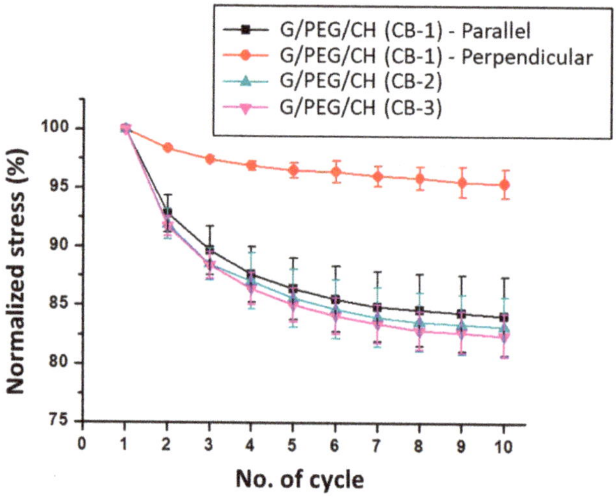

Figure 12. Amplitude of normalized stress decreases over the 10 consecutive compressive cycles of the nanocomposite hydrogels. Normalized stress was defined as the ratio of observed stress at each cycle to the maximum stress of the first cycle.

In summary, during the first cycles, the hydrogels demonstrated the Mullins effect, with a reduction in stiffness and stress at every cycle. After a few cycles, the material behavior stabilized, and the hydrogels were able to sustain a compressive strain of 50% with full strain recovery. All the nanocomposite hydrogels maintained their original shape and their load-bearing capability up to a high level of deformation. It might be possible that many microscopic flaws were created within the hydrogels, but no macroscopic cracks propagated; as a result, the hydrogels were not fractured at a macroscopic level. Overall, the G/PEG/CH (CB-1) nanocomposite hydrogel exhibited structural and mechanical anisotropy, natural tissue-like preconditioning, characteristic hysteresis, nonlinear elasticity, and energy dissipation—making it a biomimetic anisotropic hydrogel more closely resembling the native ECM of many soft tissues.

Additionally, dissipation energy calculated for each cycle also showed sharply reduced absorbed energy after the first cycle, but this was nearly constant for subsequent cycles (Figure 13a). In the case of G/PEG/CH (CB-1) nanocomposite hydrogel, when compressed in parallel to the macroporous channels, the calculated dissipation energy was double compared to those in the perpendicular direction (Figure 13a). Furthermore, the slope of this decrease was more marked along the parallel direction before the tested samples reached a constant value after a few cycles for both compression directions. This might be explained by the more stable structure in the perpendicular direction, as previously depicted. Additionally, as shown in Figure 13b, the percentage of dissipated energy was the highest for the first cycle for all hydrogels. However, the percentage of dissipation energy was found to be similar after the fifth to the subsequent tenth cycle, further confirming the achieved mechanical stability after a few cycles. Excitingly, all nanocomposite hydrogels maintained their original shapes after undergoing such a high level of deformation.

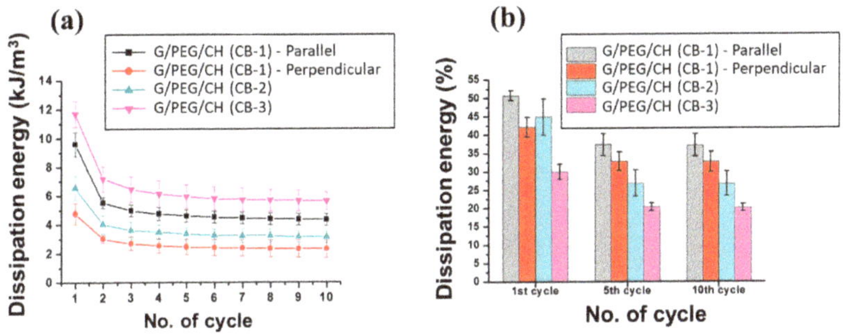

Figure 13. Energy dissipation of nanocomposite hydrogels subjected to compression consisting of 10 loading-unloading cycles at 50% strain. (**a**) Effect of number of cycles on energy dissipation and (**b**) calculated percentage energy dissipation of first, fifth, and tenth cycles.

Moreover, nanocomposite hydrogels were subjected to cyclic compression with increasing maximum strain ranging from 20 to 60%; the stress-strain curves are represented in Figure 14a–d. In addition, energy dissipation was calculated from the hysteresis loop with different strain levels (Figure 15). Figures 14 and 15 highlight that hysteresis became apparent above 30% strain and sharply increased with increasing strain levels, indicating the increased absorption of energy during higher deformation. For all cases, the hysteresis loops area became larger with increasing maximum strain. Furthermore, as presented in Figure 15a, the G/PEG/CH (CB-1) nanocomposite hydrogel showed an exponentially increasing trend in dissipation energies with an increase in maximum strain, suggesting the capability of the hydrogel to effectively dissipate energy at larger deformations, and also indicating a gradual fracture process of the gel network while reaching higher deformation levels with respect to the previous cycle. The slightly higher dissipated energies for the parallel compression of G/PEG/CH (CB-1) with respect to the perpendicular compression might be due to the migration of more pressurized water throughout the porous network. The exponential growth of dissipated energies at higher deformation levels became more apparent when the CB amount was increased to 15% and 20%, achieving values of 12 kJ/m^3 and 16 kJ/m^3 for 60% strain, respectively, which were twice and three times more than the energy dissipated at 10% CB content.

Percentage dissipation energy is often used to quantify the energy dissipation ability of a tough hydrogel. A higher level of deformation involves more dissipation due to the increased stress applied to the material (higher friction). By applying a greater force, a higher loading energy was conferred to the material and the dissipation energy increased for the water movement. As shown in Figure 15b, the G/PEG/CH (CB-1) and G/PEG/CH (CB-2) nanocomposite hydrogels exhibited similar dissipated energy capacities, which slightly increased with higher deformations. At the same time, G/PEG/CH (CB-3) demonstrated lower dissipation ability, but still increased the strain up to 60%, suggesting a less pronounced bond rupture capacity, increased elastic properties, and more deformation tolerance. It is noteworthy that, with 20% CB, the percentage dissipation energy was more constant, and it was probably able to sustain higher deformation, which led to higher values of percentage dissipation energy (not necessarily linearly increasing over 60% deformation). This CB nanofiller reinforced the hydrogel by having a greater effect on strength than modulus. This involved high compression energy during loading; as the deformation applied to the material increased, the dissipation energy increased but the compression energy increased relatively more (due to the reinforcement effect). These results clearly suggested that the nanocomposite hydrogel possessed CB concentration-dependent and strain-dependent energy dissipation behavior, and effective energy dissipation occurred at the higher strain deformation.

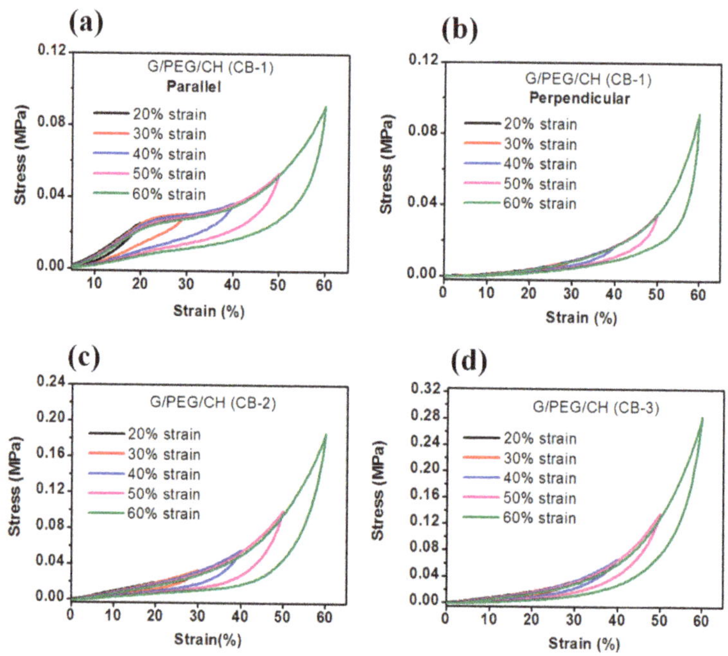

Figure 14. Stress-strain curves of the cyclic compression test with amplitudes ranging from 20 to 60% strain applied to nanocomposite hydrogels, corresponding to (**a**) G/PEG/CH (CB-1)-parallel, (**b**) (G/PEG/CH (CB-1)-perpendicular, (**c**) G/PEG/CH (CB-2), and (**d**) G/PEG/CH (CB-3).

It is generally accepted that the stiffness and toughness of conventional materials are two opposite mechanical parameters. Interestingly, our G/PEG/CH (CB) nanocomposite hydrogel could simultaneously increase both stiffness and toughness with increasing CB concentration. This could be attributed to the hybrid physical and chemical crosslinking in the nanocomposite hydrogel. The enhancement in the elastic modulus (an indicator of stiffness) of the G/PEG/CH(CB) nanocomposite hydrogels was attributed to the increase in the elastically effective G/PEG/CH chains and the reinforcing effect of the carbon black acting as a filler. The rupture (or peeling) of physically adsorbed G/PEG/CH chains from CB was accompanied by energy dissipation, which consequently improved the crack resistance of the hydrogels. In addition, as more G/PEG/CH chains were adsorbed on CB particles, more energy was dissipated. Therefore, the toughness of the G/PEG/CH (CB) nanocomposite hydrogels continued to increase with the CB concentration. Taken together, a slight decrease in the stress and no residuary strain during the consecutive loading-unloading cycles demonstrated the elasticity, excellent shape-recovery properties, and good mechanical stability of the nanocomposite hydrogels.

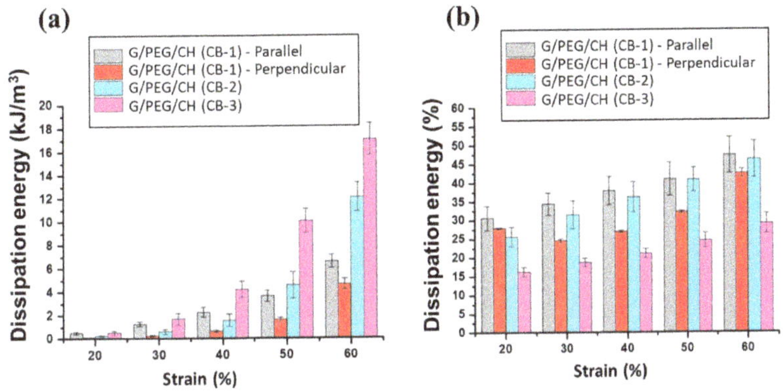

Figure 15. Energy dissipation calculated from the observed hysteresis during compression at different strains. (**a**) The hysteresis energies of nanocomposite hydrogels compressed at different maximum strains and (**b**) the percentage dissipation energy for different CB concentrations.

3.6. Electrical Impedance

To highlight the resistive and capacitive behaviors, the complex impedance associated with the samples was studied in terms of the magnitude and phase angle in dry and wet conditions, as presented in Figures 16 and 17, respectively. Considering dry testing, the results reported good optimally tunable electrical properties with a characteristic frequency of transition between resistive- and capacitive-like behaviors, depending on the CB content, despite the presence of noise at low frequency. Such a result appears to be promising in view of the application of electrical stimuli, even when compared with more complex materials. The electrical properties in wet conditions, which were more physiologically relevant, showed low magnitude at high frequencies (>1 MHz) with a decreasing phase angle, confirming a capacitive behavior (Figure 17). At lower frequencies (<1 kHz), impedance decreased with increased CB content, reaching a steady-state value in a range between 1 kHz and 1 MHz, with a near to zero phase angle confirming a resistive behavior. Overall, the impedance of the nanocomposite hydrogels decreased with increased CB concentrations; furthermore, it is worth noting that the G/PEG/CH scaffold without CB presented, within the very same frequency range, an overall value of impedance magnitude of about 10 kΩ. Additionally, a change in electrical conductivity, with and without biological cells and along with various protocols, was in progress. Since these nanocomposite hydrogels were prepared with the aim of developing novel scaffolds for tissue engineering applications, it is of utmost importance to evaluate the biocompatibility of these types of hydrogels, which is presently ongoing. However, our previous studies on G/PEG/CH hydrogels using human mesenchymal stem cells (hMSCs), assessing their tissue regeneration capability (chondrogenic and osteogenic differentiation), confirmed that this hydrogel scaffold was biocompatible and suitable for cell growth, chondrogenic and osteogenic differentiation, and mineralization [42,52].

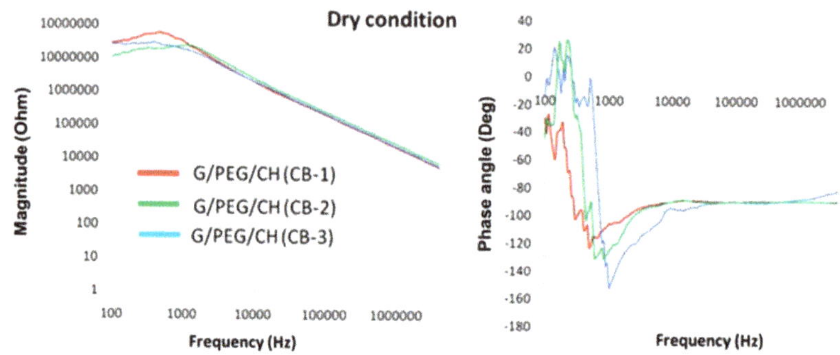

Figure 16. Complex impedance magnitude (log scale) and phase angle with respect to the frequency of nanocomposite hydrogels tested in dry conditions.

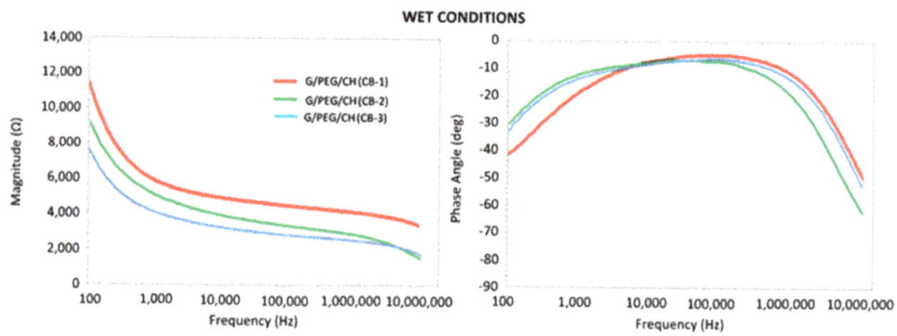

Figure 17. Complex impedance magnitude and phase angle with respect to the frequency of nanocomposite hydrogels tested in wet conditions.

4. Conclusions

Conductive nanocomposites have shown immense potential in tissue engineering due to their ability to support cell growth, promote tissue regeneration, and provide electrical stimulation to engineered tissues. These results revealed that the developed synthesis method was amenable enough to accommodate nanomaterials into macroporous hydrogels under mild conditions. The incorporation of conductive carbon black (CB) (10, 15, and 20 wt.%) significantly improved electrical conductivity and enhanced mechanical properties with the increased CB content. Interestingly, nanocomposite hydrogel with 10% CB provided both anisotropic morphology and mechanical properties, whereas anisotropic pore morphology diminished at higher CB concentrations (15 and 20%), imparting a denser texture. In general, under repeated compressive cycles at 50% deformation, all nanocomposite hydrogels showed nonlinear elasticity, toughness, preconditioning, and characteristic hysteresis. Overall, the collective mechanical behavior resembled the mechanics of soft tissues. The electrical properties of the nanocomposite hydrogels conducted in wet conditions, which is more physiologically relevant, showed a decreasing magnitude with increased CB concentrations, with a resistive-like behavior in the range of 1 kHz–1 MHz and a capacitive-like behavior for frequencies <1 kHz and >1 MHz. Overall, the impedance of the nanocomposite hydrogels decreased with increased CB concentrations. This work is still in the development and optimization stages. Finally, the combination of tissue-like morphological and mechanical behaviors, along with good electrical conductivity, could lead to their use in various applications.

Author Contributions: Conceptualization, K.D. and L.S.; methodology, K.D., E.S. (Emanuel Sandrini); N.F.L., S.T., E.S. (Emilio Sardini) and L.S.; formal analysis, K.D., E.S. (Emanuel Sandrini), A.G., G.R., N.F.L., S.T., E.S. (Emilio Sardini) and L.S.; investigation, K.D., E.S. (Emanuel Sandrini) and L.S.; data curation, K.D.; writing, E.S. (Emanuel Sandrini) and K.D.; writing—review and editing, K.D., E.S. (Emilio Sardini), N.F.L., S.T., A.G., G.R. and L.S.; visualization, K.D., A.G., G.R. and L.S.; supervision, K.D. and L.S. All authors have read and agreed to the published version of the manuscript.

Funding: This research received no external funding.

Institutional Review Board Statement: Not applicable.

Data Availability Statement: The raw/processed data required to reproduce these findings cannot be shared at this time as the data also forms part of an ongoing study.

Conflicts of Interest: The authors declare no conflict of interest.

References

1. Dey, K.; Roca, E.; Ramorino, G.; Sartore, L. Progress in the mechanical modulation of cell functions in tissue engineering. *Biomater. Sci.* **2020**, *8*, 7033–7081. [CrossRef] [PubMed]
2. Webber, M.J.; Khan, O.F.; Sydlik, S.A.; Tang, B.C.; Langer, R. A perspective on the clinical translation of scaffolds for tissue engineering. *Ann. Biomed. Eng.* **2015**, *43*, 641–656. [CrossRef] [PubMed]
3. Rahimnejad, M.; Rasouli, F.; Jahangiri, S.; Ahmadi, S.; Rabiee, N.; Ramezani Farani, M.; Akhavan, O.; Asadnia, M.; Fatahi, Y.; Hong, S.; et al. Engineered Biomimetic Membranes for Organ-on-a-Chip. *ACS Biomater. Sci. Eng.* **2022**, *8*, 5038–5059. [CrossRef]
4. Jahangirnezhad, M.; Mahmoudinezhad, S.S.; Moradi, M.; Moradi, K.; Rohani, A.; Tayebi, L. Bone scaffold materials in periodontal and tooth-supporting tissue regeneration: A review. *Curr. Stem Cell Res. Ther.* **2022**. *online ahead of print*.
5. Okamoto, M.; John, B. Synthetic biopolymer nanocomposites for tissue engineering scaffolds. *Prog. Polym. Sci.* **2013**, *38*, 1487–1503. [CrossRef]
6. Jalilinejad, N.; Rabiee, M.; Baheiraei, N.; Ghahremanzadeh, R.; Salarian, R.; Rabiee, N.; Akhavan, O.; Zarrintaj, P.; Hejna, A.; Saeb, M.R.; et al. Electrically conductive carbon-based (bio)-nanomaterials for cardiac tissue engineering. *Bioeng. Transl. Med.* **2023**, *8*, e10347. [CrossRef] [PubMed]
7. Alamdari, S.G.; Alibakhshi, A.; de la Guardia, M.; Baradaran, B.; Mohammadzadeh, R.; Amini, M.; Kesharwani, P.; Mokhtarzadeh, A.; Oroojalian, F.; Sahebkar, A. Conductive and semiconductive nanocomposite-based hydrogels for cardiac tissue engineering. *Adv. Healthc. Mater.* **2022**, *11*, 2200526. [CrossRef]
8. Xu, X.; Wang, L.; Jing, J.; Zhan, J.; Xu, C.; Xie, W.; Ye, S.; Zhao, Y.; Zhang, C.; Huang, F. Conductive collagen-based hydrogel combined with electrical stimulation to promote neural stem cell proliferation and differentiation. *Front. Bioeng. Biotechnol.* **2022**, *10*, 912497. [CrossRef]
9. Gong, B.; Zhang, X.; Zahrani, A.A.; Gao, W.; Ma, G.; Zhang, L.; Xue, J. Neural tissue engineering: From bioactive scaffolds and in situ monitoring to regeneration. *Exploration* **2022**, *2*, 20210035. [CrossRef]
10. Afjeh-Dana, E.; Naserzadeh, P.; Nazari, H.; Mottaghitalab, F.; Shabani, R.; Aminii, N.; Mehravi, B.; Rostami, F.T.; Joghataei, M.T.; Mousavizadeh, K.; et al. Gold nanorods reinforced silk fibroin nanocomposite for peripheral nerve tissue engineering applications. *Int. J. Biol. Macromol.* **2019**, *129*, 1034–1039. [CrossRef]
11. Ashtari, K.; Nazari, H.; Ko, H.; Tebon, P.; Akhshik, M.; Akbari, M.; Alhosseini, S.N.; Mozafari, M.; Mehravi, B.; Soleimani, M.; et al. Electrically conductive nanomaterials for cardiac tissue engineering. *Adv. Drug Deliv. Rev.* **2019**, *144*, 162–179. [CrossRef]
12. Esmaeili, H.; Patino-Guerrero, A.; Hasany, M.; Ansari, M.O.; Memic, A.; Dolatshahi-Pirouz, A.; Nikkhah, M. Electroconductive biomaterials for cardiac tissue engineering. *Acta Biomater.* **2022**, *139*, 118–140. [CrossRef] [PubMed]
13. Chopra, V.; Thomas, J.; Kaushik, S.; Rajput, S.; Guha, R.; Mondal, B.; Naskar, S.; Mandal, D.; Chauhan, G.; Chattopadhyay, N.; et al. Injectable bone cement reinforced with gold nanodots decorated rGO-hydroxyapatite nanocomposites, augment bone regeneration. *Small* **2023**, *19*, 2204637. [CrossRef] [PubMed]
14. Arambula-Maldonado, R.; Mequanint, K. Carbon-based electrically conductive materials for bone repair and regeneration. *Mater. Adv.* **2022**, *3*, 5186–5206.
15. Asl, M.A.; Karbasi, S.; Beigi-Boroujeni, S.; Benisi, S.Z.; Saeed, M. Polyhydroxybutyrate-starch/carbon nanotube electrospun nanocomposite: A highly potential scaffold for bone tissue engineering applications. *Int. J. Biol. Macromol.* **2022**, *223*, 524–542. [CrossRef] [PubMed]
16. Mirmusavi, M.H.; Ahmadian, M.; Karbasi, S. Polycaprolactone-chitosan/multi-walled carbon nanotube: A highly strengthened electrospun nanocomposite scaffold for cartilage tissue engineering. *Int. J. Biol. Macromol.* **2022**, *209*, 1801–1814. [CrossRef]
17. Huang, J.; Liu, F.; Su, H.; Xiong, J.; Yang, L.; Xia, J.; Liang, Y. Advanced nanocomposite hydrogels for cartilage tissue engineering. *Gels* **2022**, *8*, 138. [CrossRef]
18. Kwon, H.J.; Lee, G.S.; Chun, H. Electrical stimulation drives chondrogenesis of mesenchymal stem cells in the absence of exogenous growth factors. *Sci. Rep.* **2016**, *6*, 39302. [CrossRef]
19. Wang, Y.; Wang, Q.; Luo, S.; Chen, Z.; Zheng, X.; Kankala, R.K.; Chen, A.; Wang, S. 3D bioprinting of conductive hydrogel for enhanced myogenic differentiation. *Regen. Biomater.* **2021**, *8*, rbab035. [CrossRef]
20. Dong, R.; Ma, P.X.; Guo, B. Conductive biomaterials for muscle tissue engineering. *Biomaterials* **2020**, *229*, 119584.
21. Ma, J.; Wu, C. Bioactive inorganic particles-based biomaterials for skin tissue engineering. *Exploration* **2022**, *2*, 20210083. [CrossRef] [PubMed]

22. Shokrani, H.; Shokrani, A.; Jouyandeh, M.; Seidi, F.; Gholami, F.; Kar, S.; Munir, M.T.; Kowalkowska-Zedler, D.; Zarrintaj, P.; Rabiee, N.; et al. Green polymer nanocomposites for skin tissue engineering. *ACS App. Bio. Mater.* **2022**, *5*, 2107–2121. [CrossRef] [PubMed]
23. Elshishiny, F.; Mamdouh, W. Fabrication of nanofibrous/xerogel layer-by-layer biocomposite scaffolds for skin tissue regeneration: In Vitro study. *ACS Omega* **2020**, *5*, 2133–2147. [CrossRef] [PubMed]
24. Lim, C.; Park, C.; Sunwoo, S.H.; Kim, Y.G.; Lee, S.; Han, S.I.; Kim, D.; Kim, J.H.; Kim, D.H.; Hyeon, T. Facile and scalable synthesis of whiskered gold nanosheets for stretchable, conductive, and biocompatible nanocomposites. *ACS Nano* **2022**, *16*, 10431–10442. [CrossRef]
25. Zare, E.N.; Makvandi, P.; Ashtari, B.; Rossi, F.; Motahari, A.; Perale, G. Progress in conductive polyaniline-based nanocomposites for biomedical applications: A review. *J. Med. Chem.* **2019**, *63*, 1–22. [CrossRef]
26. Wan, X.; Zhao, Y.; Li, Z.; Li, L. Emerging polymeric electrospun fibers: From structural diversity to application in flexible bioelectronics and tissue engineering. *Exploration* **2022**, *2*, 20210029. [CrossRef]
27. Song, H.; Wang, Y.; Fei, Q.; Nguyen, D.H.; Zhang, C.; Liu, T. Cryopolymerization-enabled self-wrinkled polyaniline-based hydrogels for highly stretchable all-in-one supercapacitors. *Exploration* **2022**, *2*, 20220006. [CrossRef]
28. Chong, J.; Sung, C.; Nam, K.S.; Kang, T.; Kim, H.; Lee, H.; Park, H.; Park, S.; Kang, J. Highly conductive tissue-like hydrogel interface through template-directed assembly. *Nat. Commun.* **2023**, *14*, 2206. [CrossRef]
29. You, J.O.; Rafat, M.; Ye, G.J.; Auguste, D.T. Nanoengineering the heart: Conductive scaffolds enhance connexin 43 expression. *Nano Lett.* **2011**, *11*, 3643–3648. [CrossRef]
30. Zhao, H.; Liu, M.; Zhang, Y.; Yin, J.; Pei, R. Nanocomposite hydrogels for tissue engineering applications. *Nanoscale* **2020**, *12*, 14976–14995. [CrossRef]
31. Mehrali, M.; Thakur, A.; Pennisi, C.P.; Talebian, S.; Arpanaei, A.; Nikkhah, M.; Dolatshahi-Pirouz, A. Nanoreinforced hydrogels for tissue engineering: Biomaterials that are compatible with load-bearing and electroactive tissues. *Adv. Mater.* **2017**, *29*, 1603612. [CrossRef] [PubMed]
32. Sun, H.; Tang, J.; Mou, Y.; Zhou, J.; Qu, L.; Duval, K.; Huang, Z.; Lin, N.; Dai, R.; Liang, C.; et al. Carbon nanotube-composite hydrogels promote intercalated disc assembly in engineered cardiac tissues through β1-integrin mediated FAK and RhoA pathway. *Acta Biomater.* **2017**, *48*, 88–99. [CrossRef] [PubMed]
33. Shin, S.R.; Zihlmann, C.; Akbari, M.; Assawes, P.; Cheung, L.; Zhang, K.; Manoharan, V.; Zhang, Y.S.; Yüksekkaya, M.; Wan, K.T.; et al. Reduced graphene oxide-gelMA hybrid hydrogels as scaffolds for cardiac tissue engineering. *Small* **2016**, *12*, 3677–3689. [CrossRef]
34. Sun, Y.; Yang, Q.; Wang, H. Synthesis and characterization of nanodiamond reinforced chitosan for bone tissue engineering. *J. Funct. Biomater.* **2016**, *7*, 27. [CrossRef] [PubMed]
35. Moschou, E.A.; Peteu, S.F.; Bachas, L.G.; Madou, M.J.; Daunert, S. Artificial muscle material with fast electroactuation under neutral pH conditions. *Chem. Mater.* **2004**, *16*, 2499–2502. [CrossRef]
36. Chuang, W.J.; Chiu, W.Y.; Tai, H.J. Temperature-dependent conductive composites: Poly (N-isopropylacrylamide-co-N-methylol acrylamide) and carbon black composite films. *J. Mater. Chem.* **2012**, *22*, 20311–20318. [CrossRef]
37. Dey, K.; Agnelli, S.; Re, F.; Russo, D.; Lisignoli, G.; Manferdini, C.; Bernardi, S.; Gabusi, E.; Sartore, L. Rational design and development of anisotropic and mechanically strong gelatin-based stress relaxing hydrogels for osteogenic/chondrogenic differentiation. *Macromol. Biosci.* **2019**, *19*, 1900099. [CrossRef]
38. Dey, K.; Agnelli, S.; Sartore, L. Designing viscoelastic gelatin-PEG macroporous hybrid hydrogel with anisotropic morphology and mechanical properties for tissue engineering application. *Micro* **2023**, *3*, 434–457. [CrossRef]
39. Dey, K.; Agnelli, S.; Borsani, E.; Sartore, L. Degradation-dependent stress relaxing semi-interpenetrating networks of hydroxyethyl cellulose in gelatin-PEG hydrogel with good mechanical stability and reversibility. *Gels* **2021**, *7*, 277. [CrossRef]
40. Alimirzaei, F.; Vasheghani-Farahani, E.; Ghiaseddin, A.; Soleimani, M. pH-sensitive chitosan hydrogel with instant gelation for myocardial regeneration. *J. Tissue Sci. Eng.* **2017**, *8*, 1000212.
41. Saeedi, M.; Vahidi, O.; Moghbeli, M.R.; Ahmadi, S.; Asadnia, M.; Akhavan, O.; Seidi, F.; Rabiee, M.; Saeb, M.R.; Webster, T.J.; et al. Customizing nano-chitosan for sustainable drug delivery. *J. Control. Release* **2022**, *350*, 175–192. [CrossRef] [PubMed]
42. Manferdini, C.; Gabusi, E.; Sartore, L.; Dey, K.; Agnelli, S.; Almici, C.; Bianchetti, A.; Zini, N.; Russo, D.; Re, F.; et al. Chitosan-based scaffold counteracts hypertrophic and fibrotic markers in chondrogenic differentiated mesenchymal stromal cells. *J. Tissue Eng. Regen. Med.* **2019**, *13*, 1896–1911. [CrossRef] [PubMed]
43. Li, P.; Jia, Z.; Wang, Q.; Tang, P.; Wang, M.; Wang, K.; Fang, J.; Zhao, C.; Ren, F.; Ge, X.; et al. A resilient and flexible chitosan/silk cryogel incorporated Ag/Sr co-doped nanoscale hydroxyapatite for osteoinductivity and antibacterial properties. *J. Mater. Chem. B* **2018**, *6*, 7427–7438. [CrossRef]
44. Mazaheri, M.; Akhavan, O.; Simchi, A. Flexible bactericidal graphene oxide–chitosan layers for stem cell proliferation. *Appl. Surf. Sci.* **2014**, *301*, 456–462. [CrossRef]
45. Xing, J.; Liu, N.; Xu, N.; Chen, W.; Xing, D. Engineering complex anisotropic scaffolds beyond simply uniaxial alignment for tissue engineering. *Adv. Funct. Mater.* **2022**, *32*, 2110676. [CrossRef]
46. Hao, L.; Mao, H. Magnetically anisotropic hydrogels for tissue engineering. *Biomater. Sci.* **2023**, *11*, 6384–6402. [CrossRef]
47. Li, P.; Liu, Y.; Wang, Z.; Xiao, X.; Meng, G.; Wang, X.; Guo, H.L.; Guo, H. Dry-regulated hydrogels with anisotropic mechanical performance and ionic conductivity. *Chin. Chem. Lett.* **2022**, *33*, 871–876. [CrossRef]

48. Engelmayr Jr, G.C.; Cheng, M.; Bettinger, C.J.; Borenstein, J.T.; Langer, R.; Freed, L.E. Accordion-like honeycombs for tissue engineering of cardiac anisotropy. *Nat. Mater.* **2008**, *7*, 1003–1010. [CrossRef]
49. Wu, Y.; Wang, L.; Guo, B.; Ma, P.X. Interwoven aligned conductive nanofiber yarn/hydrogel composite scaffolds for engineered 3D cardiac anisotropy. *ACS Nano* **2017**, *11*, 5646–5659. [CrossRef]
50. Khuu, N.; Kheiri, S.; Kumacheva, E. Structurally anisotropic hydrogels for tissue engineering. *Trends Chem.* **2021**, *3*, 1002–1026. [CrossRef]
51. Diani, J.; Fayolle, B.; Gilormini, P. A review on the Mullins effect. *Eur. Polym. J.* **2009**, *45*, 601–612. [CrossRef]
52. Re, F.; Sartore, L.; Moulisova, V.; Cantini, M.; Almici, C.; Bianchetti, A.; Chinello, C.; Dey, K.; Agnelli, S.; Manferdini, C.; et al. 3D gelatin-chitosan hybrid hydrogels combined with human platelet lysate highly support human mesenchymal stem cell proliferation and osteogenic differentiation. *J. Tissue Eng.* **2019**, *10*, 2041731419845852. [CrossRef] [PubMed]

MDPI AG
Grosspeteranlage 5
4052 Basel
Switzerland
Tel.: +41 61 683 77 34

Biomimetics Editorial Office
E-mail: biomimetics@mdpi.com
www.mdpi.com/journal/biomimetics

www.ingramcontent.com/pod-product-compliance
Lightning Source LLC
Chambersburg PA
CBHW040818250926
51850CB00007BA/1198
* 9 7 8 3 7 2 5 8 1 5 7 7 7 *